Springer
Berlin
Heidelberg
New York
Barcelona
Budapest
Hong Kong
London
Milan
Paris
Santa Clara
Singapore
Tokyo

Lecture Notes in Chemistry 67

C. Pisani (Ed.)

Quantum-Mechanical Ab-initio Calculation of the Properties of Crystalline Materials

Springer

Editor

Professor Cesare Pisani
University of Torino
Department of Inorganic, Physical and Materials Chemistry
Via Giuria 5, I-10125 Torino
Italy

Library of Congress Cataloging-in-Publication Data

School of Computational Chemistry of the Italian Chemical Society (4th : 1994 : Torino, Italy)
Quantum-mechanical ab-initio calculation of the properties of crystalline materials : proceedings of the IV School of Computational Chemistry of the Italian Chemical Society / C. Pisani, editor.
p. cm.
Fourth School of Computational Chemistry of the Italian Chemical Society, held in Torino on 19-24 September 1994.--Forward
Includes bibliographical references and index.
ISBN 3-540-61645-4
1. Solid state physics--Congresses. 2. Crystals--Congresses.
3. Quantum chemistry--Congresses. I. Pisani, C. II. Title.
QC176.A1S33 1994
530.4'13--DC20 96-36109
CIP

ISSN 0342-4901
ISBN 3-540-61645-4 Springer-Verlag Berlin Heidelberg New York

Springer-Verlag Berlin Heidelberg New York
a member of BertelsmannSpringer Science+Business Media GmbH

Typesetting: Camera ready by author/editor

SPIN: 10850203 52/3142 - 54321 - Printed on acid-free paper

Foreword

Many powerful computer codes exist in the field of molecular chemistry, based on *ab initio* quantum-mechanical techniques. They can predict many properties of small- and medium-sized molecules with reasonable accuracy and at comparatively low cost. Their use has become common practice in the last ten years in many areas of activity, particularly in experimental chemical research, both in academic and in industrial environments.

The situation is much less advanced in the parallel field of condensed-matter studies. This is not due to a lack of potential interest: the development of new materials is one of the areas where the rate of progress is most rapid, and the amount of financial investment is largest. Materials for electronic and mechanical applications are in most cases crystalline structures, perfect or with controlled dosage of defects, whose properties cannot be understood without reference to an accurate description of the system at an atomic scale.

The intrinsic difficulty of handling systems of potentially infinite size explains only in part why crystalline studies are so far behind the current frontier of molecular quantum chemistry, a gap which presently is about fifteen years. Another possible reason may be that theoretical chemists have developed over many decades widely accepted interpretative schemes, a relatively standard language, a number of practical tools, good books and excellent computer codes so that having access to that technology does not require a very high-level entry point. On the other hand, making the wealth of knowledge embodied in standard solid-state theory compatible with an atomic-scale description of condensed systems is still an open problem. The question is how to transfer concepts and results from a "quantum-chemical" description of crystals to the parameterized world of solid-state physicists, and vice-versa.

Furthermore, the research and development effort which is being devoted to the implementation of powerful and user-friendly codes for the study of crystalline properties is curiously enough much less intensive than that devoted to the production of new molecular codes, and in the improvement of existing ones whose performance is already excellent. It is thus not surprising that the molecular cluster model, which can utilize standard codes with only minor modifications, is at present the favourite tool for quantum-mechanical investigations of the properties of crystals and their surfaces.

We now have the opportunity to change this state of affairs. There are a number of general-purpose, reasonably accurate and well-tested *ab initio* computer codes for crystals which are available to the scientific community. The rate of their improvement depends, in a sense, on their circulation, on the criticisms they receive and on the suggestions which are derived from their use. It is also very important that the various groups active in this field are open to unbiased comparison of the merits and drawbacks of their proposals, both as concerns basic ideas and approximations, and actual implementation of computer codes.

Schools and workshops can play a useful role to this effect, by teaching young people active in the field of Material Science how to exploit these new powerful tools. The present book contains the Proceedings of the **Fourth School of Computational Chemistry**, organized by the Interdivisional Computer Chemistry Group (GICC) of the Italian Chemical Society, and held in Torino on 19-24 September 1994, whose schedule is reported in Appendix A. Only the morning lectures (with few exceptions) are reproduced here, the afternoons being devoted to practical exercises performed by the students with use of the three codes available at the School, and described in Part 3.

The texts of the lectures can give only part of the information which can be obtained at a School through the possibility of practicing with the different programs, and from the contact with their authors. They nevertheless represent, on the whole, a useful introduction to the field, a reference for deeper study of certain specific subjects, and an objective body of information concerning the state-of-the-art in *ab-initio* simulations of the quantum-mechanical properties of crystalline materials.

Part One (Chapters 1–3) of the book provides a general introduction to the subject, addressed particularly to readers with a general knowledge in quantum chemistry, but not much confidence in solid state theory and its concepts.

Part Two (Chapters 4–7) is intended to give a deeper insight into the special algorithms and computational techniques which are currently adopted in *ab initio* computer codes for crystals.

Part Three (Chapters 8–10) presents in parallel three different programs which are available to all interested potential users on request, and based on very different approaches. These presentations may help newcomers in the field to understand the meaning hidden in the acronyms, and to choose the most suitable tool for their needs.

Finally, **Part Four (Chapters 11–16)** is an attempt to show what kind of information on the observable properties of condensed systems can be obtained from *ab initio* quantum-mechanical calculations. In particular, Resta's contribution demonstrates that important observables have become accessible to simulation, using quite unconventional new approaches. The last chapter, devoted to the hot topics of superconductivity, shows the importance of finding the connection between the results of *ab initio* calculations and high-quality theoretical schemes using parameterized Hamiltonians.

The publishing of these notes and their distribution to all the students of the above mentioned School has been possible thanks to the support of the Italian CNR (Consiglio Nazionale delle Ricerche).

A preliminary draft of these Proceedings has been distributed among attendants of the 1995 School organized under contract CHRX-CT93-0155 of the Human Capital & Mobility Programme of the European Community, which has provided additional funding for the work of revision.

I would finally like to thank Dr. Fiona Healy, who has read the manuscripts with patience and intelligence, and corrected their English, when necessary and when possible.

Cesare Pisani Torino, May 1996

Contents

Crystal Lattices and Crystal Symmetry

Davide Viterbo

Department of Inorganic, Physical and Materials Chemistry, University of Torino, via Giuria 7, I-10125 Torino, Italy

June 4, 1996

Summary. The basic concepts of the geometrical representation of crystalline solids and of their symmetry are outlined. The combination of periodic translational symmetry (describing crystal lattices) with other symmetry elements (rotation axes, mirror planes, inversion centers, etc.) is described as the basis of the space group theory.

Key words: Crystal Lattice – Translational Symmetry – Unit Cell – Crystal Structure – Crystallographic Rows and Planes – Metric Tensor – Reciprocal Lattice – Symmetry Operators – Symmetry Elements – Point Groups – Symmetry Classes – Laue Classes – Crystal Systems – Bravais Lattices – Space Groups

1. Lattice geometry

1.1. Lattices

Crystalline solids, as confirmed by evidence from several experiments (the anisotropy of their physical properties, diffraction, etc.), may be described as ordered repetitions of atoms or groups of atoms in three dimensions. Translational periodicity in crystals may be conveniently studied by focusing our attention on the geometry of the repetition rather than on the repeating motif. In an ideal crystal, all repeating units are identical and we may say that they are related by translational symmetry operations, corresponding to the set of vectors:

$$\mathbf{T} = u\,\mathbf{a} + v\,\mathbf{b} + w\,\mathbf{c} \tag{1}$$

where u, v and w are three integers ranging from minus infinity to plus infinity, zero included, and **a**, **b** and **c** are three non-coplanar vectors defining the basis of the three-dimensional space. Real crystals may present more or less marked

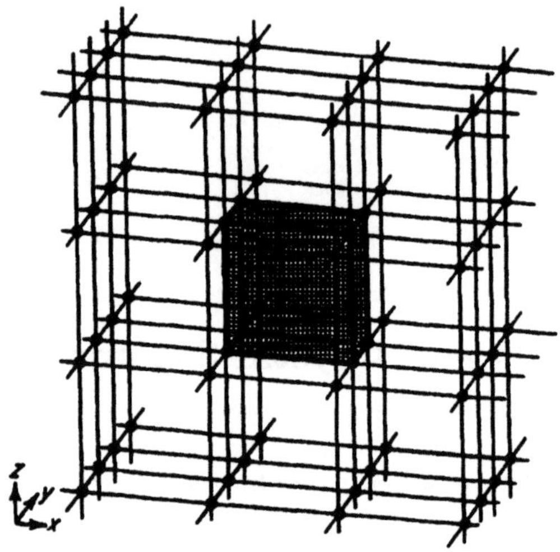

Figure 1. A three-dimensional lattice, showing a unit cell (heavy lines).

deviations from this ideal, perfect order. The set of points at the ends of all the translation vectors **T** forms a three-dimensional lattice and the points are called *lattice nodes* (Figure 1). The three integers u, v and w defining a given vector, are the corresponding coordinates of the node in the reference system defined by **a**, **b** and **c**. The parallelepiped formed by these three basis vectors is called the *unit cell* and their directions define the *crystallographic axes*: X, Y and Z. The lattice constants are the three moduli a, b and c and the three angles, α, β and γ between the vectors (α between **b** and **c**, β between **a** and **c** and γ between **a** and **b**).

A two-dimensional example will serve to illustrate these concepts. In Figure 2(a), a given "three-atom" motif is repeated at intervals **a** and **b**. If we replace each motif by a point at its centre of gravity we obtain the lattice of Figure 2(b). The same lattice is obtained if the point is located on any other position of the motif and the position of the lattice with respect to the motif is completely arbitrary. If any lattice point is chosen as the origin of the lattice, any other point in Figure 2(b) is uniquely defined by the vector:

$$\mathbf{T} = u\,\mathbf{a} + v\,\mathbf{b} \tag{2}$$

where u and v are integers and the unit cell is defined by the vectors **a** and **b**. The choice of the basis vectors is rather arbitrary, as shown in Figure 2(b), where four different choices are illustrated, all of which are consistent with relation (2) with u and v being integers. These cells contain only one lattice point, since the four points at the corners of each cell are each shared by a total of four cells. They are called *primitive* cells. Nevertheless, we are allowed to choose different types of unit cells, such as those shown in Figure 2(c), which contain two or more lattice points. Also in this case, each lattice point will satisfy (2) but u and

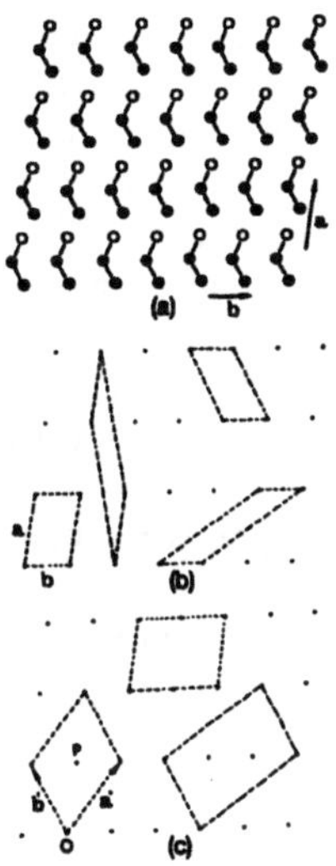

Figure 2. (a) Repetition of a graphical motif as an example of a two-dimensional crystal; (b) the corresponding lattice with some examples of primitive cells; (c) the same lattice with some examples of multiple cells.

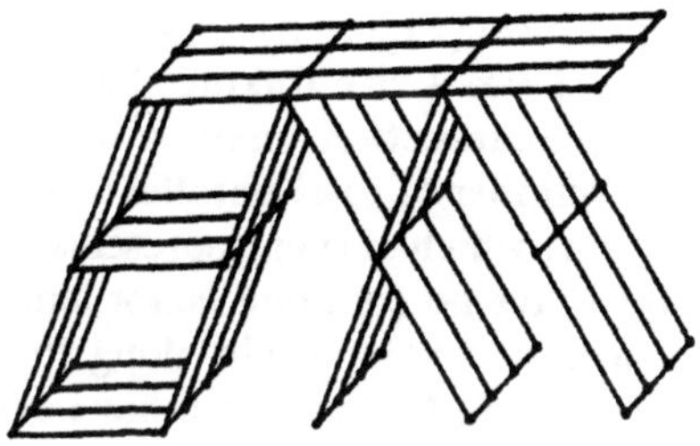

Figure 3. Lattice rows and planes.

v are no longer restricted to integer values (in Figure 2(c), point P is related to the origin and to the basis vectors $\mathbf{a}'$ and $\mathbf{b}'$ by a vector with $u = \frac{1}{2}$, $v = \frac{1}{2}$). These cells are called *multiple* or *centred* cells.

1.2. Crystal structure

The periodic repetition of the structural motif (atoms, groups of atoms or molecules) by the infinite set of vectors (1) yields the *crystal structure*, which is completely determined once the lattice constants and the coordinates, x, y and z, of all the atoms in the unit cell are known. These coordinates are the components of the vectors:

$$\mathbf{r}_j = x_j\,\mathbf{a} + y_j\,\mathbf{b} + z_j\,\mathbf{c} \qquad (j = 1, 2 \ldots N) \tag{3}$$

linking the cell origin to the nucleus of the j-th atom. For all N atoms inside the chosen unit cell, the coordinate values are in the interval 0 to 1. They are therefore called fractional coordinates and are given by:

$$x = X/a \quad y = Y/b \quad z = Z/c$$

1.3. Crystallographic rows and planes

Since crystals are anisotropic, it is necessary to identify, in a simple way, directions or planes in which specific physical properties are observed. Directions defined by two lattice nodes, and planes defined by three lattice points are called ***crystallographic directions*** and ***planes***, respectively. A ***lattice row*** is defined by two lattice points and the ***identity period***, separating the infinite number of nodes along the row is the minimum distance between two points. The lattice translations will then generate an infinite number of parallel and equally spaced, identical rows (Figure 3). Therefore, in a lattice with a primitive cell, the vector $\mathbf{T} = u\,\mathbf{a} + v\,\mathbf{b} + w\,\mathbf{c}$ linking the origin with the nearest point on a given crystallographic direction, may be used to represent a given row together with the infinite number of rows parallel to it. The three values u, v and w are the ***indices*** of the row, which is then represented by the symbol $[uvw]$. The three integers u, v and w have no common factor since they correspond to the coordinates of the point nearest to the origin along the row. The three rows corresponding to the base vectors **a**, **b**, and **c** will have symbols [100], [010] and [001], respectively.

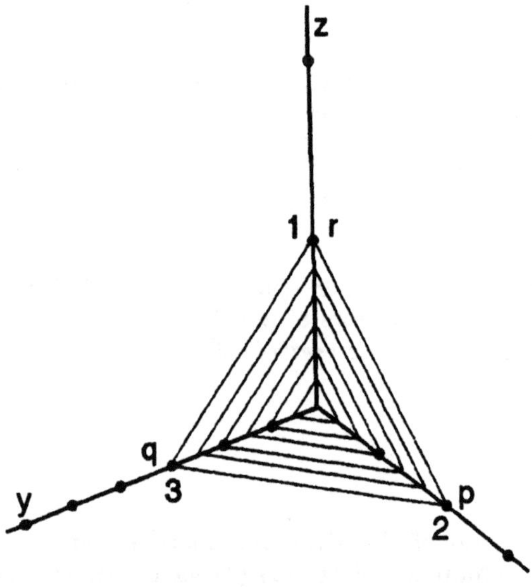

Figure 4. Some lattice planes of the set (236).

Three lattice points define a crystallographic plane. Let us consider a plane intersecting the three crystallographic axes, X, Y and Z, at the three lattice points $(p, 0, 0)$, $(0, q, 0)$ and $(0, 0, r)$ with integer values for p, q and r (Figure 4). Suppose that the largest common factor of p, q and r is 1 and that $m = pqr$ is their least common multiple. Then the equation of the plane is:

$$X/(pa) + Y/(qb) + Z/(rc) = 1 \tag{4}$$

If we introduce the fractional coordinates, $x = X/a$, $y = Y/b$ and $z = Z/c$, the equation of the plane becomes:

$$x/p + y/q + z/r = 1 \tag{5}$$

Multiplying both sides by the product pqr we obtain:

$$qrx + pry + pqz = pqr$$

which can be rewritten as:

$$hx + ky + lz = m \tag{6}$$

where $h = qr$, $k = pr$ and $l = pq$ are also integers with largest common factor equal to 1 and $m = pqr$. We can therefore generate a *family* of planes parallel to the plane (6) by varying m over all integers from $-\infty$ to $+\infty$. These will all be crystallographic planes since each of them is bound to pass through at least three lattice points. Each lattice plane is identical to any other within the family through a lattice translation. For a given triplet h, k and l, the plane with $m = 0$ will pass through the origin, while the plane with $m = 1$, that is:

$$hx + ky + lz = 1 \tag{7}$$

will be the closest to it. The intercepts of plane (7) on X, Y, Z are $1/h$, $1/k$ and $1/l$, while those of plane (6) are m/h, m/k and m/l. It is then clear that the distance of plane (6) from the origin is m times that of plane (7). The three indices, h, k and l, define the family of identical and equally spaced crystallographic planes, and are called the *Miller indices*. The symbol of a family of planes is (hkl). A simple interpretation of the Miller indices can be deduced from (6) and (7). In fact, they indicate that the planes of the family divide **a** into h parts, **b** into k parts and **c** into l parts. Crystallographic planes parallel to one of the three axes, X, Y or Z, are defined by indices $(0kl)$, $(h0l)$ or $(hk0)$ respectively. Planes parallel to the YZ (face A), XZ (face B) and XY (face C) planes are of type $(h00)$, $(0k0)$ or $(00l)$ respectively.

1.4. The metric tensor

In matrix notation, the set of basis vectors chosen to define a lattice may be written as:

$$|\mathbf{e}\rangle = \begin{pmatrix} \mathbf{a} \\ \mathbf{b} \\ \mathbf{c} \end{pmatrix} \tag{8}$$

and a lattice vector with indices u, v and w, is written as:

$$|\mathbf{T}\rangle = \langle \mathbf{u}|\mathbf{e}\rangle \quad \text{with} \quad \langle \mathbf{u}| = (u\ v\ w) \tag{9}$$

The modulus of this vector is obtained from the scalar product of the vector by its transpose:

$$\langle \mathbf{T}|\mathbf{T}\rangle = \langle \mathbf{u}|\mathbf{e}\rangle\langle \mathbf{e}|\mathbf{u}\rangle = \langle \mathbf{u}|\mathbf{G}|\mathbf{u}\rangle \tag{10}$$

where $\mathbf{G} = |\mathbf{e}\rangle\langle\mathbf{e}|$ is a square matrix:

$$\mathbf{G} = \begin{pmatrix} \mathbf{a}\cdot\mathbf{a} & \mathbf{a}\cdot\mathbf{b} & \mathbf{a}\cdot\mathbf{c} \\ \mathbf{b}\cdot\mathbf{a} & \mathbf{b}\cdot\mathbf{b} & \mathbf{b}\cdot\mathbf{c} \\ \mathbf{c}\cdot\mathbf{a} & \mathbf{c}\cdot\mathbf{b} & \mathbf{c}\cdot\mathbf{c} \end{pmatrix} \tag{11}$$

which is known as the *metric tensor* or *matrix* and allows the calculation of all geometrical features of the lattice.

1.5. The reciprocal lattice

In metric calculations and in the interpretation of many physical properties of crystals (diffraction, conductivity, etc.), it is very useful to introduce the so-called *reciprocal lattice*. Given the basis of a lattice (called here a *direct lattice*) as defined by (8), we can associate with it the basis $|\mathbf{e}^\star\rangle$ of a new lattice, defined by the vectors $\mathbf{a}^\star$, $\mathbf{b}^\star$, $\mathbf{c}^\star$, satisfying the condition $\mathbf{I} = |\mathbf{e}^\star\rangle\langle\mathbf{e}|$ or explicitly:

$$\begin{pmatrix} 1 & 0 & 0 \\ 0 & 1 & 0 \\ 0 & 0 & 1 \end{pmatrix} = \begin{pmatrix} \mathbf{a}^\star\cdot\mathbf{a} & \mathbf{a}^\star\cdot\mathbf{b} & \mathbf{a}^\star\cdot\mathbf{c} \\ \mathbf{b}^\star\cdot\mathbf{a} & \mathbf{b}^\star\cdot\mathbf{b} & \mathbf{b}^\star\cdot\mathbf{c} \\ \mathbf{c}^\star\cdot\mathbf{a} & \mathbf{c}^\star\cdot\mathbf{b} & \mathbf{c}^\star\cdot\mathbf{c} \end{pmatrix} \tag{12}$$

from which:

$$\begin{array}{lll} \mathbf{a}^\star\cdot\mathbf{a}=1 & \mathbf{b}^\star\cdot\mathbf{b}=1 & \mathbf{c}^\star\cdot\mathbf{c}=1 \\ \mathbf{a}^\star\cdot\mathbf{b}=0 & \mathbf{a}^\star\cdot\mathbf{c}=0 & \\ \mathbf{b}^\star\cdot\mathbf{a}=0 & \mathbf{b}^\star\cdot\mathbf{c}=0 & \\ \mathbf{c}^\star\cdot\mathbf{a}=0 & \mathbf{c}^\star\cdot\mathbf{b}=0 & \end{array} \tag{13}$$

The last three lines of equations (13) suggest that $\mathbf{a}^\star$ is normal to the plane $(\mathbf{b},\mathbf{c})$, $\mathbf{b}^\star$ to the plane $(\mathbf{a},\mathbf{c})$ and $\mathbf{c}^\star$ to the plane $(\mathbf{a},\mathbf{b})$. The modulus and direction of $\mathbf{a}^\star$, $\mathbf{b}^\star$, $\mathbf{c}^\star$ are fixed by relations (13). According to the second row of (13), we can write:

$$\mathbf{a}^\star = p\,(\mathbf{b}\wedge\mathbf{c}) \tag{14}$$

where p is a constant. Scalar multiplication of both sides of (14) by $\mathbf{a}$, will give:

$$\mathbf{a}^\star\cdot\mathbf{a} = 1 = p\,(\mathbf{b}\wedge\mathbf{c})\cdot\mathbf{a} = p\,V \tag{15}$$

from which $p = 1/V$ (V is the volume of the unit cell). Equation (14) and its analogues obtained from the third and fourth row of (13) may then be written as:

$$\mathbf{a}^\star = (\mathbf{b}\wedge\mathbf{c})/V \;\; ; \;\; \mathbf{b}^\star = (\mathbf{c}\wedge\mathbf{a})/V \;\; ; \;\; \mathbf{c}^\star = (\mathbf{a}\wedge\mathbf{b})/V \tag{16}$$

giving the reciprocal basis vectors in terms of the direct unit cell. Equations (13) also suggest that the roles of direct and reciprocal space may be interchanged: i.e. the reciprocal of the reciprocal lattice is the direct lattice. It may be easily seen that a reciprocal lattice vector:

$$\mathbf{T}^\star = h\mathbf{a}^\star + k\mathbf{b}^\star + l\mathbf{c}^\star \quad (h,k,l \text{ integers}) \tag{17}$$

is perpendicular to the family of direct lattice planes (hkl). For the reciprocal lattice, we can also define a metric matrix $\mathbf{G}^\star$:

$$\mathbf{G}^\star = \begin{pmatrix} \mathbf{a}^\star\cdot\mathbf{a}^\star & \mathbf{a}^\star\cdot\mathbf{b}^\star & \mathbf{a}^\star\cdot\mathbf{c}^\star \\ \mathbf{b}^\star\cdot\mathbf{a}^\star & \mathbf{b}^\star\cdot\mathbf{b}^\star & \mathbf{b}^\star\cdot\mathbf{c}^\star \\ \mathbf{c}^\star\cdot\mathbf{a}^\star & \mathbf{c}^\star\cdot\mathbf{b}^\star & \mathbf{c}^\star\cdot\mathbf{c}^\star \end{pmatrix} \tag{18}$$

and it may be verified that:

$$\mathbf{G}^\star = \mathbf{G}^{-1} \tag{19}$$

Finally, it is worth pointing out that the "1's" in equations (12) and (13) are sometimes substituted by a constant, such as 2π: the resulting reciprocal lattice is simply scaled by a factor equal to the constant.

2. Symmetry operations and symmetry elements

2.1. Isometric transformations

In order to understand the periodic and ordered nature of crystals, it is also necessary to know all other operations, besides translation, by which the repetition of the basic motif may be obtained. Given two identical objects, placed in random positions and orientations, we must know which operations should be performed to superimpose one object onto the other. The existence of enantiomeric molecules also requires that we know how to superimpose two objects of opposite handedness (***enantiomorphic objects***). The objects are said to be congruent if each point on one object corresponds to a point on the other and if the distance between two points on one object is equal to that between the corresponding points on the other. As a consequence, the angles formed by three related points will also be equal in absolute value and the correspondence is called ***isometric***. The congruence may either be ***direct*** or ***opposite*** according to whether the corresponding angles have the same or opposite signs. The study of operations relating congruent objects is the goal of the theory of ***isometric transformations***, of which we will only consider the most useful results (more detailed accounts may be found in references [1]-[9]). For a direct congruence, the ***movements*** by which two objects are brought to coincidence, may be: a) a ***translation***, which we have already considered; b) a ***rotation*** of $2\pi/n$ around an axis (an axis of order n or an n-axis); c) a ***rototranslation*** or ***screw*** movement, which may be considered as the combination (product) of a rotation around an axis and a translation along the axial direction. For an opposite congruence, which relates enantiomorphic objects, these may be brought to coincidence by the following operations: a) an ***inversion*** with respect to a point; b) a ***reflection*** with respect to a plane; c) a ***rotoinversion***, the product of a rotation around an axis by an inversion with respect to a point on that axis (an $\bar{n}$-axis); d) the product of a reflection by a translation parallel to the reflection plane: the plane is then called a ***glide plane***; e) a ***rotoreflection***, the product of a rotation by a reflection with respect to a plane perpendicular to the axis. As is well known, the description of the chemical and physical properties of molecules may be greatly simplified when their symmetry is taken into account and the symmetry operations which are considered are among those listed above.

2.2. Symmetry restrictions imposed by lattice periodicity

Suppose now that the isometric operations not only bring to coincidence a pair of congruent objects, but act on the entire space. If all the properties of the space remain unchanged after the operation has been performed, the operation will be a ***symmetry operation*** and ***symmetry elements*** are the points, axes or planes with respect to which symmetry operations are carried out.

When, as in the case of ideal crystals, the space is filled by the periodic repetition of a motif, we have seen that it may be described by a lattice. In this case, not all symmetry elements are compatible with the periodic nature of the space, which requires the repetition of these elements also. Without losing generality, let us assume that a rotation axis of order n passes through the origin O of the lattice. Since all lattice points are identical, there will be an

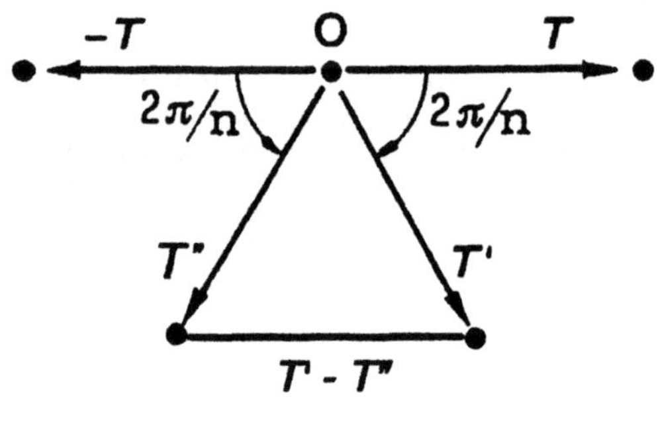

Figure 5. Lattice points in a plane normal to the symmetry axis n passing through O.

n axis passing through each and every lattice point, parallel to that passing through the origin. Each symmetry axis will lie along a lattice row and will be perpendicular to a crystallographic plane. In Figure 5, $\mathbf{T}$ is the period vector of a row passing through O and normal to n. Lattice points at $-\mathbf{T}$, $\mathbf{T}'$ and $\mathbf{T}''$ will also be present, and the vector $\mathbf{T}' - \mathbf{T}''$, parallel to $\mathbf{T}$, will have to be a lattice vector. This imposes the condition that $\mathbf{T}' - \mathbf{T}'' = m\mathbf{T}$, m being an integer; in a scalar form:

$$2\cos(2\pi/n) = m \quad (m \text{ integer}) \tag{20}$$

which is only verified for $n = 1, 2, 3, 4, 6$. Axes with $n = 5$ or $n > 6$ are not allowed in crystals. This is the reason why it is impossible to pave a room only with pentagonal tiles. The same restrictions apply to rotoinversion axes ($\overline{n}$ axes) and to screw axes. For the latter, the possible values of the translational component $\mathbf{t}$ are also restricted by periodicity: for an axis of order n, the translational component is repeated n times, and, because of lattice periodicity, we must have $n\mathbf{t} = p\mathbf{T}$, with p integer and $0 \le p < n$, or:

$$\mathbf{t} = (p/n)\mathbf{T} \tag{21}$$

Thus, for a screw axis of order 4, the allowed translational components will be $(0/4)\mathbf{T}$, $(1/4)\mathbf{T}$, $(2/4)\mathbf{T}$ and $(3/4)\mathbf{T}$. A normal n axis may be thought of as a special screw with $\mathbf{t} = 0$ and the properties of a screw axis are completely defined by the symbol n_p.

Reflection planes are indicated by the symbol m and must be parallel to the crystallographic planes. For glide planes, the periodicity requires that the translational component $\mathbf{t}$ must be either $\mathbf{t} = 0$ (mirror plane) or $\mathbf{t} = \frac{1}{2}\mathbf{T}$. Glide planes with translational components equal to $\mathbf{a}/2$, $\mathbf{b}/2$ and $\mathbf{c}/2$ will be indicated by the symbols a, b and c respectively, while the diagonal glides with translational components: $(\mathbf{a}+\mathbf{b})/2$, $(\mathbf{a}+\mathbf{c})/2$, $(\mathbf{b}+\mathbf{c})/2$ and $(\mathbf{a}+\mathbf{b}+\mathbf{c})/2$, are indicated by the symbol n. As we have seen, in a non-primitive cell, $\mathbf{T}$ is a vector with rational components and $\mathbf{t}$ may assume other values: in this case, the symbol d is used for the glide plane.

Table 1. Graphical symbols for symmetry elements: (a) axes normal to the plane of projection; (b) axes 2 and 2_1 parallel to the plane of projection; (c) axes parallel or inclined to the plane of projection; (d) symmetry planes normal to the plane of projection; (e) symmetry planes parallel to the plane of projection.

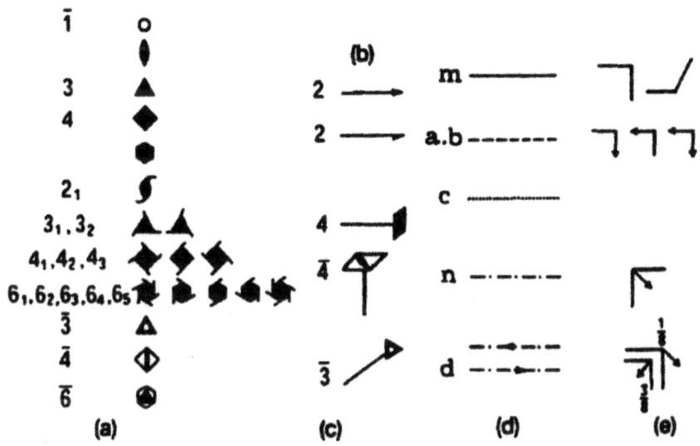

Of the five rotoinversion axes, only $\bar{3}$, $\bar{4}$, and $\bar{6}$ are independent, while $\bar{1}$ and $\bar{2}$ are equivalent to an inversion point and to a mirror plane, m, perpendicular to the axis, respectively. Finally, it can be shown that all the rotoreflection axes are equivalent to rotoinversion axes and these symmetry operations will no longer be considered separately.

The graphic symbols of all independent symmetry elements compatible with a crystal lattice, used in the *International Tables for Crystallography* [1], are shown in Table 1, while their effects on the space are illustrated in Figure 6, where, following international notation, an object is represented by a circle with a + or a - sign next to it indicating whether it is above or below the plane of the page; its enantiomorph is indicated by a comma inside the circle and, when both object and enantiomorph fall one on top of each other in the drawing plane, they are represented by a single circle divided into two halves (each with its appropriate sign), one of which contains a comma.

Symmetry operations relating objects through a direct congruence are called *proper*, while those relating objects through an opposite congruence are called *improper*.

3. Point groups and symmetry classes

In crystals more symmetry axes, both proper and improper, with or without translational components, may coexist. Let us first consider the combinations of symmetry operators which do not imply translations, i.e. the combinations of proper and improper axes intersecting in a point. These are called *point groups*, since the operators form a mathematical *group* and leave one point fixed. The set of crystals having the same point group is called *crystal class* and its symbol

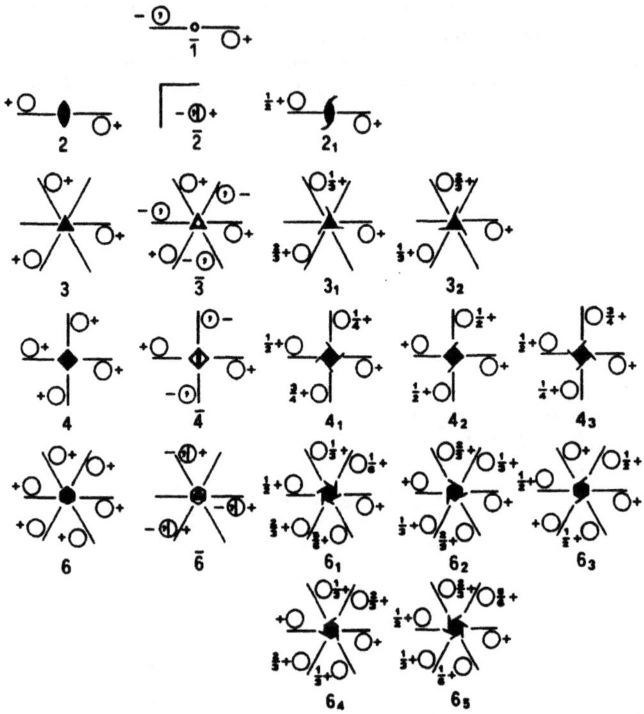

Figure 6. Arrangements of symmetry-equivalent objects, showing the effect of rotation, inversion and screw axes.

is that of the point group. In three dimensions, there are 32 possible crystallographic point groups. The simplest groups are those with only one symmetry axis. As we have seen, there are 5 possible proper (P) axes and 5 improper (I) axes giving rise to 10 point groups. We must also consider the cases where both a proper and an improper axis are simultaneously present along the same direction; these give rise to 3 new groups indicated by the symbol $n/\bar{n}$. In the remaining 19 groups, more axes coexist. We will not consider the rules governing the combinations of different axes (see references [1]-[9]), but simply list, in Table 2, the conventional symbols of the resulting groups, together with those of the 13 single-axis point groups. It may be noted that crystals with inversion symmetry operators which have an equal number of 'left' and 'right' moieties, each being the enantiomorph of the other. It is important to note that there is no direct relation between the crystal symmetry classes and the molecular symmetry of the individual molecules forming the crystals, in the sense that molecules which have a molecular symmetry not allowed in crystals (for instance with a 5 axis) may well form crystals. Of course, the symmetry of the crystal will belong to one of the 32 point groups.

3.1. Point groups in one and two dimensions

In a one-dimensional lattice, there are only two possible point groups: 1 and $\bar{1}$. In two dimensions, instead of reflection with respect to a plane, we have reflection with respect to a line and rotoinversion axes are not used. The following 10 point groups result: 1, 2, 3, 4, 6, *m*, 2*mm*, 3*m*, 4*mm* and 6*mm*.

4. Laue classes and crystal systems

4.1. The eleven Laue classes

It is also very important to understand how the symmetry of the physical properties of a crystal relates to its point group. Of basic relevance to this is a postulate of crystal physics, known as the *Neumann principle*: "the symmetry elements of any physical property must include the symmetry elements of the crystal point group". In keeping with this principle, the physical properties may present a higher, but not a lower, symmetry than the point group. As a consequence physical experiments do not normally reveal the true symmetry of the crystal. Some of them, such as diffraction, always add an inversion centre and point groups differing only by the presence of such a centre will not be differentiated by these experiments. When these groups are collected in classes they form the 11 *Laue classes* also listed in Table 2.

Table 2. List of the 32 point groups.

Crystal Systems	Point groups			Laue Classes	Lattice point groups
	Non-centrosymmetric		Centrosymmetric		
Triclinic	1		$\bar{1}$	$\bar{1}$	$\bar{1}$
Monoclinic	2	m	2/m	2/m	2/m
Orthorhombic	222	mm2	mmm	mmm	mmm
Tetragonal	4	$\bar{4}$	4/m	4/m	4/mmm
	422	4mm, $\bar{4}2m$	4/mmm	4/mmm	
Trigonal	3		$\bar{3}$	$\bar{3}$	$\bar{3}m$
	32	3m	$\bar{3}m$	$\bar{3}m$	
Hexagonal	6	$\bar{6}$	6/m	6/m	6/mmm
	622	6mm, $\bar{6}2m$	6/mmm	6/mmm	
Cubic	23		$m\bar{3}$	$m\bar{3}$	$m\bar{3}m$
	432	$\bar{4}3m$	$m\bar{3}m$	$m\bar{3}m$	

4.2. The seven crystal systems

If the crystal periodicity is only compatible with axes of order 1, 2, 3, 4 and 6, the presence of these different types of axes will in turn impose some restrictions on the geometry of the lattice. It is therefore convenient to group together the symmetry classes with common lattice features which may be described by unit cells of the same type. In turn, the cells will be chosen in the most suitable way to show the symmetry actually present. Seven different groups may be identified, which are listed in the first column of Table 2. They are known as the *seven crystal systems*.

Triclinic system - Point groups 1 and $\bar{1}$ have no symmetry axes and therefore there are no constraints on the unit cell: the ratios $a : b : c$ and the angles α, β, γ can assume any value.

Monoclinic system - Point groups 2, $\bar{2} \equiv m$ and $2/m$ all present a two-fold axis. Conventionally, this axis is assumed to coincide with the b axis of the unit cell (y-axis *unique*) and then **a** and **c** can be chosen on the lattice plane normal to **b**. We will then have $\alpha = \beta = 90°$ and γ unrestricted and the axial ratio $a : b : c$ also unrestricted.

Orthorhombic system - Classes 222, $mm2$ and mmm all have three mutually orthogonal two-fold axes. If we assume these as reference axes we will obtain a unit cell with angles $\alpha = \beta = \gamma = 90°$ and with unrestricted $a : b : c$ ratio.

Tetragonal system - The 7 point groups: 4, $\bar{4}$, $4/m$, 422, $\bar{4}2m$, $4mm$ and $4/mmm$, all have a single four-fold axis which is chosen to coincide with the **c** axis. Then the **a** and **b** axes become symmetry equivalent on the plane normal to **c**. The cell angles will be $\alpha = \beta = \gamma = 90°$ and the axial ratio $a : b : c =$ 1:1:c.

Trigonal and hexagonal systems - 3, $\bar{3}$, 32, $3m$, $\bar{3}m$, 6, $\bar{6}$, $6/m$, 622, $6mm$, $\bar{6}2m$ and $6/mmm$ all have a single three-fold or six-fold axis which is assumed to coincide with the **c** axis. Axes **a** and **b** become symmetry equivalent on the plane normal to **c**. Both systems are characterized by a unit cell with angles $\alpha = \beta = 90°$ and $\gamma = 120°$ and ratio $a : b : c =$ 1:1:c.

Cubic system - 23, $m\bar{3}$, 432, $\bar{4}3m$ and $m\bar{3}m$ all have four three-fold axes directed along the diagonals of a cube and can be referred to orthogonal unit cell axes coinciding with the cube edges. The presence of the three-fold axes ensures that these directions are symmetry equivalent. The chosen unit cell will have $\alpha = \beta = \gamma = 90°$ and axial ratio $a : b : c = 1 : 1 : 1$.

5. Bravais Lattices

As we have seen, each crystal system can be associated with a primitive cell compatible with the point groups belonging to that system. Each of these primitive cells defines a lattice type, but there are other types of lattices, based on non-primitive cells, which are also compatible with some of the crystal systems but cannot be related to those already seen. In this section, we will describe the 5 possible plane lattices and the 14 possible space lattices based on both primitive and non-primitive cells. These are called *Bravais lattices*, after Auguste Bravais who first listed them in 1850.

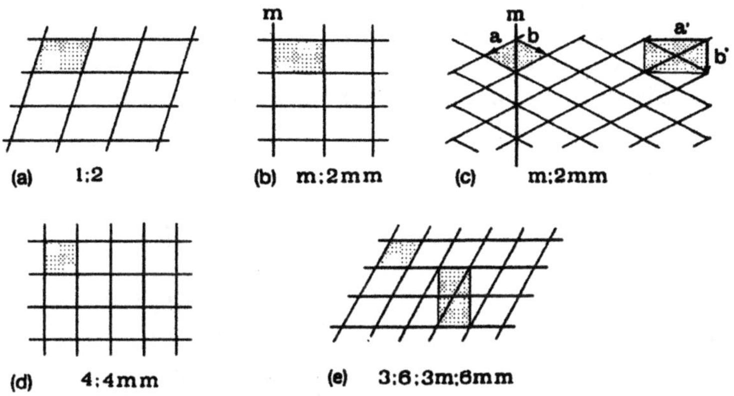

Figure 7. The five plane lattices and the corresponding two-dimensional point groups.

5.1. Plane lattices

As may be seen in Figure 7(a), an oblique cell is compatible with the presence of axes 1 and 2 normal to the cell. This cell is primitive and has point group 2. If the row indicated by m in Figure 7(b) is a reflection line, the cell must be rectangular. It is primitive and compatible with the point groups, m and $2mm$. The lattice shown in Figure 7(c) with a = b and $\gamma \neq 90°$, is also compatible with m.

Table 3. The five plane lattices.

Cell	Type of cell	Point group of the net	Lattice parameters
Oblique	p	2	a, b, γ
Rectangular	p, c	2mm	a, b, $\gamma = 90°$
Square	p	4mm	a = b, $\gamma = 90°$
Hexagonal	p	6mm	a = b, $\gamma = 120°$

Even though this lattice has an oblique primitive cell, each lattice point has $2mm$ symmetry and therefore, the lattice must be compatible with a rectangular system. This can be seen by choosing the rectangular centred cell, defined by the unit vectors: $\mathbf{a}'$ and $\mathbf{b}'$. The choice of an orthogonal cell is more convenient because it is more simply related to the symmetry of the lattice. It is important to note that the two lattices shown in Figures 7(b) and (c) are of different types, even though they are compatible with the same point groups. In Figure 7(d), a plane lattice is represented, which is compatible with the presence of a 4 axis. The square cell is primitive and compatible with the point groups 4 and $4mm$. Finally, in Figure 7(e), a plane lattice is shown, compatible with the presence of a 3 or a 6 axis. A primitive unit cell with a rhombus shape and angles of

60° and 120° (an hexagonal cell) may be chosen. This is compatible with the point groups 3, 6, $3m$ and $6mm$. A centred rectangular cell can also be selected, but such a cell, which is not as simply related to the lattice symmetry, is less convenient. The basic features of the five plane lattices are listed in Table 3.

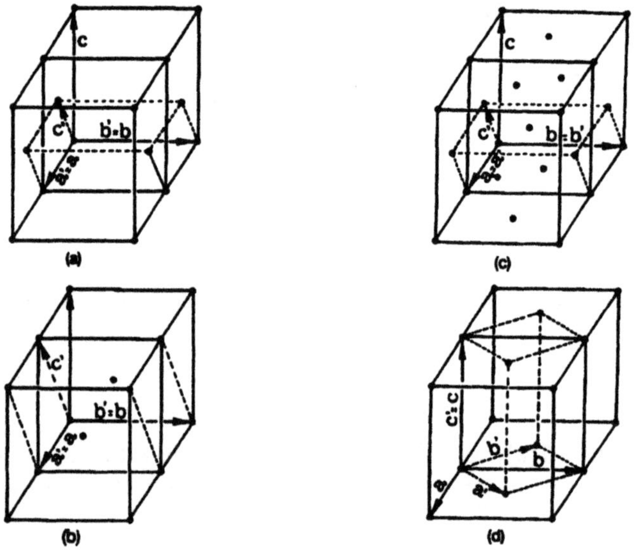

Figure 8. Monoclinic lattices: (a) reduction of a B-centred cell to a P cell; (b) reduction of an I-centred to an A-centred cell: (c) reduction of an F-centred to a C-centred cell; (d) reduction of a C-centred to a P non-monoclinic cell.

5.2. Space lattices

In Table 4, the different types of cells are described. The positions of the additional lattice points define the minimal translational components which will move an object onto an equivalent one. For instance, in an A-type cell, an object at (x, y, z) is repeated by translation into $(x, y + v/2, z + w/2)$, with v and w integers; the shortest translation from the original position will be (0, 1/2, 1/2).

Let us now examine the different types of three-dimensional lattices grouped in the appropriate crystal systems.

Triclinic lattices - Even though non-primitive cells can always be chosen, the absence of axes of order greater than one makes such choices inconvenient. The simplest choice is that of the primitive cell described in section 4.2.

Monoclinic lattices - In the same section, we have described the conventional primitive monoclinic cell as having $\alpha = \beta = 90°$ with γ and a:b:c unrestricted. Figure 8(a) shows a B-centred monoclinic cell with unit vectors **a**, **b** and **c**. Choosing unit vectors, so that $\mathbf{a}' = \mathbf{a}$, $\mathbf{b}' = \mathbf{b}$, $\mathbf{c}' = (\mathbf{a}+\mathbf{c})/2$, gives the primitive cell shown by the dotted line, which is also monoclinic ($\mathbf{c}'$ lies on the **ac** plane

Table 4. The conventional types of unit cell.

Symbol	Type	Positions of additional lattice points	Number of lattice points per cell
P	primitive	-	1
I	body-centred	(1/2, 1/2, 1/2)	2
A	A face-centred	(0, 1/2, 1/2)	2
B	B face-centred	(1/2, 0, 1/2)	2
C	C face-centred	(1/2 1/2, 0)	2
F	all faces centred	(1/2, 1/2, 0), (1/2, 0, 1/2)	2
		(0, 1/2, 1/2)	4
R	rhombohedrally centred (description with "hexagonal axes")	(1/3, 2/3, 2/3),(2/3, 1/3, 1/3)	3

and only the value of β will change). A B-type monoclinic cell can therefore always be reduced to the conventional primitive cell. With similar arguments, it may be shown that an I cell can be reduced to an A cell [Figure 8(b)] (since the **a** and **c** axes can always be interchanged, an A cell is equivalent to a C cell), and an F cell may become a C cell [Figure 8(c)]. However, a C cell is not reducible to a monoclinic P cell, as may be seen in Figure 8(d), where the primitive cell, shown by the dotted line with: $\mathbf{a}' = (\mathbf{a}+\mathbf{b})/2$, $\mathbf{b}' = (-\mathbf{a}+\mathbf{b})/2$, $\mathbf{c}' = \mathbf{c}$, no longer has the features of a monoclinic cell, since $\gamma' \neq 90°$, $|\mathbf{a}'| = |\mathbf{b}'| \neq |\mathbf{c}'|$ and the 2 axis lies along the diagonal of a face. It is worth noting that the choice of the C cell becomes necessary if we want to preserve the symmetry features of the monoclinic lattice in the cell geometry. The P cell shown by the dotted line is an allowed, but less convenient, reference system, which is seldom used to describe crystal structures. Nevertheless, in some cases, the choice of a P cell may become more convenient from a computational point of view and then an appropriate transformation must be performed from the conventional C cell to the corresponding primitive one. All transformation matrices, which allow passage from a centred cell to the corresponding primitive one and vice versa, are given in Table 5. It can therefore be concluded that there are two distinct monoclinic lattices, conventionally described by P and C cells, which are not interchangeable.

Orthorhombic lattices - With arguments similar to those used for monoclinic lattices, it can be verified that there are four different types of orthorhombic lattice, which are conventionally described in terms of P, C, I and F cells.

Tetragonal lattices - As a result of the four-fold axis, an A cell will always be equivalent to a B cell and therefore, to an F cell; indeed, the combination of the two lattice translations (0, 1/2, 1/2) and (1/2, 0, 1/2) gives the third translation (1/2, 1/2, 0) of the F cell (see Table 4). The latter is then transformable to a tetragonal I cell. A C cell is interchangeable with another tetragonal P cell. Thus only two different tetragonal lattices, P and I, are found.

Table 5. Transformation matrices, **M**, conventionally used to generate centered from primitive lattices, and vice versa, according to the relationship: **A' = M A**.

$$\mathrm{I} \rightarrow \mathrm{P}: \begin{pmatrix} -\frac{1}{2} & \frac{1}{2} & \frac{1}{2} \\ \frac{1}{2} & -\frac{1}{2} & \frac{1}{2} \\ \frac{1}{2} & \frac{1}{2} & -\frac{1}{2} \end{pmatrix} \quad \mathrm{P} \rightarrow \mathrm{I}: \begin{pmatrix} 0 & 1 & 1 \\ 1 & 0 & 1 \\ 1 & 1 & 0 \end{pmatrix} \quad \mathrm{R_h} \rightarrow \mathrm{R_{obv}}: \begin{pmatrix} \frac{2}{3} & \frac{1}{3} & \frac{1}{3} \\ -\frac{1}{3} & \frac{2}{3} & \frac{1}{3} \\ -\frac{1}{3} & -\frac{2}{3} & \frac{1}{3} \end{pmatrix} \quad \mathrm{R_{obv}} \rightarrow \mathrm{R_h}: \begin{pmatrix} 1 & -1 & 0 \\ 0 & 1 & -1 \\ 1 & 1 & 1 \end{pmatrix}$$

$$\mathrm{R_h} \rightarrow \mathrm{R_{rev}}: \begin{pmatrix} \frac{1}{3} & -\frac{1}{3} & \frac{1}{3} \\ \frac{1}{3} & \frac{2}{3} & \frac{1}{3} \\ -\frac{2}{3} & -\frac{1}{3} & \frac{1}{3} \end{pmatrix} \quad \mathrm{R_{rev}} \rightarrow \mathrm{R_h}: \begin{pmatrix} 1 & 0 & -1 \\ -1 & 1 & 0 \\ 1 & 1 & 1 \end{pmatrix} \quad \mathrm{F} \rightarrow \mathrm{P}: \begin{pmatrix} 0 & \frac{1}{2} & \frac{1}{2} \\ \frac{1}{2} & 0 & \frac{1}{2} \\ \frac{1}{2} & \frac{1}{2} & 0 \end{pmatrix} \quad \mathrm{P} \rightarrow \mathrm{F}: \begin{pmatrix} -1 & 1 & 1 \\ 1 & -1 & 1 \\ 1 & 1 & -1 \end{pmatrix}$$

$$\mathrm{A} \rightarrow \mathrm{P}: \begin{pmatrix} -1 & 0 & 0 \\ 0 & -\frac{1}{2} & \frac{1}{2} \\ 0 & \frac{1}{2} & \frac{1}{2} \end{pmatrix} \quad \mathrm{P} \rightarrow \mathrm{A}: \begin{pmatrix} -1 & 0 & 0 \\ 0 & -1 & 1 \\ 0 & 1 & 1 \end{pmatrix} \quad \mathrm{B} \rightarrow \mathrm{P}: \begin{pmatrix} -\frac{1}{2} & 0 & \frac{1}{2} \\ 0 & -1 & 0 \\ \frac{1}{2} & 0 & \frac{1}{2} \end{pmatrix} \quad \mathrm{P} \rightarrow \mathrm{B}: \begin{pmatrix} -1 & 0 & 1 \\ 0 & -1 & 0 \\ 1 & 0 & 1 \end{pmatrix}$$

$$\mathrm{C} \rightarrow \mathrm{P}: \begin{pmatrix} \frac{1}{2} & \frac{1}{2} & 0 \\ \frac{1}{2} & -\frac{1}{2} & 0 \\ 0 & 0 & -1 \end{pmatrix} \quad \mathrm{P} \rightarrow \mathrm{C}: \begin{pmatrix} 1 & 1 & 0 \\ 1 & -1 & 0 \\ 0 & 0 & -1 \end{pmatrix}$$

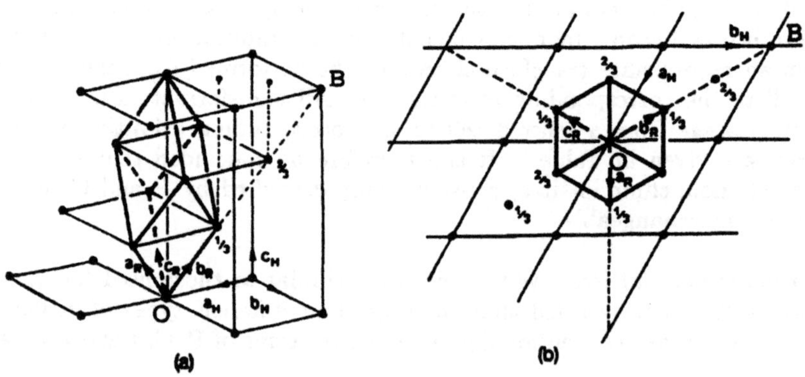

Figure 9. *The rhombohedral lattice.* The basis of the rhombohedral cell is labelled a_h, b_h and c_h (numerical fractions are calculated in terms of the c_h axis). (a) Obverse setting; (b) the same figure as in (a), projected along c_h.

Cubic lattices - For cubic lattices, symmetry requires that an A-type (or B or C) cell is also an F cell. There are three independent cubic lattices, P, I and F.

Hexagonal lattices - Only the P hexagonal Bravais lattice exists and is described in terms of the conventional primitive cell ($\alpha = \beta = 90°$ and $\gamma = 120°$ and ratio $a : b : c = 1 : 1 : c$) described in section 4.2.

Trigonal lattices - As in the case of the hexagonal lattices, all centred cells are transformable to the conventional P trigonal cell. As may be seen in Figure 9, due to the presence of a three-fold axis, some lattices can exist which may be described by a primitive cell of rhombohedral shape, with unit vectors $\mathbf{a}_R$, $\mathbf{b}_R$, $\mathbf{c}_R$, such that $a_R = b_R = c_R$, $\alpha_R = \beta_R = \gamma_R$, and the three-fold axis lies along the $\mathbf{a}_R + \mathbf{b}_R + \mathbf{c}_R$ direction. Such lattices may also be described by the three triple hexagonal cells with basis vectors:

$$\mathbf{a}_H = \mathbf{a}_R - \mathbf{b}_R \; ; \; \mathbf{b}_H = \mathbf{b}_R - \mathbf{c}_R \; ; \; \mathbf{c}_H = \mathbf{a}_R + \mathbf{b}_R + \mathbf{c}_R \quad \text{or}$$

$$\mathbf{a}_H = \mathbf{b}_R - \mathbf{c}_R \; ; \; \mathbf{b}_H = \mathbf{c}_R - \mathbf{a}_R \; ; \; \mathbf{c}_H = \mathbf{a}_R + \mathbf{b}_R + \mathbf{c}_R \quad \text{or}$$

$$\mathbf{a}_H = \mathbf{c}_R - \mathbf{a}_R \; ; \; \mathbf{b}_H = \mathbf{a}_R - \mathbf{b}_R \; ; \; \mathbf{c}_H = \mathbf{a}_R + \mathbf{b}_R + \mathbf{c}_R$$

which are said to be in *obverse setting*. Three further triple hexagonal cells, said to be in *reverse setting*, can be obtained by changing $\mathbf{a}_H$ and $\mathbf{b}_H$ into -$\mathbf{a}_H$ and -$\mathbf{b}_H$. In conclusion, some trigonal lattices may be described by a hexagonal P cell, others by a triple hexagonal cell. In the first case, the nodes lying on the different lattice planes normal to the three-fold axis will lie exactly on top of each other. In the second case, these planes are shifted with respect to each other, in such a way that the nth plane will be superimposed on the $(n + 3)$th plane (see Figure 9).

The 14 Bravais lattices are illustrated in Figure 10 by means of their conventional unit cells. Each of these lattices will have a corresponding reciprocal lattice. It may be easily verified that the reciprocal of each type of crystal system will be a lattice of the same crystal system: the reciprocal of an F lattice is an I lattice and vice versa.

5.3. Wigner-Seitz cells and Brillouin zones

Other cells may also be used to describe the 14 Bravais lattices. If we regard the conventional cells, which are parallelepipeds, as particular types of polyhedra, we can find several families of polyhedra with which we can fill up space by translation. A very important type of space filling is obtained through the *Dirichlet construction*. Each lattice point is connected with a line to its nearest neighbours and then, through the mid-points of all these segments, the planes perpendicular to them are traced. These intersecting planes will delimit a region of space which is called the *Dirichlet region*, *Wigner-Seitz cell* or *Voronoi cell*. An example in two dimensions is given in Figure 11(a), and two three-dimensional examples are shown in Figure 11(b). The Wigner-Seitz cell is always primitive and coincides with the Bravais cell when this is both rectangular and primitive. A construction identical to the Wigner-Seitz cell delimits in reciprocal space a

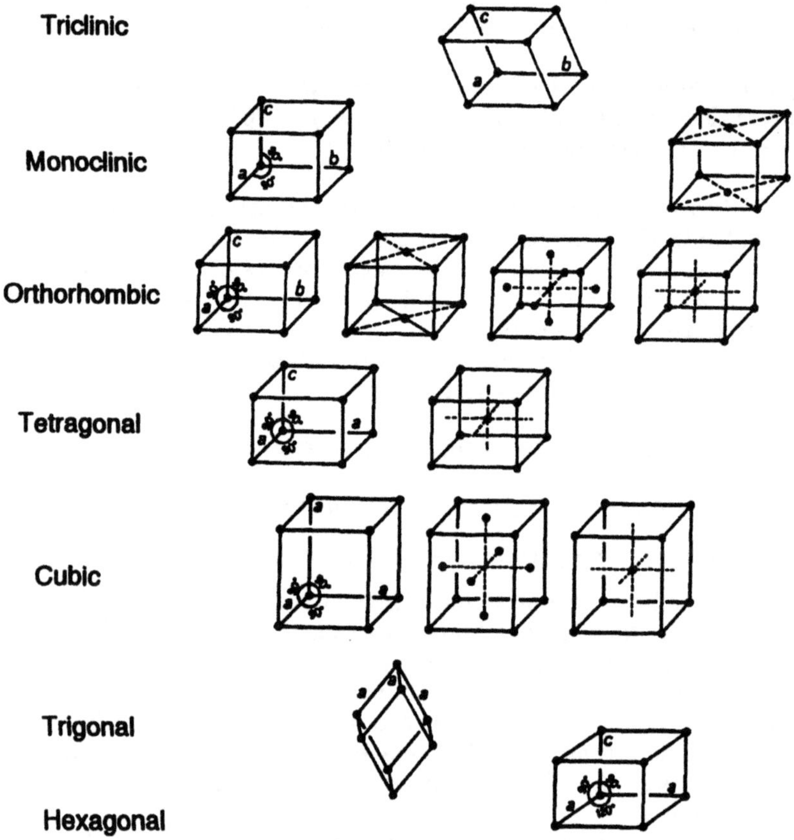

Figure 10. The 14 three-dimensional Bravais lattices.

cell conventionally known as the *first Brillouin zone*. There will be 14 first Brillouin zones corresponding to the 14 reciprocal Bravais lattices and that of an I lattice will look like a Wigner-Seitz cell of an F lattice and vice versa. As will be shown in the rest of this book, the Brillouin zones are very important in the study of lattice dynamics and in the band theory of solids.

6. Space groups

In section 3, we have seen that 32 combinations of either simple rotation or rotoinversion axes are compatible with the periodic nature of crystals. By combining the 32 point groups with the 14 Bravais lattices, 73 space groups are obtained. Furthermore, if we also consider symmetry operations with translational components, such as screw axes and glide planes, the total number of possible combinations, i.e. of *crystallographic space groups*, becomes 230. Note, however, that when such combinations have more than one axis, the restriction

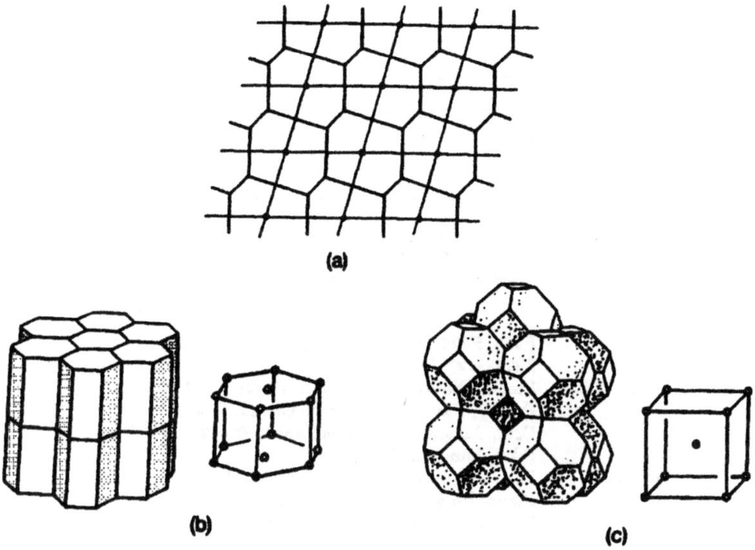

Figure 11. Examples of Wigner-Seitz cells.

that all symmetry elements must intersect in a point no longer applies. As a consequence of the presence of symmetry elements, several symmetry-equivalent objects (groups of atoms or molecules) will coexist within the unit cell. We will call the smallest part of the unit cell an ***asymmetric unit***, which will generate the whole cell contents when the symmetry operations of the space group are applied to it. The asymmetric unit is not usually uniquely defined and can be chosen with some degree of freedom. It is, nevertheless, obvious that when rotation or inversion axes are present, they must lie at the borders of the asymmetric unit. We can, therefore, define as a crystallographic space group, the set of symmetry operations which, in a three-dimensional periodic medium, convert a given asymmetric unit into all its equivalent units. The space groups were first independently derived by the mathematicians Fedorov (1890) and Schoenflies (1891) and are listed in Table 6.

Here, we will simply introduce the main terms of the space group language, following the international notation of Herman-Mauguin, used in the *International Tables for Crystallography* [1]. Indeed, the aim of this section is to introduce the reader to the use of these tables, which are an essential tool for the study of solid-state physics and chemistry. Each space group symbol consists of a letter (P, A, B, C, I, F, R), which indicates the type of centring in the conventional cell, followed by a set of characters indicating the symmetry elements. Such a set is organized according to the following rules:

1. For triclinic groups, no symmetry directions are needed and only two space groups exist: P1 and $P\bar{1}$.
2. For monoclinic groups, only one symbol is needed, which gives the nature of the unique dyad (two-fold) axis (proper and/or inversion). Two *settings* are used: y-axis unique or z-axis unique.

Table 6. The 230 three-dimensional space groups arranged by crystal systems and point groups.

Crystal system	Point group	Space groups
Triclinic	1	P1
	1	P$\bar{1}$
Monoclinic	2	P2, P2$_1$, C2
	m	Pm, Pc, Cm, Cc
	2/m	P2/m, P2$_1$/m, C2/m, P2/c, P2$_1$/c, C2/c
Orthorhombic	222	P222, P222$_1$, P2$_1$2$_1$2, P2$_1$2$_1$2$_1$, C222$_1$, C222, F222, I222, I2$_1$2$_1$2$_1$
	mm2	Pmm2, Pmc2$_1$, Pcc2, Pma2$_1$, Pca2$_1$, Pnc2$_1$, Pmn2$_1$, Pba2, Pna2$_1$, Pnn2, Cmm2, Cmc2$_1$, Ccc2, Amm2, Abm2, Ama2,Aba2, Fmm2, Fdd2, Imm2, Iba2, Ima2
	mmm	Pmmm, Pnnn, Pccm, Pban, Pmma, Pnna, Pmna, Pcca, Pbam, Pccn, Pbcm, Pnnm, Pmmn, Pbcn, Pbca, Pnma, Cmcm, Cmca, Cmmm, Cccm, Cmma, Ccca, Fmmm, Fddd, Immm, Ibam, Ibca, Imma
Tetragonal	4	P4, P4$_1$, P4$_2$, P4$_3$, I4, I4$_1$
	$\bar{4}$	P$\bar{4}$, I$\bar{4}$
	4/m	P4/m, P44$_2$/m, P4/n, P4$_2$/n, I4/m, I4$_1$/a
	422	P422, P42$_1$2, P4$_1$22, P4$_1$2$_1$2, P4$_2$22, P4$_2$2$_1$2, P4$_3$22, P4$_3$2$_1$2, I422, I4$_1$22
	4mm	P4mm, P4bm, P4$_2$cm, P4$_2$nm, P4cc, P4nc, P4$_2$mc, P4$_2$bc, I4mm, I4cm, I4$_1$md, I4$_1$cd
	$\bar{4}$m	P$\bar{4}$2m, P$\bar{4}$2c, P$\bar{4}$2$_1$m, P$\bar{4}$2$_1$c, P$\bar{4}$m2, P$\bar{4}$c2, P$\bar{4}$b2, P$\bar{4}$n2, I$\bar{4}$m2, I$\bar{4}$c2, I$\bar{4}$2m, I$\bar{4}$2d
	4/mmm	P4/mmm, P4/mcc, P4/nbm, P4nnc, P4/mbm, P4/mnc, P4nmm, P4/ncc, P4$_2$/mmc, P4$_2$/mcm, P4$_2$/nbc, P4$_2$/nnm, P4$_2$/mbc, P4$_2$/mnm, P4$_2$/nmc, P4$_2$/ncm, I4/mmm, I4mcm, I4$_1$/amd, I4$_1$/acd
Trigonal-hexagonal	3	P3, P3$_1$, P3$_1$, R3
	$\bar{3}$	P$\bar{3}$, R$\bar{3}$
	32	P312, P321, P3$_1$12, P3$_1$21, P3$_2$12, P3$_2$21, R32
	3m	P3m1, P31m, P3c1, P31c, R3m, R3c
	$\bar{3}$m	P$\bar{3}$2m, P$\bar{3}$2c, P$\bar{3}$m1, P$\bar{3}$c1, R$\bar{3}$m, R$\bar{3}$c
	6	P6, P6$_1$, P6$_5$, P6$_3$, P6$_2$, P6$_4$
	$\bar{6}$	P$\bar{6}$
	6/m	P6/m, P6$_3$/m
	622	P622, P6$_1$22, P6$_5$22, P6$_2$22, P6$_4$22, P6$_3$22
	6mm	P6mm, P6cc, P6$_3$cm, P6$_3$mc
	$\bar{6}$m	P$\bar{6}$m2, P$\bar{6}$c2, P$\bar{6}$2m, P$\bar{6}$2c
	6/mmm	P6mmm, P6mcc, P6$_3$/mcm, P6$_3$/mmc
Cubic	23	P23, F23, I23, P2$_1$3, I2$_1$3
	m$\bar{3}$	Pm$\bar{3}$, Pn$\bar{3}$, Fm$\bar{3}$, Fd$\bar{3}$, Im$\bar{3}$, Pa$\bar{3}$, Ia$\bar{3}$
	432	P432, P4$_2$32, F432, F4$_1$32, I432, P4$_3$32, P4$_1$32, I4$_1$32
	$\bar{4}$3m	P$\bar{4}$3m, F$\bar{4}$3m, I$\bar{4}$3, P$\bar{4}$3n, F$\bar{4}$3c, I$\bar{4}$3d
	m$\bar{3}$m	Pm$\bar{3}$m, Pn$\bar{3}$n, Pm$\bar{3}$n, Pn$\bar{3}$m, Fm$\bar{3}$m, Fd$\bar{3}$m, Fd$\bar{3}$c, Fm$\bar{3}$c, Im$\bar{3}$m, Ia$\bar{3}$d

3. For orthorhombic groups, the three dyads (proper and/or inversion) along x, y and z are given in sequential order. Thus Pca21 means: primitive cell, glide plane of type c (translational component along z) normal to the x-axis, glide plane of type a (translational component along x) normal to the y-axis with a two-fold screw axis along z.
4. For tetragonal groups, first the tetrad (proper and/or inversion) axis along z is specified, then the dyad (proper and/or inversion) along x is given, and after that the dyad along [110] is specified. For instance, P$4_2/nbc$ denotes a space group with primitive cell, a 4_2 screw axis along z, to which a diagonal (n) glide plane is perpendicular, a glide plane of type b normal to x, and a c glide plane normal to [110]. Because of the tetragonal symmetry, the x and y axes are equivalent and there is no need to specify the symmetry along y.
5. For trigonal and hexagonal groups, the triad or hexad (proper and/or inversion) along the z-axis is first given, then the dyad along x and after that the dyad along $[1\bar{1}0]$ is specified. For example, P6_3mc has a primitive cell, a six-fold screw axis 6_3 along z, a reflection plane normal to x and a c glide normal to $[1\bar{1}0]$.
6. For cubic groups, the order is: dyads or tetrads (proper and/or inversion) along x, followed by triads along [111] and dyads along [110]. Thus F$4_1$32 has a face centred cell, a 4_1 axis along x, a three-fold rotation axis along [111] and a two-fold rotation axis along [110].

For a given space group, lattice symmetry directions that carry no symmetry elements are represented by the symbol '1', as in P3m1 and P31m. In general, the '1' at the end of the space group symbol is omitted, as in P6 (instead of P611) or F23 (instead of F231). We note that:

- The 73 space groups obtained from combinations of the Bravais lattices with symmetry elements which have no translational components are called *symmorphic* space groups. Examples are: Pm, P222, Cmm2, F23, etc.
- Many compounds, such as biological molecules, are enantiomorphic and will then crystallize in space groups with no inversion centres or mirror planes; there are 65 groups of this type.
- Among the 230 space groups, there are 11 enantiomorphic pairs: P3_1-P3_2, P$3_1$12-P$3_2$12, P$3_1$21-P$3_2$21, P4_1-P4_3, P$4_1$22-P$4_3$22, P$4_12_1$2-P$4_32_1$2, P6_1-P6_5, P6_2-P6_4, P$6_1$22-P$6_5$22, P$6_2$22-P$6_4$22, P$4_1$32-P$4_3$32. If one chiral form of an optically active compound crystallizes in one of the two enantiomorphic space groups, the other form will crystallize in the other.
- The point group to which a given space group belongs may be obtained from its symbol by omitting the lattice indication and by replacing the screw axes and the glide planes with their corresponding symmorphic symmetry elements. For instance, the space groups P$4_2/mmc$, P$4/nnc$, I$4_1/acd$ all belong to the point group $4/mmm$.
- The frequency of the different space groups is not uniform. Organic compounds tend to crystallize in those space groups which permit close packing of objects of irregular shape. It may be understood intuitively that this close packing is better achieved when screw axes and glide planes are present. Indeed, these symmetry elements, through their translational components, facilitate the approach of concave and convex parts of an irregular molecule in crystal formation.

Ccca D_{2h}^{22} *mmm* Orthorhombic

No. 68 *C* 2/*c* 2/*c* 2/*a* Patterson symmetry *Cmmm*

ORIGIN CHOICE 1

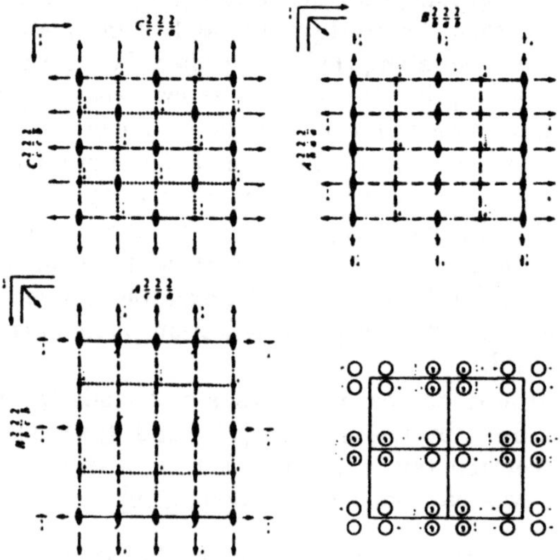

Origin at 222 at 2/*n* 2/*n* 2, at 0,¼,¼ from $\bar{1}$

Asymmetric unit 0≤*x*≤¼; 0≤*y*≤½; 0≤*z*≤½

Symmetry operations

For (0,0,0)+ set

(1) 1 (2) 2 ¼,¼,*z* (3) 2 0,*y*,0 (4) 2(½,0,0) *x*,¼,0
(5) $\bar{1}$ 0,¼,¼ (6) *a* *x*,*y*,¼ (7) *c* *x*,¼,*z* (8) *c* ¼,*y*,*z*

For (½,½,0)+ set

(1) *t*(½,½,0) (2) 2 0,0,*z* (3) 2(0,½,0) ¼,*y*,0 (4) 2 *x*,0,0
(5) $\bar{1}$ ¼,0,¼ (6) *b* *x*,*y*,¼ (7) *n*(½,0,½) *x*,0,*z* (8) *n*(0,½,½) 0,*y*,*z*

Generators selected (1); *t*(1,0,0); *t*(0,1,0); *t*(0,0,1); *t*(½,½,0); (2); (3); (5)

Positions

Multiplicity, Wyckoff letter, Site symmetry			Coordinates (0,0,0)+ (½,½,0)+				Reflection conditions
							General:
16	*i*	1	(1) *x*,*y*,*z* (5) $\bar{x}$,$\bar{y}$+½,$\bar{z}$+½	(2) $\bar{x}$+½,$\bar{y}$+½,*z* (6) *x*+½,*y*,$\bar{z}$+½	(3) $\bar{x}$,*y*,$\bar{z}$ (7) *x*,$\bar{y}$+½,*z*+½	(4) *x*+½,$\bar{y}$+½,$\bar{z}$ (8) $\bar{x}$+½,*y*,*z*+½	*hkl* : *h*+*k*=2*n*; 0*kl* : *k*,*l*=2*n*; *h*0*l*: *h*,*l*=2*n*; *hk*0: *h*,*k*=2*n*; *h*00: *h*=2*n*; 0*k*0: *k*=2*n*; 00*l* : *l*=2*n*
							Special: as above, plus
8	*h*	..2	¼,¼,*z*	¾,¼,$\bar{z}$	¾,¼,$\bar{z}$+½	¼,¼,*z*+½	*hkl* : *l*=2*n*
8	*g*	..2	0,0,*z*	0,0,$\bar{z}$	0,½,$\bar{z}$+½	0,½,*z*+½	*hkl* : *k*+*l*=2*n*
8	*f*	.2.	0,*y*,0	½,$\bar{y}$+½,0	0,$\bar{y}$+½,½	½,*y*,½	*hkl* : *k*+*l*=2*n*
8	*e*	2..	*x*,0,0	$\bar{x}$+½,½,0	$\bar{x}$,½,½	*x*+½,0,½	*hkl* : *k*+*l*=2*n*
8	*d*	$\bar{1}$	0,¼,¼	½,¼,¼	0,¼,¾	½,¼,¾	*hkl* : *k*,*l*=2*n*
8	*c*	$\bar{1}$	¼,0,¼	¼,½,¼	¾,0,¾	¾,½,¾	*hkl* : *k*,*l*=2*n*
4	*b*	222	0,0,½	0,½,0			*hkl* : *k*+*l*=2*n*
4	*a*	222	0,0,0	0,½,½			*hkl* : *k*+*l*=2*n*

Figure 12. (a): The space group *Ccca*, as described in Volume A of the *International Tables of Crystallography* [1]: Origin Choice 1.

Ccca	D_{2h}^{22}	*mmm*	**Orthorhombic**
No. 68	$C\,2/c\,2/c\,2/a$		Patterson symmetry *Cmmm*

ORIGIN CHOICE 2

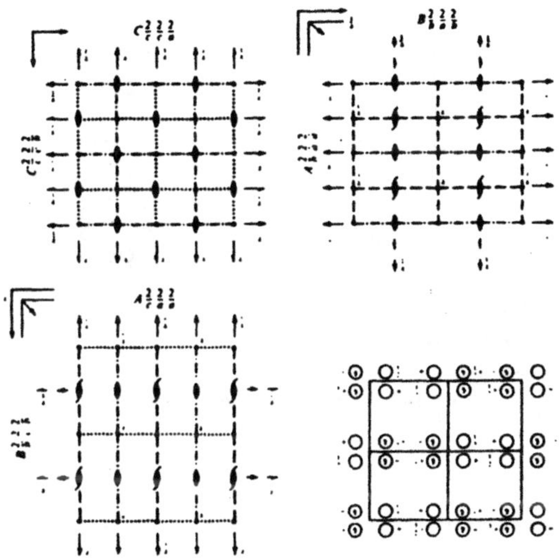

Origin at $\bar{1}$ at *nca*, at $0,-\frac{1}{4},-\frac{1}{4}$ from 222

Asymmetric unit $0\le x\le \frac{1}{4};\ 0\le y\le \frac{1}{2};\ 0\le z\le \frac{1}{2}$

Symmetry operations

For (0,0,0)+ set

(1) 1	(2) 2 $\frac{1}{4},0,z$	(3) 2 $0,y,\frac{1}{4}$	(4) 2($\frac{1}{2}$,0,0) $x,0,\frac{1}{4}$
(5) $\bar{1}$ 0,0,0	(6) *a* $x,y,0$	(7) *c* $x,0,z$	(8) *c* $\frac{1}{4},y,z$

For ($\frac{1}{2},\frac{1}{2}$,0)+ set

(1) *t*($\frac{1}{2},\frac{1}{2}$,0)	(2) 2 $0,\frac{1}{4},z$	(3) 2(0,$\frac{1}{2}$,0) $\frac{1}{4},y,\frac{1}{4}$	(4) 2 $x,\frac{1}{4},\frac{1}{4}$
(5) $\bar{1}$ $\frac{1}{4},\frac{1}{4},0$	(6) *b* $x,y,0$	(7) *n*($\frac{1}{2}$,0,$\frac{1}{2}$) $x,\frac{1}{4},z$	(8) *n*(0,$\frac{1}{2},\frac{1}{2}$) $0,y,z$

Generators selected (1); *t*(1,0,0); *t*(0,1,0); *t*(0,0,1); *t*($\frac{1}{2},\frac{1}{2}$,0); (2); (3); (5)

Positions

Multiplicity, Wyckoff letter, Site symmetry — Coordinates (0,0,0)+ ($\frac{1}{2},\frac{1}{2}$,0)+ — Reflection conditions

Mult.	Wyckoff	Site symm.	Coordinates				Reflection conditions
							General:
16	*i*	1	(1) x,y,z	(2) $\bar{x}+\frac{1}{2},\bar{y},z$	(3) $\bar{x},y,\bar{z}+\frac{1}{2}$	(4) $x+\frac{1}{2},\bar{y},\bar{z}+\frac{1}{2}$	*hkl* : $h+k=2n$
			(5) $\bar{x},\bar{y},\bar{z}$	(6) $x+\frac{1}{2},y,\bar{z}$	(7) $x,\bar{y},z+\frac{1}{2}$	(8) $\bar{x}+\frac{1}{2},y,z+\frac{1}{2}$	0*kl* : $k,l=2n$
							*h*0*l* : $h,l=2n$
							*hk*0 : $h,k=2n$
							*h*00 : $h=2n$
							0*k*0 : $k=2n$
							00*l* : $l=2n$
							Special: as above, plus
8	*h*	..2	$\frac{1}{4},0,z$	$\frac{1}{4},0,\bar{z}+\frac{1}{2}$	$\frac{3}{4},0,\bar{z}$	$\frac{3}{4},0,z+\frac{1}{2}$	*hkl* : $l=2n$
8	*g*	..2	$0,\frac{1}{4},z$	$0,\frac{1}{4},\bar{z}+\frac{1}{2}$	$0,\frac{3}{4},\bar{z}$	$0,\frac{3}{4},z+\frac{1}{2}$	*hkl* : $k+l=2n$
8	*f*	.2.	$0,y,\frac{1}{4}$	$\frac{1}{2},\bar{y},\frac{1}{4}$	$0,\bar{y},\frac{3}{4}$	$\frac{1}{2},y,\frac{3}{4}$	*hkl* : $k+l=2n$
8	*e*	2..	$x,\frac{1}{4},\frac{1}{4}$	$\bar{x}+\frac{1}{2},\frac{3}{4},\frac{1}{4}$	$\bar{x},\frac{3}{4},\frac{3}{4}$	$x+\frac{1}{2},\frac{1}{4},\frac{3}{4}$	*hkl* : $k+l=2n$
8	*d*	$\bar{1}$	0,0,0	$\frac{1}{2}$,0,0	0,0,$\frac{1}{2}$	$\frac{1}{2}$,0,$\frac{1}{2}$	*hkl* : $k,l=2n$
8	*c*	$\bar{1}$	$\frac{1}{4},\frac{3}{4},0$	$\frac{1}{4},\frac{1}{4},0$	$\frac{3}{4},\frac{3}{4},\frac{1}{2}$	$\frac{3}{4},\frac{1}{4},\frac{1}{2}$	*hkl* : $k,l=2n$
4	*b*	222	$0,\frac{1}{4},\frac{3}{4}$	$0,\frac{3}{4},\frac{1}{4}$			*hkl* : $k+l=2n$
4	*a*	222	$0,\frac{1}{4},\frac{1}{4}$	$0,\frac{3}{4},\frac{3}{4}$			*hkl* : $k+l=2n$

Figure 12. (b): The space group C*ca*, as described in Volume A of the *International Tables of Crystallography* [1]: Origin Choice 2.

$P4_222$ D_4^5 422 Tetragonal

No. 93 $P4_222$

Patterson symmetry $P4/mmm$

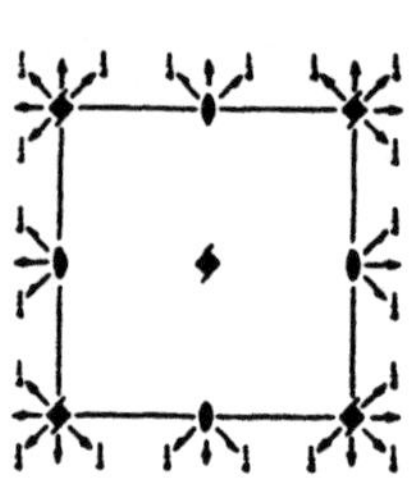

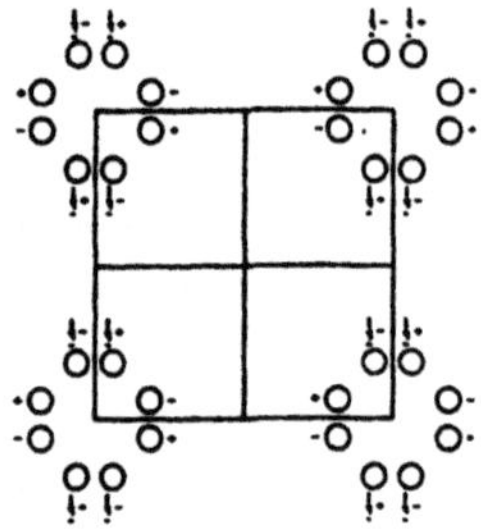

Origin at 222 at 4_221

Asymmetric unit $0\le x\le \frac{1}{2}$; $0\le y\le 1$; $0\le z\le \frac{1}{4}$

Symmetry operations

(1) 1	(2) 2 0,0,z	(3) $4^+(0,0,\frac{1}{2})$ 0,0,z	(4) $4^-(0,0,\frac{1}{2})$ 0,0,z
(5) 2 0,y,0	(6) 2 x,0,0	(7) 2 $x,x,\frac{1}{4}$	(8) 2 $x,\bar{x},\frac{1}{4}$

Generators selected (1); $t(1,0,0)$; $t(0,1,0)$; $t(0,0,1)$; (2); (3); (5)

Positions

Multiplicity, Wyckoff letter, Site symmetry			Coordinates				Reflection conditions
							General:
8	p	1	(1) x,y,z (5) $\bar{x},y,\bar{z}$	(2) $\bar{x},\bar{y},z$ (6) $x,\bar{y},\bar{z}$	(3) $\bar{y},x,z+\frac{1}{2}$ (7) $y,x,\bar{z}+\frac{1}{2}$	(4) $y,\bar{x},z+\frac{1}{2}$ (8) $\bar{y},\bar{x},\bar{z}+\frac{1}{2}$	$00l$: $l=2n$
							Special: as above, plus
4	o	..2	$x,x,\frac{3}{4}$	$\bar{x},\bar{x},\frac{3}{4}$	$\bar{x},x,\frac{1}{4}$	$x,\bar{x},\frac{1}{4}$	$0kl$: $l=2n$
4	n	..2	$x,x,\frac{1}{4}$	$\bar{x},\bar{x},\frac{1}{4}$	$\bar{x},x,\frac{3}{4}$	$x,\bar{x},\frac{3}{4}$	$0kl$: $l=2n$
4	m	.2.	$x,\frac{1}{2},0$	$\bar{x},\frac{1}{2},0$	$\frac{1}{2},x,\frac{1}{2}$	$\frac{1}{2},\bar{x},\frac{1}{2}$	hhl : $l=2n$
4	l	.2.	$x,0,\frac{1}{2}$	$\bar{x},0,\frac{1}{2}$	$0,x,0$	$0,\bar{x},0$	hhl : $l=2n$
4	k	.2.	$x,\frac{1}{2},\frac{1}{2}$	$\bar{x},\frac{1}{2},\frac{1}{2}$	$\frac{1}{2},x,0$	$\frac{1}{2},\bar{x},0$	hhl : $l=2n$
4	j	.2.	$x,0,0$	$\bar{x},0,0$	$0,x,\frac{1}{2}$	$0,\bar{x},\frac{1}{2}$	hhl : $l=2n$
4	i	2..	$0,\frac{1}{2},z$	$\frac{1}{2},0,z+\frac{1}{2}$	$0,\frac{1}{2},\bar{z}$	$\frac{1}{2},0,\bar{z}+\frac{1}{2}$	hkl : $h+k+l=2n$
4	h	2..	$\frac{1}{2},\frac{1}{2},z$	$\frac{1}{2},\frac{1}{2},z+\frac{1}{2}$	$\frac{1}{2},\frac{1}{2},\bar{z}$	$\frac{1}{2},\frac{1}{2},\bar{z}+\frac{1}{2}$	hkl : $l=2n$
4	g	2..	$0,0,z$	$0,0,z+\frac{1}{2}$	$0,0,\bar{z}$	$0,0,\bar{z}+\frac{1}{2}$	hkl : $l=2n$
2	f	2.22	$\frac{1}{2},\frac{1}{2},\frac{1}{4}$	$\frac{1}{2},\frac{1}{2},\frac{3}{4}$			hkl : $l=2n$
2	e	2.22	$0,0,\frac{1}{4}$	$0,0,\frac{3}{4}$			hkl : $l=2n$
2	d	222.	$0,\frac{1}{2},\frac{1}{2}$	$\frac{1}{2},0,0$			hkl : $h+k+l=2n$
2	c	222.	$0,\frac{1}{2},0$	$\frac{1}{2},0,\frac{1}{2}$			hkl : $h+k+l=2n$
2	b	222.	$\frac{1}{2},\frac{1}{2},0$	$\frac{1}{2},\frac{1}{2},\frac{1}{2}$			hkl : $l=2n$
2	a	222.	$0,0,0$	$0,0,\frac{1}{2}$			hkl : $l=2n$

Figure 13. The space group $P4_222$, as described in Volume A of the *International Tables of Crystallography* [1].

As we have already mentioned, the standard description of the crystallographic space groups is contained in Volume A of the *International Tables of Crystallography* [1]. For each space group the *Tables* include an extensive set of descriptors. In Figures 12 and 13, the space groups *Ccca* and $P4_222$ are shown as examples. Here is an outline of the information most relevant to this course.

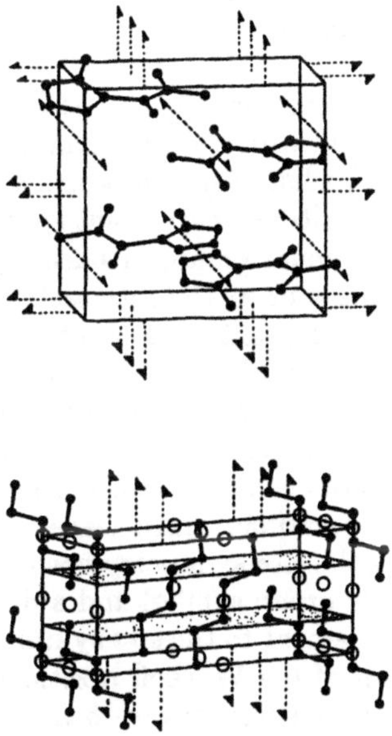

Figure 14. (a) A $P2_12_12_1$ crystal structure and its symmetry elements (G. Chiari, D. Viterbo, A. Gaetani Manfredotti and C. Guastini (1975), *Cryst. Struct. Commun.*, **4**, 561) and (b) a $P2_1/c$ crystal structure and its symmetry elements. Glide planes are emphasised by the shading (M. Calleri, G. Ferraris and D. Viterbo (1966), *Acta Cryst.*, **20**, 73). For clarity, hydrogen atoms are not shown in either figure.

1. The first line gives the abbreviated international Herman-Mauguin and Schoenflies symbols of the space group, the point group symbol and the crystal system.

2. The second line gives the sequential number of the space group, the full international Herman-Mauguin symbol and the so called Patterson symmetry (this corresponds to the point group symbol, preceded by the lattice symbol and we shall not consider it further). The short and full versions of the symbols differ only in some space groups. While in the short symbols symmetry axes are suppressed as much as possible, in the full symbols, axes and planes are listed for each direction.

3. When needed, the third line gives the chosen unique axis, settings, origin choices or reference axes.

4. Two types of space group diagrams (as orthogonal projections along a cell axis) are given: one shows the arrangement of the symmetry elements and the other, the position of a set of symmetry equivalent points. Close to the graphical symbols of a symmetry plane or axis parallel to the projection plane, the 'height' is printed as h, a fraction of the shortest lattice translation normal to the projection plane (unless $h = 0$). For some space groups, the diagrams illustrating the arrangement of the symmetry elements are shown in more

than one projection. In the most recent edition of the *Tables*, for the cubic groups, the second type of diagram is shown as a stereo pair.

5. Information is next given concerning the chosen origin position (see later), the definition of the asymmetric unit, the list of symmetry operations with their positions and the symmetry generators (the minimum set of symmetry operations of the group from which all the others are derived).

6. The block *positions* (called also *Wyckoff positions*) contain: (a) the *general position*, a set of symmetry equivalent points, each of which is left invariant only by application of the identity operation and (b) a list of *special positions*. A set of symmetry equivalent points is in a special position if each point is left invariant by at least two symmetry operations of the space group; this usually happens when the points are on a symmetry element or at the intersection of more than one symmetry element. For instance, a point on a three-fold rotation axis is left invariant not only by the identity, but also by the rotation itself. The first three columns of the block give the *multiplicity* (number of equivalent points per unit cell), the *Wyckoff letter* (a code scheme starting with *a* at the bottom position and continuing upwards in alphabetical order) and the *site symmetry* (the group of symmetry operations which leave the site invariant). The symbol adopted for describing the site symmetry has the same sequence of symmetry directions as the space group symbol and a dot marks the directions which do not contribute any element to the site symmetry. In the next column, the fractional coordinates of the Wyckoff positions are given. We note that, by definition, the general position always has site symmetry 1. For this position, each fractional coordinate is preceded by a number between parentheses, which is the same number preceding the corresponding symmetry operation listed in a previous block. For structures, such as molecular crystals, in which a formula unit is well defined, crystallographic papers give the number of 'molecules' in the unit cell (usually denoted by the letter Z). If the molecules, as in the case of the structure illustrated in Figure 6 (a), are in a general position, Z is equal to the multiplicity of the general position (in $P2_12_12_1$, Z = 4). In the structure shown in Figure 6 (b), the centre of the molecule lies on an inversion centre and the asymmetric unit is formed by half a molecule; Z is then equal to the multiplicity of the special position at site $\bar{1}$ (in $P2_1/c$, Z = 2), which is half that of the general position. We shall not consider the final column which contains the *reflection conditions*, which are important for diffraction experiments, nor describe the information given in the remaining blocks, which have not been reproduced in Figures 12 and 13.

6.1. Choice of origin

The characterization and description of crystal structures are greatly simplified by the choice of a suitable *origin*. In the *International Tables*, the following conventions are used.

- For centrosymmetric space groups, the origin is normally chosen on an inversion centre. An alternative choice is given if the space group possesses points of high site symmetry not coincident with the inversion centre. Thus for $Pn\bar{3}n$, two origin choices are considered: at $-\frac{1}{4}-\frac{1}{4}-\frac{1}{4}$ from the centre of $\bar{4}3m$ and at the centre of $\bar{3}m$.

- For non-centrosymmetric space groups, the origin is chosen at the point of highest site symmetry (e.g. the origin for $P\bar{4}c$ is chosen to be at $\bar{4}1c$) or at a point conveniently placed with respect to the symmetry elements. For example, on the 2_1 axis in $P2_1$, on the glide plane in Pc, at $1a2_1$ in $Pca2_1$ or at a point symmetrically surrounded by the three screw axes in $P2_12_12_1$.

6.2. The plane groups

In two dimensions, 17 plane groups may be defined. They are listed in Table 7.

Table 7. The 17 plane groups.

Oblique cell	p1, p2
Rectangular cell	pm, pg, cm, p3mm, p3mg, p2gg, c2mm
Square cell	p4, p4mm, p4gm
Hexagonal cell	p3, p3m1, p32m, p6, p6mm

In the symbols, the lattice letter is lower case and *g* is a glide reflection line. The 17 plane group diagrams are shown in Figure 15 and the ways by which they repeat a given motif are shown in Figure 16. It should be noted that any periodic decoration of a plane surface, such as a floor, wall or fabric, will belong to one of these plane groups. Some very beautiful examples are given in reference [10], where many drawings and paintings of the Dutch painter M.C. Escher are reproduced.

6.3. Matrix representation of symmetry operators

A symmetry operation acts on the fractional coordinates (x, y, z) of a point P to obtain the coordinates (x', y', z') of a symmetry equivalent point P'. In matrix notation, this transformation is written as

$$\mathbf{X}' = \begin{pmatrix} x' \\ y' \\ z' \end{pmatrix} = \begin{pmatrix} R_{11} & R_{12} & R_{13} \\ R_{21} & R_{22} & R_{23} \\ R_{31} & R_{32} & R_{33} \end{pmatrix} \begin{pmatrix} x \\ y \\ z \end{pmatrix} + \begin{pmatrix} T_1 \\ T_2 \\ T_3 \end{pmatrix} = \mathbf{RX} + \mathbf{T} \quad (22)$$

The matrix $\mathbf{R}$ is the *rotational component* (proper or improper) of the symmetry operation, while $\mathbf{T}$ is the matrix of the *translational component*. As an example, let us consider the space group *Ccca* (origin choice 1) shown in Figure 12. In the block 'positions', we can see that under the operation number (2) (a 2-axis)

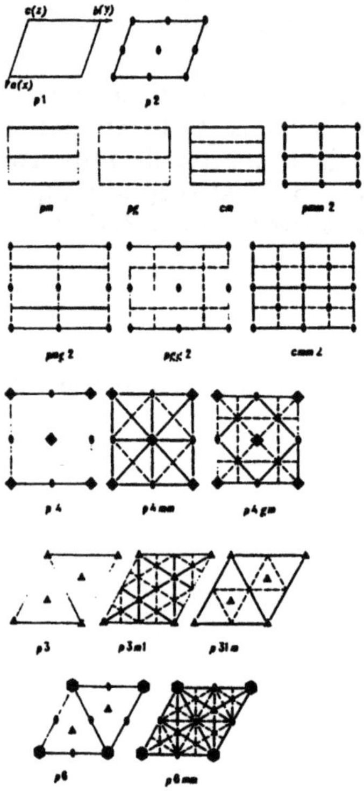

Figure 15. The 17 plane group diagrams.

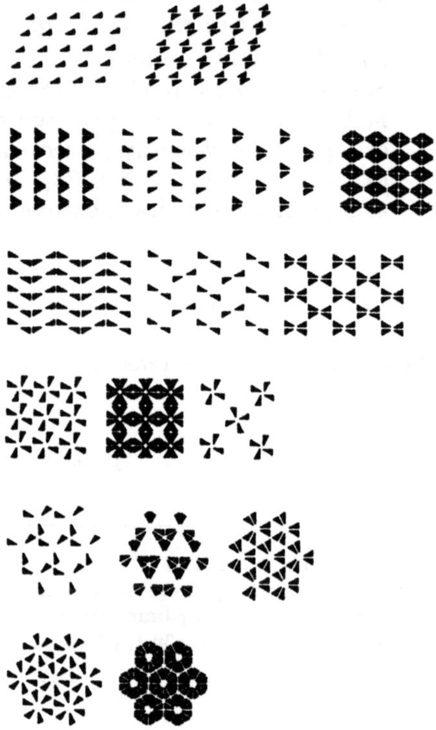

Figure 16. A simple motif repeated through the 17 plane groups.

a point at (x, y, z) is moved to $(-x + \frac{1}{2}, -y + \frac{1}{2}, z)$. Using matrix notation, this operation is represented as

$$\begin{pmatrix} -x + \frac{1}{2} \\ -y + \frac{1}{2} \\ z \end{pmatrix} = \begin{pmatrix} -1 & 0 & 0 \\ 0 & -1 & 0 \\ 0 & 0 & 1 \end{pmatrix} \begin{pmatrix} x \\ y \\ z \end{pmatrix} + \begin{pmatrix} \frac{1}{2} \\ \frac{1}{2} \\ 0 \end{pmatrix}$$

It may easily be seen that all rotation matrices have a determinant equal to ± 1 (the + sign corresponds to a proper rotation and the - sign to an improper rotation) and their elements have values 0, 1 and -1.

Acknowledgements
I would like to express my thanks to Prof. Piero Ugliengo for the helpful discussions and the critical reading of these notes.

References

1. *International Tables for Crystallography, Vol. A, Space group symmetry,* (T. Hahn, ed.), Kluwer, Dordrecht (1992).
2. C.J. Bradley and A.P. Cracknell, *The Mathematical Theory of Symmetry in Solids. Representation theory for Point and Space groups,* Clarendon Press, Oxford (1972).
3. E.H. Lockwood and R.H. MacMillan, *Geometric Symmetry,* Cambridge University Press (1978).
4. B.K. Vainshtein, *Modern Crystallography I: Symmetry of Crystals , Methods of Structural Crystallography,* Springer, Berlin (1981).
5. C. Giacovazzo, H.L. Monaco, D. Viterbo, F. Scordari, G. Gilli, G. Zanotti and M. Catti, *Fundamentals of Crystallography,* Oxford University Press (1992).
6. G. Rigault, *Introduzione alla Cristallografia,* Levrotto-Bella, Torino (1983).
7. E. Prince, *Mathematical Techniques in Crystallography and Materials Science,* Springer-Verlag, Berlin (1994).
8. D.E. Sands, *Vectors and Tensors in Crystallography,* Addison-Wesley, Reading MA (1982).
9. T. Hahn and H. Wondratschek, *Symmetry of Crystals,* Heron Press, Sofia (1994)
10. D. Schattschneider, *Visions of symmetry. Notebooks, periodic drawings, and related works of M.C. Escher,* W.H. Freeman, New York (1990).

References

[illegible]

The Language of Band Theory

Roberto Dovesi

Department of Inorganic, Physical and Materials Chemistry, University of Torino, via Giuria 5, I-10125 Torino, Italy

May 17, 1996

Summary. Reciprocal space vectors, Bloch Functions and periodic boundary conditions are introduced and discussed. The π bands of two-dimensional graphite are obtained with a model hamiltonian and are used to illustrate the effect of translational symmetry on the wave function.

Key words: Reciprocal Lattice – Brillouin Zone – Bloch Theorem – Bloch functions – Periodic Boundary conditions – Band Structure – Density of states

1. Introduction

In this chapter, some concepts of band theory are introduced with reference to the electronic states of a crystalline compound. Most of these concepts, which are related to the periodic nature of the solid, can be applied to the single particle spectrum obtained from other "parts" of the hamiltonian: for example, there is a vibrational *band structure* of the solid (phonons; see Chapter 12) and the magnetic properties of a crystal can be also described in terms of a spin hamiltonian whose eigenvalues are organized in *bands* [1, 2].

For a chemist, probably the most immediate intuition of a solid is as a *large molecule*, and actually many chemical properties can be understood with this model. Quantum mechanical approaches exist (see Chapter 13) for the investigation of *local* properties of solids (local defects, d spectrum of a transition metal in an ionic compound, core electron properties), which treat the solid as a molecule or a cluster of atoms perturbed by the *rest* of the infinite system. However, for most of the crystalline properties, it is necessary to take into account, explicitly, the infinite nature of the system. In a brute force extension of the familiar LCAO (Linear Combination of Atomic Orbitals) scheme to solids, one would need to consider a very large number of basis functions, and work with unmanageably large matrices in the solution of the eigenvalue problem. However, if translational symmetry is taken into account and Bloch functions (see below) are used to form representative sets, which are the bases for the irreducible representations (IRs) of the translation group (TG), the problem is factorized into n problems of dimension m, where n is the number of IRs of the TG, and m the number of basis functions within the unit cell. In fact, a general theorem of quantum mechanics [3] states that matrix elements of totally symmetric operators (as the hamiltonian is) between functions which are bases for different IRs (or different rows of the same IR) of the group are zero.

The calculated electronic band structure is the eigenvalue spectrum resulting from a single particle approach (within the Hartree-Fock, density functional theory or other formalisms) to the problem of the electronic wavefunction of a periodic solid. The difference with respect to the molecular case is due to the fact that symmetry plays such a fundamental role in solids that it is convenient (as discussed above) to identify each *crystalline orbital* CO (the equivalent of the *molecular orbital*, MO) with two labels, one ($\mathbf{k}$) for the IR of the translation group to which the CO belongs, and another (n) ordering the eigenvalues, $\varepsilon_n^{\mathbf{k}}$, related to the COs at given $\mathbf{k}$: $\psi_n^{\mathbf{k}}$.

In this chapter, we will discuss the relationship between the translational invariance of the hamiltonian and $\mathbf{k}$ vectors through the Bloch theorem, and explain why $\varepsilon_n^{\mathbf{k}}$ and $\psi_n^{\mathbf{k}}$ are continuous functions of $\mathbf{k}$, while remaining discrete with respect to n, as in the molecular case. The continuous nature of $\varepsilon_n^{\mathbf{k}}$ entails the problem of counting the number of states in a given energy interval; to this purpose, we will introduce the definition of Density of States (DOS); the vector nature of $\mathbf{k}$ requires a careful definition of the surface, separating, according to the Aufbau principle, empty and occupied states; the Fermi energy and the Fermi surface need to be defined. The infinite order of the translation group, which entails an infinite number of eigenvalues and eigenvectors, does not create unsolvable problems, because the above mentioned continuity in $\mathbf{k}$ space of $\varepsilon_n^{\mathbf{k}}$ and $\psi_n^{\mathbf{k}}(\mathbf{r})$ can be exploited by solving the secular equation for a suitable finite number of $\mathbf{k}$ points, and then interpolating the $\mathbf{k}$ dependent functions of interest (see Chapter 4 for details).

2. Reciprocal lattice

In the previous Chapter, the unit vectors of the direct ($\mathbf{a}$, $\mathbf{b}$, $\mathbf{c}$) and reciprocal ($\mathbf{a}^*$, $\mathbf{b}^*$, $\mathbf{c}^*$) lattices were introduced. We will use from now on the notation $\mathbf{a}_1$, $\mathbf{a}_2$, $\mathbf{a}_3$ instead of $\mathbf{a}$, $\mathbf{b}$, $\mathbf{c}$, and $\mathbf{b}_1$, $\mathbf{b}_2$, $\mathbf{b}_3$ for $2\pi\mathbf{a}^*$,$2\pi\mathbf{b}^*$,$2\pi\mathbf{c}^*$. We then have (see Chapter 1, eq. (13)):

$$\mathbf{b}_i \cdot \mathbf{a}_j = 2\pi\delta_{ij} \qquad \text{for} \qquad i,j = 1,3 \tag{1}$$

The scalar product of two vectors, $\mathbf{k} = \sum_i k_i \mathbf{b}_i$ and $\mathbf{r} = \sum_i r_i \mathbf{a}_i$, becomes

$$\mathbf{k} \cdot \mathbf{r} = 2\pi \sum_{i=1}^{3} k_i r_i \tag{2}$$

The reciprocal lattice can be generated by analogy with the direct lattice through the relation:

$$\mathbf{K}_m = \sum_{i=1}^{3} m_i \mathbf{b}_i \quad (3)$$

where the m_i are integers (positive, negative or zero). The difference between $\mathbf{k}$ and $\mathbf{K}_m$ must be noted: k_i coefficients are not required to be integers.

With our choice of the length of the reciprocal unit vectors, it may be immediately verified that:

$$exp(i\mathbf{K} \cdot \mathbf{T}) = 1 \quad (4)$$

for all couples of reciprocal ($\mathbf{K}$) and direct ($\mathbf{T}$) lattice vectors.

When the vectors $\mathbf{a}_i$ are orthogonal, equation (1) is satisfied if each $\mathbf{b}_i$ is parallel to $\mathbf{a}_i$, and has a length $2\pi / \mid \mathbf{a}_i \mid$, whence the name *reciprocal vector.* The relation between $\mathbf{a}_i$ and $\mathbf{b}_i$ vectors in the case of two-dimensional hexagonal graphite is illustrated in figure 1. Suppose we are now interested in finding all

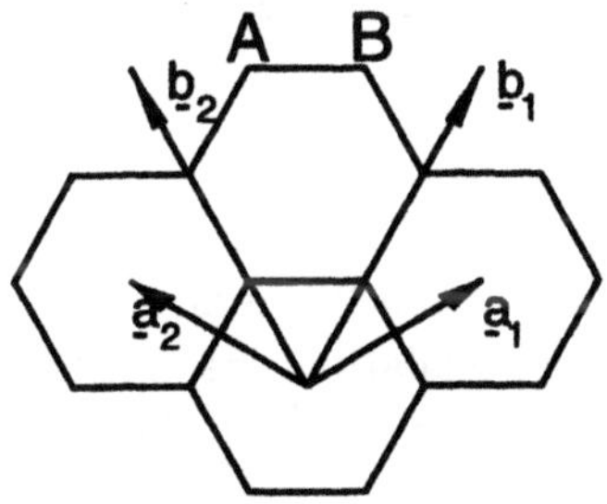

Figure 1. Direct ($\mathbf{a}_1$ and $\mathbf{a}_2$) and reciprocal ($\mathbf{b}_1$ and $\mathbf{b}_2$) lattice vectors of two-dimensional graphite.

the $\mathbf{k}$ vectors which generate different values for the function $\exp(i\mathbf{k} \cdot \mathbf{T})$, where $\mathbf{T}$ is any direct lattice vector. It is easily seen from equation (4) that vectors $\mathbf{k}'$ and $\mathbf{k}$ which differ by a reciprocal lattice vector $\mathbf{K}$ give the same value for the exponential function, that is, they are *equivalent.* Only the points in the *zero* (or *reference*) reciprocal cell (or any other cell obtained from it by translation) need then to be considered. An alternative and totally equivalent choice which is frequently used refers to the so-called *First Brillouin Zone*, or simply the Brillouin Zone (BZ), whose points, $\boldsymbol{\kappa}$, have the property of being at the shortest distance from the origin among the infinite equivalent points $\mathbf{k}=\boldsymbol{\kappa}+\mathbf{K}$, with $\mathbf{K}$ running over all the reciprocal lattice vectors (see Chapter 1, Section 5.3). The BZ of the two-dimensional graphite lattice is shown in figure 2.

3. The Bloch theorem

Let $\widehat{T}$ be a translation operator, associated with the Bravais lattice vector $\mathbf{T}$, that shifts the argument of any function, $F(\mathbf{r})$, by $\mathbf{T}$:

$$\widehat{T}F(\mathbf{r}) = F(\mathbf{r} + \mathbf{T}) \quad (5)$$

If $F(\mathbf{r})$ is the product of the Hamiltonian, $\widehat{H}(\mathbf{r})$, times a generic function, $\psi(\mathbf{r})$, we obtain:

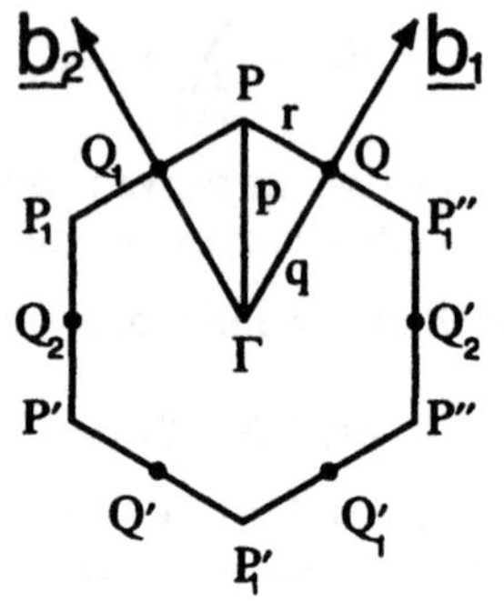

Figure 2. First Brillouin Zone (BZ) of two-dimensional graphite. Symbols for the special points (capital letters) and lines (lower case letter) are indicated. The irreducible part of the BZ (IBZ) is the triangle Γ-P-Q (or any other triangle obtained from it using one of the point symmetry operators). Q′, Q_1', Q_2' are obtained by translating Q, Q_1 and Q_2; Q_1 and Q_2 are obtained from Q by a rotation of $\pi/3$ and $2\pi/3$.

$$\widehat{T}\widehat{H}(\mathbf{r})\psi(\mathbf{r}) = \widehat{H}(\mathbf{r}+\mathbf{T})\psi(\mathbf{r}+\mathbf{T}) = \widehat{H}(\mathbf{r})\psi(\mathbf{r}+\mathbf{T}) = \widehat{H}(\mathbf{r})\widehat{T}\psi(\mathbf{r}) \tag{6}$$

where the second step follows from the periodic nature of $\widehat{H}$ (the kinetic energy operator is invariant to any translation; the potential operator, $\widehat{V}(\mathbf{r})$, is periodic by hypothesis). Equation (6) must hold for *any* ψ function at any $\mathbf{r}$ point; which implies that:

$$\widehat{T}\widehat{H} = \widehat{H}\widehat{T} \tag{7}$$

that is, $\widehat{H}$ commutes with the space (as well as the point) symmetry operators of the space group. A fundamental theorem of quantum mechanics [4] states that, when two operators commute, it is possible to choose the eigenvectors of the first to be also eigenvectors of the second (if the levels are not degenerate, the theorem simply states that the eigenvectors of the first operator are eigenvectors also of the second):

$$\widehat{H}\psi = E\psi \tag{8}$$

$$\widehat{T}\psi = \gamma(\mathbf{T})\psi \tag{9}$$

Let us consider now the application of two translation operators to the function ψ :

$$\widehat{T}\widehat{T'}\psi(\mathbf{r}) = \widehat{T}\psi(\mathbf{r}+\mathbf{T}') = \psi(\mathbf{r}+\mathbf{T}+\mathbf{T}') = \widehat{T'}\widehat{T}\psi(\mathbf{r}) = \widehat{T''}\psi(\mathbf{r}) \tag{10}$$

where $\widehat{T''}$ is the translation operator associated with the lattice vector: $\mathbf{T}'' = \mathbf{T} + \mathbf{T}'$. Therefore, the translation operators commute and the group is Abelian [4]:

$$\widehat{T}\widehat{T'} = \widehat{T'}\widehat{T} = \widehat{T''} \tag{11}$$

and then, combining (9) and (11), we get:

$$\widehat{T}\widehat{T'}\psi(\mathbf{r}) = \gamma(\mathbf{T})\gamma(\mathbf{T}')\psi(\mathbf{r}) = \gamma(\mathbf{T}'')\psi(\mathbf{r}) = \gamma(\mathbf{T}+\mathbf{T}')\psi(\mathbf{r}) \tag{12}$$

whence

$$\gamma(\mathbf{T}+\mathbf{T}') = \gamma(\mathbf{T})\gamma(\mathbf{T}') \tag{13}$$

The general Bravais lattice vector $\mathbf{T_n}$ is a linear combination of the primitive vectors of the Bravais lattice with integer coefficients:

$$\mathbf{T_n} = n_1\mathbf{a}_1 + n_2\mathbf{a}_2 + n_3\mathbf{a}_3 \tag{14}$$

Equation (13) then takes the form:

$$\gamma(\mathbf{T_n}) = \gamma(\mathbf{a}_1)^{n_1}\gamma(\mathbf{a}_2)^{n_2}\gamma(\mathbf{a}_3)^{n_3} \tag{15}$$

With no loss of generality, the number $\gamma(\mathbf{a}_j)$ can be written in exponential form:

$$\gamma(\mathbf{a}_j) = \exp(2\pi i k_j) \tag{16}$$

Equation (15) then becomes:

$$\gamma(\mathbf{T_n}) = \exp[2\pi i(n_1k_1 + n_2k_2 + n_3k_3)] \tag{17}$$

If we define a reciprocal vector $\mathbf{k}$ with components k_i:

$$\mathbf{k} = k_1\mathbf{b}_1 + k_2\mathbf{b}_2 + k_3\mathbf{b}_3 \tag{18}$$

and remember the property given in eq. (1), then eq. (17) can be written in a more compact form:

$$\gamma(\mathbf{T}) = \exp(i\mathbf{k}\cdot\mathbf{T}) \tag{19}$$

Substituting back (19) into (9), we obtain:

$$\widehat{T}\psi^{\mathbf{k}}(\mathbf{r}) = \psi^{\mathbf{k}}(\mathbf{r}+\mathbf{T}) = \exp(i\mathbf{k}\cdot\mathbf{T})\psi^{\mathbf{k}}(\mathbf{r}) \tag{20}$$

which is the *Bloch theorem*. It must be noticed that eq (20) has been obtained for a one-electron hamiltonian of the form $\widehat{H} = \frac{-\nabla^2}{2} + V(\mathbf{r})$, under only one assumption: *the periodic nature of the potential.* It is easily generalized to any one-electron hamiltonian which includes, in some way, electron-electron interactions, such as the Fock or the Kohn-Sham hamiltonians (see Chapter 3), provided that the electronic structure reflects the symmetry of the lattice, as is normally the case for the ground state. Exceptions may be found in magnetic crystals, where it may be necessary to consider unit cells larger than the crystalline cell (see Chapter 14).

Each eigenfunction of the periodic Hamiltonian $\widehat{H}$ can then be labelled with $\mathbf{k}$, according to its translational properties (eq. 20). The $\mathbf{k}$ vector can be chosen in the BZ, because any $\mathbf{k}'$ external to the BZ can be written as $\mathbf{k}' = \boldsymbol{\kappa} + \mathbf{K}$, where $\boldsymbol{\kappa}$ is within the BZ and $\mathbf{K}$ is a reciprocal lattice vector;

$$\exp(i\mathbf{k}'\cdot\mathbf{T}) = \exp[i(\boldsymbol{\kappa}+\mathbf{K})\cdot\mathbf{T}] = \exp(i\boldsymbol{\kappa}\cdot\mathbf{T}) \tag{21}$$

The Bloch theorem can be cast in a different way; we can write:

$$\psi^{\mathbf{k}}(\mathbf{r}) = \exp(i\mathbf{k}\cdot\mathbf{r})w(\mathbf{r}) \tag{22}$$

where

$$w(\mathbf{r}) = \exp(-i\mathbf{k}\cdot\mathbf{r})\psi^{\mathbf{k}}(\mathbf{r}) \tag{23}$$

is immediately shown from eq. (20) to be a periodic function of $\mathbf{r}$:

$$\begin{aligned} w(\mathbf{r}+\mathbf{T}) &= \exp[-i\mathbf{k}\cdot(\mathbf{r}+\mathbf{T})]\psi^{\mathbf{k}}(\mathbf{r}+\mathbf{T}) = \\ &= \exp(-i\mathbf{k}\cdot\mathbf{r})\exp(-i\mathbf{k}\cdot\mathbf{T})\exp(i\mathbf{k}\cdot\mathbf{T})\psi^{\mathbf{k}}(\mathbf{r}) = w(\mathbf{r}) \end{aligned} \tag{24}$$

A function defined as in eq. (22) is called *Bloch function* (BF).

4. The periodic boundary conditions

Our model of the solid is as follows: the system is finite, but macroscopic; for simplicity, we can suppose that it is a parallelepiped containing $N = N_1 \cdot N_2 \cdot N_3$ unit cells, whose sides are $N_i \cdot \mathbf{a}_i$. The electrostatic potential, the atomic positions and the electron charge distribution close to the surface are different from those at the center of the finite crystal. However, due to the macroscopic nature of our system, the number of atoms perturbed by surface effects is a small fraction of the total number of atoms (in a cubic box of side a, if surface effects penetrate for Δx, the fraction of perturbed volume is $6a^2\Delta x/a^3$, which tends to zero when a tends to infinity). This argument requires further comment and a deeper discussion is given in ref.s [10, 11]; in fact, unscreened electric charges at the surface would create an electrostatic potential decaying very slowly ($1/r$) within the bulk; in this case, bulk properties would be influenced by the nature of the surface (charge and shape).

Under normal conditions, *we can expect the bulk properties* (electron charge distribution, energy per atom, single particle spectrum, vibrational frequencies, etc) *to be insensitive to the surface, and also to the boundary conditions imposed on the wavefunctions.* Among the many possible choices for the boundary conditions, the most convenient ones from a computational point of view are then chosen. The usual assumption is that the finite crystal is part of an infinite crystal, and it is delimited in a purely formal way. It is then assumed that the (single particle) wavefunction obeys the following condition *(Born-Von Karman or periodic boundary conditions)*:

$$\psi^{\mathbf{k}}(\mathbf{r} + N_j\mathbf{a}_j) = \psi^{\mathbf{k}}(\mathbf{r}) \qquad \text{for} \qquad j = 1,3 \tag{25}$$

If this condition is applied to the Bloch function, one gets (eq. 20):

$$\psi^{\mathbf{k}}(\mathbf{r} + N_j\mathbf{a}_j) = \exp(iN_j\mathbf{k} \cdot \mathbf{a}_j)\psi^{\mathbf{k}}(\mathbf{r}) = \psi(\mathbf{r}) \tag{26}$$

which implies:

$$\exp(iN_j\mathbf{k} \cdot \mathbf{a}_j) = 1 \tag{27}$$

This may be written, remembering the definition (18) and the property (1), as:

$$\exp(2\pi i N_j k_j) = 1 \tag{28}$$

which is satisfied if:

$$k_j = n_j/N_j \tag{29}$$

with n_j integer. The above equation shows that, *because of the periodic boundary conditions, vectors* **k** *are real*; the general $\boldsymbol{\kappa}$ vector inside the unit reciprocal cell is defined as:

$$\boldsymbol{\kappa} = n_1\mathbf{b}_1/N_1 + n_2\mathbf{b}_2/N_2 + n_3\mathbf{b}_3/N_3 \tag{30}$$

From equation (30), each $\boldsymbol{\kappa}$ point can be attributed a small parallelepiped with edges $\mathbf{b}_j/N_j$ and volume $\nu_{\boldsymbol{\kappa}}$:

$$\nu_{\boldsymbol{\kappa}} = (\mathbf{b}_1/N_1) \cdot (\mathbf{b}_2/N_2) \times (\mathbf{b}_3/N_3) = [\mathbf{b}_1 \cdot (\mathbf{b}_2 \times \mathbf{b}_3)]/N = V_{BZ}/N \tag{31}$$

The quantity in square brackets is the volume of the reciprocal lattice cell (or of the BZ, V_{BZ}), which is equal to $(2\pi)^3/V$ (where V is the volume of the direct lattice cell). This results from equation (1) and is obvious in the case of orthogonal $\mathbf{a}_j$ vectors. Equation (31) can therefore be recast as:

$$\nu_{\boldsymbol{\kappa}} = (2\pi)^3/(VN) \tag{32}$$

which shows that the number of $\boldsymbol{\kappa}$ points in the reciprocal primitive cell or in the BZ is equal to N, the number of unit cells in the crystal.

We can show, at this point, that $\varepsilon_n^{\boldsymbol{\kappa}}$ is continuous in $\boldsymbol{\kappa}$ and discrete in n. If we substitute eq. (22) in the Schrödinger equation (8) and eliminate the exponential factor from both sides, we obtain the following equation for the periodic function $w(\mathbf{r})$:

$$\left[\frac{1}{2}(-i\nabla + \boldsymbol{\kappa})^2 + U(\mathbf{r})\right] w^{\boldsymbol{\kappa}}(\mathbf{r}) = \varepsilon^{\boldsymbol{\kappa}} w^{\boldsymbol{\kappa}}(\mathbf{r}) \tag{33}$$

where $U(\mathbf{r})$ (the periodic potential) and $w^{\boldsymbol{\kappa}}(\mathbf{r})$ must have the same periodicity, which allows us to look at the above equation as referring to the unit cell volume. That is to say, eq. (33) is an hermitian eigenvalue problem referring to a ***finite volume***. On general grounds, such an eigenvalue problem is expected to have an infinite set of solutions, $w_n^{\boldsymbol{\kappa}}(\boldsymbol{r})$, with discretely spaced eigenvalues, $\varepsilon_n^{\boldsymbol{\kappa}}(\boldsymbol{r})$, labelled by n. On the other hand, the hamiltonian, in the above equation, contains $\boldsymbol{\kappa}$ as a continuous parameter, and therefore, again on the basis of general considerations, the solutions $w_n^{\boldsymbol{\kappa}}(\mathbf{r})$ and the corresponding eigenvalues vary with continuity as a function of $\boldsymbol{\kappa}$.

In many situations, $\mathbf{k}$–dependent functions must be summed over all the $\boldsymbol{\kappa}$ values within a portion of the BZ; in the limit of infinite N, the sum is replaced by an integral as follows:

$$\frac{1}{N}\sum_{\boldsymbol{\kappa}}{}' f(\boldsymbol{\kappa}) = \frac{1}{(N\nu_{\boldsymbol{\kappa}})}\sum_{\boldsymbol{\kappa}}{}' f(\boldsymbol{\kappa})\nu_{\boldsymbol{\kappa}} \longrightarrow \frac{1}{V_{BZ}}\int' f(\boldsymbol{\kappa})d\boldsymbol{\kappa} \tag{34}$$

where the prime indicates that the sum (integral) extends only to the points in the selected region of the BZ, and the weighting factor results from the discussion following equation (32).

5. Bloch functions

We can summarize the results of the two previous sections as follows:

a) the eigenvectors of the periodic hamiltonian must satisfy the Bloch theorem (eq. 20);
b) different eigenfunctions can satisfy the Bloch theorem for the same $\boldsymbol{\kappa}$ value; this implies that the eigenvectors of our hamiltonian are also labelled with an n index: $\psi_n^{\boldsymbol{\kappa}}(\mathbf{r})$. In group theory language, $\boldsymbol{\kappa}$ labels one of the (infinite, one-dimensional) IRs of the abelian translation group; $\psi_n^{\boldsymbol{\kappa}}(\mathbf{r})$ is a basis function of that representation;
c) there are as many solutions as there are non-equivalent $\boldsymbol{\kappa}$ points; the number of $\boldsymbol{\kappa}$ points is N, the number of cells in the crystal. As the crystal size increases, the $\boldsymbol{\kappa}$ points get closer and closer; at the limit of an infinite (and therefore, also for a finite but macroscopic) lattice, $\boldsymbol{\kappa}$ becomes a continuous

variable that can take on all the possible values within the BZ, and expression (34) can be used to perform all the summations that are needed.

In solid state computational chemistry and physics, the unknown single-particle crystalline wavefunction, $\psi_n^{\mathbf{k}}(\mathbf{r})$, is expanded in a finite set of BFs, $\phi_\mu^{\boldsymbol{\kappa}}(\mathbf{r})$:

$$\psi_n^{\mathbf{k}}(\mathbf{r}) = \sum_\mu c_{\mu n}^{\boldsymbol{\kappa}} \phi_\mu^{\boldsymbol{\kappa}}(\mathbf{r}) \tag{35}$$

The coefficients, $c_{\mu n}^{\boldsymbol{\kappa}}$, are determined variationally by solving the set of coupled matrix equations (see eq. 26 in Chapter 3):

$$H^{\boldsymbol{\kappa}} C^{\boldsymbol{\kappa}} = S^{\boldsymbol{\kappa}} C^{\boldsymbol{\kappa}} E^{\boldsymbol{\kappa}} \tag{36}$$

$$(C^{\boldsymbol{\kappa}})^+ S^{\boldsymbol{\kappa}} C^{\boldsymbol{\kappa}} = I \tag{37}$$

where $H^{\boldsymbol{\kappa}}$ is the hamiltonian matrix in the basis set of the $\phi_i^{\boldsymbol{\kappa}}(\mathbf{r})$ functions; $S^{\boldsymbol{\kappa}}$ is the overlap matrix among these functions ($S^{\boldsymbol{\kappa}} = I$ if the basis functions are orthogonal); $C^{\boldsymbol{\kappa}}$ is the matrix of the variational coefficients, $c_{\mu n}^{\boldsymbol{\kappa}}$; and $E^{\boldsymbol{\kappa}}$ is the diagonal matrix of the single particle eigenvalues, $\varepsilon_n^{\boldsymbol{\kappa}}$. Two basic types of BFs are used for the expansion (35):

a) localized functions, or *atomic orbital* (AO) based BFs:

$$\phi_\mu^{\boldsymbol{\kappa}}(\mathbf{r}) = \sum_{\mathbf{T}} \exp(i\boldsymbol{\kappa} \cdot \mathbf{T}) \varphi_\mu(\mathbf{r} - \mathbf{A}_\varphi - \mathbf{T}) \tag{38}$$

where μ labels the AOs in the unit cell, and A_φ is the position (in the reference or zero cell) of the atom on which the atomic orbital φ is centered. The AOs are usually linear combinations of the products of Gaussian functions by real, solid harmonics [10, 12, 13]; Gaussian functions are usually preferred to Slater functions because of their analytical properties. It is easy to see that the function defined in equation (38) satisfies Bloch's theorem:

$$\begin{aligned} \phi_\mu^{\boldsymbol{\kappa}}(\mathbf{r}+\mathbf{T}) &= \sum_{T''} \exp(i\boldsymbol{\kappa} \cdot \mathbf{T}'') \varphi_\mu(\mathbf{r} - \mathbf{A}_\varphi - \mathbf{T}'' + \mathbf{T}) \\ &= \sum_{\mathbf{T}'} \exp[i\boldsymbol{\kappa} \cdot (\mathbf{T}' + \mathbf{T})] \varphi_\mu(\mathbf{r} - \mathbf{A}_\varphi - \mathbf{T}') \\ &= \exp(i\boldsymbol{\kappa} \cdot \mathbf{T}) \phi_\mu^{\boldsymbol{\kappa}}(\mathbf{r}) \end{aligned} \tag{39}$$

where $\mathbf{T}' = \mathbf{T}'' - \mathbf{T}$; the summation index has been changed from $\mathbf{T}''$ to $\mathbf{T}'$ because all the lattice vectors are spanned in both cases.

b) Plane waves:

$$\phi_{\mathbf{K}}^{\boldsymbol{\kappa}}(\mathbf{r}) = \exp[i\mathbf{r} \cdot (\boldsymbol{\kappa} + \mathbf{K})] \tag{40}$$

The plane waves also satisfy Bloch's theorem:

$$\begin{aligned} \phi_{\mathbf{K}}^{\boldsymbol{\kappa}}(\mathbf{r}+\mathbf{T}) &= \exp[i(\mathbf{r}+\mathbf{T}) \cdot (\boldsymbol{\kappa} + \mathbf{K})] = \exp(i\boldsymbol{\kappa} \cdot \mathbf{T}) \exp[i\mathbf{r} \cdot (\boldsymbol{\kappa} + \mathbf{K})] \\ &= \exp(i\boldsymbol{\kappa} \cdot \mathbf{T}) \phi_{\mathbf{K}}^{\boldsymbol{\kappa}}(\mathbf{r}) \end{aligned} \tag{41}$$

In both equations (38) and (41), normalization factors have been disregarded. It should be appreciated that there is a very large difference between the two kinds of basis functions; for example, the number of AO-based BFs that can be built does not depend on the size of the unit cell, but rather on the number of atoms in it; the number of PW–based BFs, below a given kinetic energy cutoff (see Chapters 4 and 6), on the contrary, depends on the unit cell size and shape, but not (explicitly) on the number or type of atoms in the unit cell. Another important difference is that the plane waves $\phi_{\mathbf{K}}^{\boldsymbol{\kappa}}$ and $\phi_{\mathbf{K}'}^{\boldsymbol{\kappa}}$ are orthogonal, whereas AO-based BFs are not.

6. Band Structure: general comments

Examples of the electronic band structure of a metal (aluminum), a semiconductor (silicon) and an ionic compound (NaCl) obtained with the Local Density Approximation (LDA) approach are reported in Chapter 10. In the NaCl case, the bands are very flat (very small *dispersion*, that is, very small dependence on the $\boldsymbol{\kappa}$ vector), and the *band gap* (energy difference between the highest occupied and lowest empty eigenstate) very large; at the opposite extreme, in the metallic case, the dispersion is large and there is continuity between valence and conduction (*lowest non-occupied*) bands. The Si band structure is intermediate (*semiconductor*). The reported band structures refer to two lines only in the First Brillouin Zone, $\Gamma - X$ and $\Gamma - L$ (see fig. 3), which contain, however, the most significant points; for each band, the eigenvalue at a generic $\boldsymbol{\kappa}$ point will fall within the eigenvalues at the special points. The notation for the special points

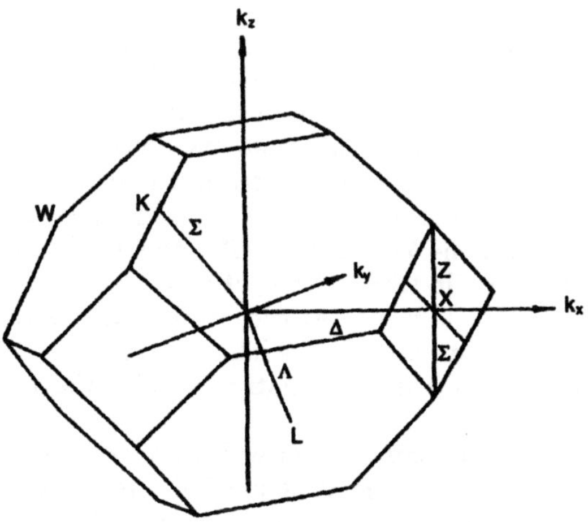

Figure 3. First Brillouin Zone for the face-centered cubic lattice. Points and lines of symmetry are indicated. Note in particular that X and L are at the center of the square and hexagonal faces, respectively.

is very old [5]; for the simplest lattices it can be found, for example in Slater's [6] and Callaway's [7] books. At Γ and L the band structure presents degeneracies, which are a consequence of point symmetry (see, for example, refs. [8, 9] for a discussion of these aspects).

The general features of the bands obtained with the Hartree-Fock (HF) and LDA hamiltonians are very similar, the most evident difference being the energy scale: the HF bandwidths and bandgaps are always larger (by a factor 2 to 4) than the LDA ones. This difference is not related to the use of a finite basis set or to some numerical inaccuracy, but depends in an essential way on the differences between the HF and Kohn-Sham hamiltonians. The question "Which is the more correct: the HF or LDA band structure?" entails another question: "Can we compare the band structure to some physical observable, in particular to the optical spectrum?". For a long time, the quality of a periodic *ab initio* calculation has been judged on the basis of the agreement with the *experimental* band structure. However, the HF and LDA eigenvalues can be related only very vaguely (see comments on this point in Chapter 3) to this quantity, which requires an appropriate description of both the excited and the ground states. In spite of these limitations, the calculated band structure is a useful qualitative tool for the description of the electronic structure of crystalline systems. In Chapter 13, for example, the band structure is used to describe the modifications resulting from cutting the infinite crystalline system and creating a surface.

In many situations, the HF and LDA approaches give very similar ground state properties, in spite of large quantitative differences in the band structure. The reasons for this are:

a) The ground state properties involve only the occupied single particle states.
b) In the sum over the occupied crystalline orbitals, to be performed for calculating any ground state property, the numerical value of the eigenvalue is not involved.
c) The total energy of the system is NOT equal to the sum of the occupied $\varepsilon_n^{\boldsymbol{\kappa}}$.

In other situations, the two methods provide a qualitatively different description of the occupied manifold that entails also large differences in the ground state electronic properties: systems which are insulators at the HF level and metallic at the LDA level; transition metal insulators which show a different energy ordering in the d states, and then differences in the magnetic properties. As a last comment, we remember that the *ab initio* techniques described in Chapters 8, 9, 10 and on which the WIEN, PWSCF and CRYSTAL codes are based, are self-consistent schemes, in which the solution at cycle n is used to build the effective hamiltonian at cycle $n+1$. At each cycle, the effective hamiltonian is then diagonalized at a given set of $\boldsymbol{\kappa}$ points, the number of occupied energy levels is determined, the new density matrix is built, the new hamiltonian is calculated, and so on.

There are two aspects in this process which are related to the present discussion: a) how the $\boldsymbol{\kappa}$ points in the Brillouin Zone are chosen; and b) how the eigenvalues at a given $\boldsymbol{\kappa}$ point are occupied. As regards the first point, which will be discussed at length later on (in Chapter 4, in particular), we simply recall here that, in general, a commensurate net of non-equivalent $\boldsymbol{\kappa}$ points in the BZ is used; the integration over the BZ is then replaced by a weighted sum; the number of terms in the sum necessary to give reasonable results is determined by using nets with increasing numbers of points, and monitoring the convergence of the most important properties, *in primis*, the total energy. As regards the second point, we recall once again that the total energy of the system is not simply the sum of the eigenvalues: at each cycle of the SCF process, the total energy resulting from filling the eigenvalues with lowest energy, according to the Aufbau principle, is in general lower than that resulting from a different filling scheme; this is true, in particular, when the energy gap is large; when, however, the gap reduces to a fraction of an eV, it could happen that ***anti-Aufbau*** filling would produce a lower energy than the ***canonical*** Aufbau scheme.

7. Band Structure: a simple example

In this section, we consider a very simple example, the π bands of two-dimensional graphite, in order to illustrate some of the features of the band structure. We will consider BFs built on atomic orbitals (s,p) centered on the two atoms of the unit cell (see figure 1). The s, p_x and p_y AOs have the appropriate symmetry for building states symmetric with respect to the graphite plane (σ states), whereas p_z AOs contribute to π states. At any given κ point, π states are a linear combination of BFs with the appropriate symmetry, that is, in our case, built on the p_z AOs. In a minimal basis set scheme, at each κ point we can build two BFs of π symmetry, one with the p_z AOs centered on atom A, the other on atom B (see figure 1). As discussed in the previous section, it is sufficient to focus on special points which, in the graphite case (see figure 2), are Γ, Q and P.

$$\kappa = \Gamma = (0,0)$$

At this special point, BFs reduce to a sum of p_z functions, each centered on atom A (ϕ_A^Γ) or B (ϕ_B^Γ). In figure 4, the structure of ϕ_B^Γ is shown.
If we adopt a simple Extended Hückel type hamiltonian, we can also estimate the

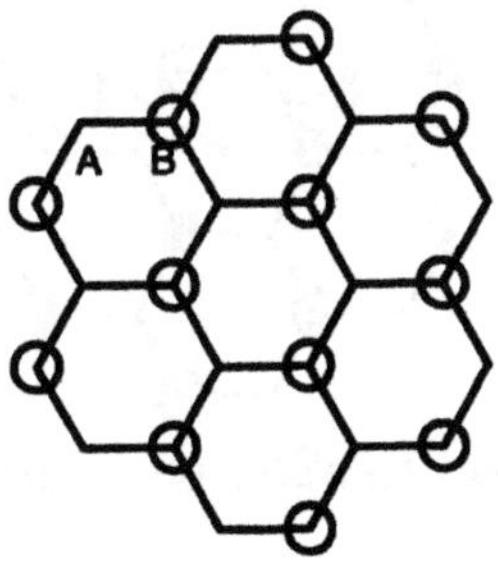

Figure 4. Top view of the graphite two dimensional lattice and of $\phi_B^\Gamma(\mathbf{r})$, the BF at $\kappa = \Gamma$ built from the p_z atomic orbitals centered on atoms of type B.

relative strength of the interaction between the two BFs at the different κ points. According to this model, in the AO basis, only diagonal and first neighbour matrix elements must be taken into account (the values being indicated as α and β respectively); next-nearest neighbours and farther interactions are disregarded. The other assumption of the method is that basis functions are orthogonal, so that equation (36) simplifies to the case S=I. We must then evaluate the 2×2 Hückel matrix, with the above prescriptions. Let us begin with H_{AA}:

$$H_{AA}^\Gamma = \frac{1}{N}\sum_{\mathbf{T}}\sum_{\mathbf{T}'}\langle\varphi_A(\mathbf{r}-\mathbf{S_A}-\mathbf{T}')|\hat{\mathbf{h}}|\varphi_A(\mathbf{r}-\mathbf{S_A}-\mathbf{T})\rangle$$

$$= \sum_{\mathbf{T}} \langle \varphi_A(\mathbf{r} - \mathbf{S_A}) | \hat{\mathbf{h}} | \varphi_A(\mathbf{r} - \mathbf{S_A} - \mathbf{T}) \rangle = \alpha$$

where the N^{-1} factor comes from the $N^{-1/2}$ normalization factor of each BF, the second step is justified because the $\mathbf{T}$ summations extend to all lattice vectors; the last step is due to the fact that the first neighbours of the central A atom are of type B, meaning that only the diagonal ($\mathbf{T} = 0$) contribution is different from zero. Operating in the same way for the off-diagonal elements, we obtain $H^{\Gamma}_{AB} = 3\beta$ (each atom A has three B neighbours). We need then to diagonalize the following matrix:

$$\begin{pmatrix} \alpha & 3\beta \\ 3\beta & \alpha \end{pmatrix}$$

whose eigenvalues are $\alpha \pm 3\beta$. The splitting of the two bands is 6β. Since, in this case, we have one π electron per atom, only the lowest state is occupied.
At the point Q:

$$\boldsymbol{\kappa} = Q = (1/2, 0)$$

The exponential factor in eq. 38 becomes ($\mathbf{T} = m\mathbf{a}_1 + n\mathbf{a}_2$):

$$\exp(im\pi) = \left\{ \begin{array}{rl} 1 & \text{if } m \text{ is even} \\ -1 & \text{if } m \text{ is odd} \end{array} \right\}$$

Note that at Q, as at Γ, the BFs are real. The BF centered on atom B is shown in figure 5. The BF on atom A, or those for the point $Q_1 = (0, 1/2)$ (see figure 2)

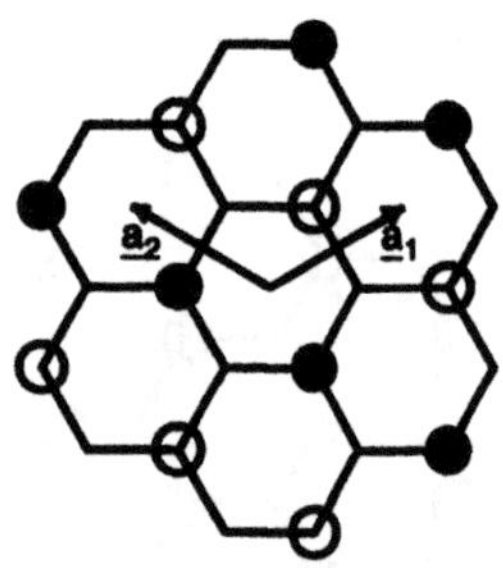

Figure 5. Same as figure 4 for $\phi^Q_B(\mathbf{r})$; Black circles indicate negative values.

can be easily obtained by translation and a rotation through $\pi/3$. The matrix elements of the Hückel matrix are easily evaluated: $H^Q_{AA} = \alpha$, because, as in the previous case, only the $\mathbf{T} = \mathbf{0}$ term is different from zero. For H^Q_{AB}, the situation is similar to that at Γ, but, as figure 5 shows, we have two positive and one negative contribution, so that the resulting value is β. The two eigenvalues are then $\alpha \pm \beta$. If β is a negative number (simulating attractive terms), at Q the first eigenvalue is higher and the second is lower than at Γ.
At P:

$$\boldsymbol{\kappa} = P = (1/3, 1/3)$$

Here, the exponential factor does not reduce to a real function, and the BF is now complex, so that in figure 6 both the real and the imaginary parts are represented. The exponential factor:

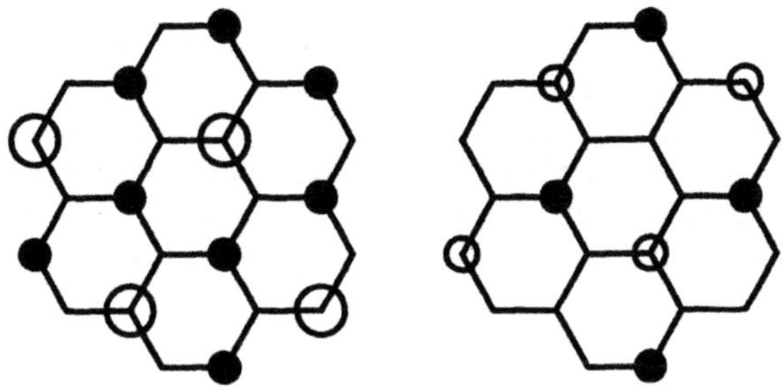

Figure 6. Top view of the real (left) and imaginary (right) parts of $\phi_B^P(\mathbf{r})$, the BF at κ=P built from the p_z atomic orbitals centered on atoms of type B.

$$\exp[2\pi i(m+n)/3) = \cos[2\pi(m+n)/3] + i\sin[2\pi(m+n)/3]$$

multiplies the p_z functions by $-0.5, -0.5$ and 1.0 (the real part) and by $t/2, -t/2$ and 0.0 (the imaginary part, where t is the square root of three), for $m+n$ equal to $3s+1$, $3s+2$ and $3s$ respectively. In the figure, small circles indicate AOs bearing the 0.5 or $t/2$ factors; black circles indicate negative values. As regards the matrix elements: in $H_{AA}^P = \langle\varphi_A^P(\mathbf{r})|\widehat{\mathbf{h}}|\varphi_A^P(\mathbf{r})\rangle$, the exponential factor cancels for the $\mathbf{T}=\mathbf{0}$ terms (note that the function on the left is the complex conjugate of that on the right); for $\mathbf{T} \neq \mathbf{0}$ the matrix elements between AOs are zero, as at Q and Γ, so that also in this case $H_{AA}^P = \alpha$. For H_{AB}^P there are three contributions to the real part and to the imaginary part, each of which sum to zero: $-0.5\beta + (-0.5\beta) + \beta = 0$ and $-0.5\beta + 0.5\beta + 0\beta = 0$, respectively; at P, the matrix is then already diagonal, and therefore, the eigenvalues are degenerate and equal to α.

The Hartree-Fock band structure of graphite, resulting from an *all electron* calculation (1s, 2s and 2p AOs in the basis set), is shown on the left of figure 7; π bands are qualitatively similar to those discussed for our simple model where the amount of dispersion depends on the β value; small β values give flat bands. In order to remove the degeneracy at the P point, which is not an artifact of

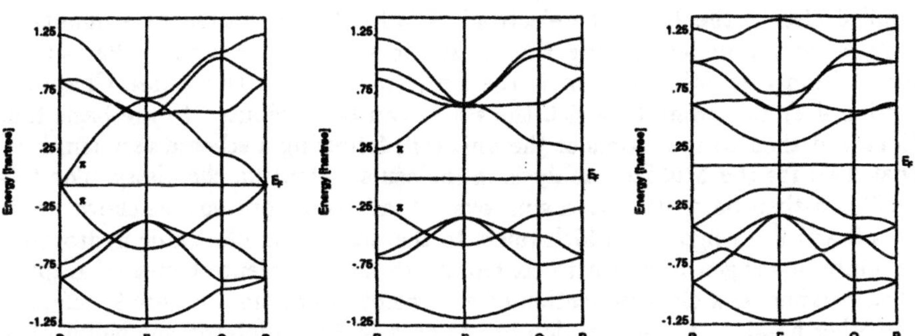

Figure 7. Hartree–Fock band structure of graphite (left), hexagonal BN (center) and the system H+graphite (right), where hydrogen is adsorbed on top of each carbon atom.

the Hückel or Hartree-Fock models nor of the limited basis set, but is a result of symmetry constraints, one of the following two perturbations, with respect to the above situation, are necessary: a) two different atoms are used, for example N and B instead of the two C atoms, as shown in figure 7 (center), where the HF band structure of hexagonal BN is reported; b) atoms are added above and/or below the layer, as for example in three-dimensional graphite or when an atom (H, for example) is adsorbed on top of the C atoms; in this case, positive and negative interactions are allowed and the degeneracy is removed, as shown on the right in figure 7.

8. Density of States

We must often calculate weighted sums over the electronic single particle levels:

$$S = 2\sum_{n,\boldsymbol{\kappa}} f(n;\boldsymbol{\kappa}) \tag{42}$$

If $f(n;\boldsymbol{\kappa})$ depends on n and $\boldsymbol{\kappa}$ only through the eigenvalues, $\varepsilon_n^{\boldsymbol{\kappa}}$, we can write:

$$S = 2\sum_{n,\boldsymbol{\kappa}} f(\varepsilon_n^{\boldsymbol{\kappa}}) = 2\sum_{n,\boldsymbol{\kappa}} \int f(\varepsilon)\delta(\varepsilon - \varepsilon_n^{\boldsymbol{\kappa}})d\varepsilon = \int f(\varepsilon)n(\varepsilon)d\varepsilon \tag{43}$$

where

$$n(\varepsilon) = 2\sum_{n,\boldsymbol{\kappa}} \delta(\varepsilon - \varepsilon_n^{\boldsymbol{\kappa}}) = (2/V_{BZ})\sum_{n} \int' \delta(\varepsilon - \varepsilon_n^{\boldsymbol{\kappa}})d\boldsymbol{\kappa} \tag{44}$$

is the density of states (DOS), δ is the Dirac *delta* function, and the $\boldsymbol{\kappa}$ integral extends over the BZ; the energy integral often extends to the energy interval $-\infty < \varepsilon < \varepsilon_F$. The number of electrons in the unit cell is given by:

$$\int_{-\infty}^{\varepsilon_F} n(\varepsilon)d(\varepsilon) \tag{45}$$

The comments concerning the physical meaning of the band structure can be applied also to the DOS; the electronic DOS is to be considered a qualitative instrument for understanding the electronic structure of a solid. Particularly useful from this point of view is the Projected DOS (PDOS) (see Chapter 3, section 4.2), according to which the states can be attributed to the basis functions and then to the atoms of the unit cell, following a scheme very similar to that used for the Mulliken analysis in molecular quantum chemistry. The total DOS can then be written as a sum over atomic contributions, as shown in figures 8 and 9. In figure 8, which refers to the adsorption of atomic hydrogen on graphite, hydrogen is shown to contribute to all the valence bands of graphite; this "mixing" is a clear indication of a covalent interaction. Figure 9, refers, on the other hand, to the adsorption of CO on a MgO monolayer; the bands of the adsorbate and of the substrate are completely separate; the dispersion of both is very small; the only difference between the superposition of the bands of the non-interacting and of the interacting systems is a modest shift of the adsorbate bands. This is a very clear example of a purely electrostatic interaction.

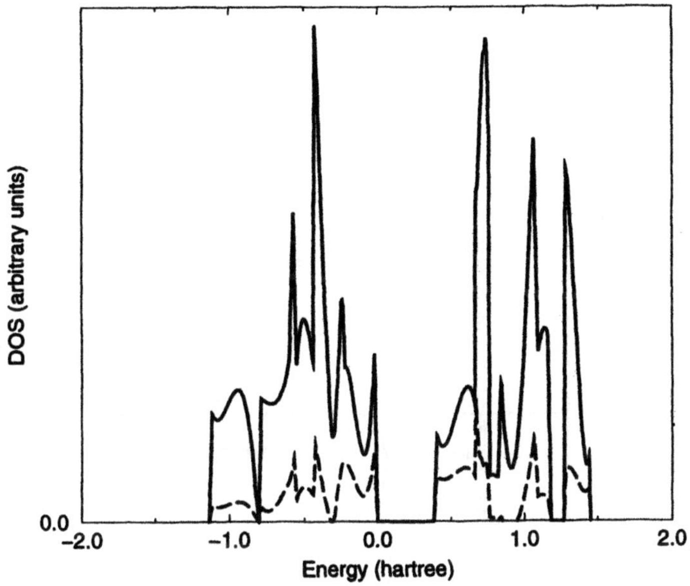

Figure 8. Density of states spectra for H on graphite (on top 1:1 adsorption). The higher contour corresponds to the total density, the area below the broken line to the hydrogen contribution. Valence and conduction bands are at negative and positive energies, respectively.

9. Fermi Energy and Fermi surface

In a closed shell model, each energy band can allocate $2 \times N$ (N is the number of cells) electrons; if in the unit cell there are n electrons, and the bands do not cross, the lowest $n/2$ bands are occupied and are separated from the empty bands; however, if n is odd, or if the valence bands cross, more than $n/2$ bands are partially occupied. At each cycle of the SCF process, an energy ε_F (the Fermi energy) must be determined, such that the number of one electron levels with energy below ε_F is equal to the number of electrons (or, in other words, the number of filled bands below ε_F is equal to half the number of electrons in the unit cell). The Fermi surface is the surface in reciprocal space, which satisfies the condition $\varepsilon_n^{\boldsymbol{\kappa}} = \varepsilon_F$ (see Chapter 4).

References

1. N.W. Ashcroft and N.D. Mermin, *Solid State Physics*, Holt, Rinehart and Winston, ch. 33 (1976).
2. H. Ibach and H.Lüth, *Solid State Physics*, Springer Verlag, ch. 8 (1991).
3. L. Landau and E. Lifchitz, *Mécanique Quantique*, Editions Mir, Moscou, p. 424 (1966).
4. N.W. Ashcroft and N.D. Mermin, *Solid State Physics*, Holt, Rinehart and Winston, p. 47 (1976).
5. L.P. Bouckaert, R. Smoluchowski and E. Wigner, Phys. Rev. **50**, 58 (1936).
6. J.C. Slater, *Quantum Theory of Molecules and Solids*, Vol. 2, McGraw-Hill (1965).
7. J. Callaway, *Quantum Theory of the Solid State*, Part A, Academic Press (1974).
8. M. Lax in *Symmetry Principles in Solid State and Molecular Physycs* Wiley (1974).

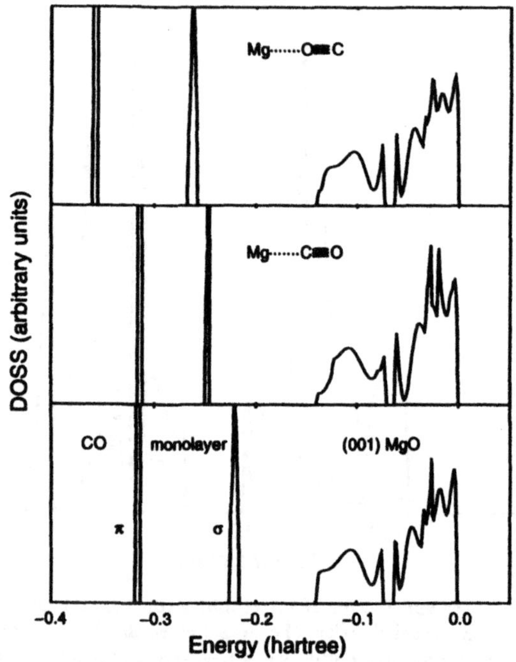

Figure 9. Valence density of states of the $CO + MgO$ system. In the bottom figure, the superposition of the density of states of the two non-interacting systems is reported; the other two figures refer to the vertical adsorption on Mg^{2+} in the two possible vertical orientations.

9. C. J. Bradley and A. P. Cracknell in *The Mathematical Theory of Symmetry in Solids*, Clarendon Press, Oxford (1972).
10. V. R. Saunders, C. Freyria Fava, R. Dovesi, L. Salasco and C. Roetti, Mol. Phys. **77**, 629-665 (1992).
11. F.E. Harris in *Theoretical Chemistry: Advances and Perspectives* , edited by H. Eyring and D. Henderson, Vol. 1 (Academic Press), p. 147.
12. V.R. Saunders in *Computational Techniques in Quantum Chemistry*, edited by G.H.F. Diercksen, B.T. Sutcliffe and A. Veillard (Reidel) p. 347 (1975).
13. C. Pisani, R. Dovesi and C. Roetti, *Lecture Notes in Chemistry* , vol 48, Springer, Heidelberg (1988).

Ab-Initio Approaches to the Quantum-Mechanical Treatment of periodic Systems

Cesare Pisani

Department of Inorganic, Physical and Materials Chemistry, University of Torino, via Giuria 5, I-10125 Torino, Italy

May 17, 1996

Summary. A general overview is proposed of the different strategies that are currently adopted for the *ab initio* study of the electronic structure of periodic systems.

Key words: Electron gas – Hartree-Fock (HF) – Electron Correlation – Relativistic Effects – Density-Functional Theory (DFT) – Kohn-Sham (KS) – Local Density Approximation (LDA) – Gradient Corrections (GC) – Self-Interaction Correction (SIC) – Pseudopotentials (PP) – Plane Waves (PW) – Gaussian Type Orbitals (GTO) – Lattice Sums – Self-Consistent-Field (SCF) – Size-consistency – Car-Parrinello (CP) – Montecarlo – KKR – Recursion Method – N-scaling

1. Introduction

The purpose of this Chapter is to provide an overview of the methods and computational tools currently adopted in the *ab initio* quantum-mechanical (QM) study of the chemical and physical properties of crystals, and more specifically: a) to make it easier for non-specialists to find their way in a disconcerting variety of proposals, each characterized by a different language, different algorithms, different topical problems; b) to give references for those who want to go deeper in certain subjects; c) to serve as an introduction to subsequent chapters of this book.

Given the chemical composition and the crystalline structure of a periodic system, the aim of *ab initio* computational methods is to calculate its chemical and physical properties as accurately as possible at a reasonable cost, without the need for empirical *a priori* information (although such information can be useful, see section 3.1). In principle, a QM description should be provided for all involved particles, that is, for all electrons and nuclei. In fact, the Born-Oppenheimer separation is universally adopted: the potential energy surface is first determined, corresponding to the fundamental level of the electronic subsystem for each given nuclear configuration; nuclear motion is studied *a posteriori*, by considering the electronic energy as an external potential [1]. In this Chapter and in most of the subsequent ones, we shall be concerned only with electronic properties, that is, those that depend on the electronic structure of the crystal. This, in turn, is a function of the geometry and the electric charge of the nuclei, which will be taken as fixed and arranged according to a crystalline lattice. The problem of nuclear motion is one of great relevance, because thermodynamic, acoustic and in part, transport properties of the system depend precisely on nuclear vibration or migration. Chapter 12 will discuss some aspects of this problem (see also section 5 of this Chapter).

Preliminary to any QM study of the electronic structure is the choice of the electronic hamiltonian, that is, one must decide which energetic contributions should be included and which expression should be adopted for them. We shall make reference in the following to the non-relativistic electrostatic Hamiltonian $\hat{H}^{nr}$. It includes the kinetic energy of electrons, their Coulomb interaction with nuclei and between themselves, as well as the electrostatic interaction between nuclei as a constant term:

$$\hat{H}^{nr} = -\sum_i (\nabla_i^2/2) - \sum_{i,A} Z_A/r_{iA} + \sum_{i>j} 1/r_{ij} + \sum_{A>B} Z_A Z_B/r_{AB} \qquad (1)$$

Here and in the following, use is made of atomic units, a.u., see table 1.

Table 1. Atomic units. e and m_0 are the electron charge and rest mass, respectively; h is the Planck constant.

Quantity	Atomic unit (name)	SI Equivalent
Mass	m_0	9.109610^{-31} kg
Charge	e	1.602210^{-18} C
Angular momentum	$\hbar$	1.054610^{-34} Js
Length	$a_0 = \hbar^2/(me^2)(Bohr)$	5.291810^{-11} m
Energy	$E_h = e^2/a_0$ (Hartree)	4.359810^{-18} J
Time	$t_0 = \hbar/E_h$	2.418910^{-17} s
Velocity	$a_0/t_0 = E_h/(me^2) \approx c/137$	2.157910^{6} ms^{-1}

Neglecting relativistic effects may be a poor approximation in some instances. The mean square velocity v of an electron in a $1s$ orbital is Z a.u.: for heavy atoms it may be an appreciable fraction of the velocity of light ($c = 137.036 a.u.$). In mercury ($Z = 80$), $v/c \approx 0.58$, and the corresponding relativistic mass is $m = 1.23\, m_0$. Since the electron mass appears in the denominator of the Bohr radius, the $1s$ orbital is contracted by about the same amount. Such important relativistic effects concerning core electrons, indirectly affect valence electrons. In heavy atoms, valence s and p orbitals are contracted and stabilized with respect to the non-relativistic case, due to the orthogonality constraints with the orbitals of the innermost shells; on the other hand, valence orbitals of d and f type are significantly expanded and destabilized, because of the more effective screening of the nuclear charge by the *s* and *p* electrons. These effects can be partially taken into account by using a relativistic treatment for core electrons (see Chapter 9), or by adopting relativistic pseudo-potentials in valence-only calculations (see section 2.4). In addition, the relativistic Dirac hamiltonian contains the spin variable explicitly (for instance, in the so-called spin-orbit coupling term) [2] while in non-relativistic quantum mechanics, spin effects are important only because of the exclusion principle. A recent study of mercury (atom, clusters and bulk) provides an instructive example of the importance of relativistic effects in solid state theory [3]. Neglected relativistic terms may be taken into account *a posteriori* by a perturbative treatment or some other corrective scheme.

The knowledge of the Hamiltonian, and of the total number of electrons contains *in nuce* all information about the electronic structure of the system, both concerning static and time-dependent features. In the following, we shall take the system as neutral, that is, the number of electrons per crystalline unit cell, n, will equal the sum of the nuclear charges of the translationally inequivalent nuclei. To make progress, in particular to be able to exploit the essential simplification introduced by the periodic nature of the external field, we must specify which aspects of the electronic structure we are mainly interested in. We shall see that the ground state electronic structure (GSES) plays a fundamental rôle in solid state physics; its determination, and the associated properties will be the main object of our interest. A straightforward, though unfeasible means for determining the GSES would be to solve the corresponding Schrödinger equation:

$$\widehat{H}^{nr}\Psi_0 = E_0(R)\,\Psi_0 \tag{2}$$

where the ground state energy, $E_0(R)$, corresponds to the lowest eigenvalue for a given nuclear configuration, R. In fact, it is not only impossible to determine, but also to manipulate such a cumbersome object as Ψ_0 (an antisymmetric function of $3nN$ spatial coordinates and nN spin co-ordinates, where N is the number of crystalline cells, see Chapter 2). It is better to consider simpler functions, which contain as much useful information as possible. One possible choice is to look for simplified forms of Ψ_0: in particular, in the Hartree-Fock (HF) approach, we approximate Ψ_0 by a Slater determinant, corresponding to the antisymmetrized product of nN one-electron functions. Other quantities can be chosen to describe the GSES. The one-electron density matrix (or one-matrix):

$$\begin{aligned}\gamma_0(\mathbf{r},\mathbf{r}') \;&=\; nN \int\cdots\int d\mathbf{r}_2\cdots d\mathbf{r}_{nN}\, d\sigma_1\cdots d\sigma_{nN} \;\times \\ &\times\;\; [\Psi_0(\mathbf{r},\sigma_1;\mathbf{r}_2,\sigma_2;\cdots;\mathbf{r}_{nN},\sigma_{nN})\;\; \Psi_0^*(\mathbf{r}',\sigma_1;\mathbf{r}_2,\sigma_2;\cdots;\mathbf{r}_{nN},\sigma_{nN})]\end{aligned} \tag{3}$$

is a very simple function which contains a lot of important information (see Chapter 14) and is the primary objective of some procedures (see sections 5 and 6). In density functional theory (DFT), the objective is still simpler, namely, the determination of the electron density: $\rho_0(\mathbf{r}) \equiv \gamma_0(\mathbf{r},\mathbf{r})$; from the knowledge of this function of three spatial coordinates, all ground state properties can be derived, in principle, by means of suitable integrals (see sections 2.2 and 6). The dream of solid state physicists is to determine the self-energy function, $\Sigma(\mathbf{r},\mathbf{r}',\varepsilon)$, which would permit all ground state one-particle properties to be obtained, including one-electron excitation energies [4]; $\Sigma(\mathbf{r},\mathbf{r}',\varepsilon)$ can be estimated, for instance, through Hedin's GW-approximation [5].

The main obstacle to the solution of equation 2, or to the determination of some intermediate quantity which contains the desired information, lies in the inter-electronic interaction term, $\sum_{i>j} 1/r_{ij}$, which prevents the problem from being factorized into nN problems for the individual electrons. An essential simplification, therefore, consists in introducing an effective potential for each electron in the field of the others, and substituting a sum of one-electron Hamiltonians ($\widehat{H}^{nr} \approx \sum_i \widehat{h}_i$) for the full Hamiltonian, at the cost of neglecting some higher order effects to be specified below. This allows us to achieve two advantages simultaneously: firstly, to de-couple the electronic motions; secondly, to introduce, quite naturally, the translational symmetry of the problem. In fact, since we are interested in the GSES, any effective potential must reflect the symmetry of the external potential and therefore be translationally periodic: the methods, language and techniques of band theory can be used (Chapter 2). The following scheme summarizes the concepts introduced up to now (in the lines below, we list the approximations introduced and the effects that are disregarded):

$$\begin{array}{cccccccc} \Rightarrow & \text{(fixed nuclei)} & \Rightarrow & \widehat{H}_{nr} & \Rightarrow & \{\widehat{h}_i\} & \Rightarrow & \text{GSES} \\ \Downarrow & & \Downarrow & & \Downarrow & & \Downarrow & \\ \text{nuclear} & & \text{relativistic} & & \text{higher order} & & \text{excited} & \\ \text{motion} & & \text{effects} & & \text{effects} & & \text{states} & \end{array} \qquad (4)$$

In the rest of this Chapter, we will concentrate our attention on the last two parts of this scheme. Section 2 is concerned with the formulation of one-electron approximations for the Hamiltonian, starting from the consideration of the prototypical periodic system, the homogeneous electron gas. Two *ab initio* schemes are currently adopted. The first is based on DFT, and its fundamental objective is to derive the correct ground state electron density, $\rho_0(\mathbf{r})$, and the exact ground state energy, E_0; the second consists of adopting the HF approximation, and aims at obtaining the best single-determinant representation of the ground state wave-function in a variational sense. The fundamental limitation of DF theory is that one must use approximate expressions for the effective potential, since an exact formulation is not available. On the other hand, with the HF approach one is always faced with an electronic correlation problem [1, 6], as is discussed below. The use of pseudo-potentials (PP) to dispense with the treatment of core electrons in one-electron hamiltonians is well established in solid state theory and is summarily treated here.

The intermediate object of our study (the GSES, described either by $\gamma_0(\mathbf{r}, \mathbf{r}')$, or by a simplified expression for Ψ_0) is some function of the electron coordinates which is usually expressed as a linear combination of a finite set of one-electron functions, $\phi_\mu(\mathbf{r})$ ($\mu = 1, \ldots, p$) or of the products thereof. The choice of the basis set is one of the critical aspects of most computational schemes: the algorithms, the efficiency, the very structure of the program are often intimately related with this choice. A general introduction to the basis set problem in *ab initio* crystalline studies is provided in section 3. More detailed analyses, with reference to specific choices, are contained in Chapters 9, 10 and 11.

Section 4 outlines the standard methods for the determination of the GSES, which are based on the solution of one-electron equations and on the use of band theory. The key features and the critical computational issues are similar for DF and HF schemes and are exposed briefly in parallel.

Non-standard methods which have relevance to solid state studies and are not considered in this book, are summarily treated in section 5 for information. The general problem of the scaling of the computational cost with the size of the irreducible part of the crystal is also considered.

Section 6 serves as an introduction to the last part of this book, concerning the calculation of crystalline properties.

2. One-Electron Hamiltonians

2.1. Introduction: The electron gas problem

The problem of the homogeneous electron gas (a system of interacting electrons in a uniform sea of compensating positive charge) has traditionally played a crucial rôle in solid state theory, but its importance has grown dramatically with the advent and the extraordinary success of DF theory. The discussion of the merits and limitations of one-electron Hamiltonians in solid state physics can begin with some of the essential results concerning this ideal system. A concise account of the subject is given in chapter 2 of Fulde's book [6]; a clear and detailed treatment (except for the more recent studies) can be found in Raimes' book [7].

The only parameter which characterizes the electron gas is its uniform density, ρ (number of electrons per unit volume), or equivalently, the radius of the Wigner-Seitz sphere, $r_s = [(4/3)\pi\rho]^{-1/3}$. One often speaks of *low density* or *high density* electron gas according to whether $r_s > 1$ or $r_s < 1$ (for instance, for valence electrons in sodium $r_s = 3.9$, for core electrons in carbon $r_s = 0.22$; as always, a.u. are used). Two one-electron approximations are applied to solve the non-relativistic problem of equation 2. The *free-electron* approach takes into account the fact that classically, the electrostatic self-interactions of the electron gas and of the positive background and their mutual interaction cancel exactly. The only term which is left is the sum of the kinetic energy terms for the individual electrons. In addition, an exchange term appears in the HF approximation, which takes into account the antisymmetry of the wave-function (see section 2.3). The problem then becomes an elementary one for the individual electrons:

$$[-\frac{1}{2}\nabla^2 + (v_x)_i]\psi_i(\mathbf{r}) = \varepsilon_i\psi_i(\mathbf{r}) \tag{5}$$

In this equation, the exchange term, $(v_x)_i$, is zero for the free-electron gas approach; in the HF scheme, it is a constant which may depend on i but is independent of $\mathbf{r}$ for symmetry reasons. The eigenvectors are then plane waves: $\psi_i(\mathbf{r}) \equiv \psi(\mathbf{r};\mathbf{k}) = \Omega^{-\frac{1}{2}}\exp(\imath\,\mathbf{k}\cdot\mathbf{r})$, with eigenvalues $\varepsilon_i \equiv \varepsilon(k) = \frac{1}{2}\,k^2 + v_x(k)$ (for symmetry reasons, ε_i cannot depend on the direction of $\mathbf{k}$). According to the Aufbau principle, the occupied orbitals fill a sphere in $\mathbf{k}$-space, centered in $\mathbf{k} = 0$ of radius $k_F = 1/(\alpha r_s)$, with $\alpha = [4/(9\pi)]^{1/3} = 0.521$. Since the expression of the occupied orbitals is known, the HF value of $v_x(k)$ can be calculated and it turns out to be:

$$v_x(k) = \frac{k_F}{2\pi}[2 + \frac{k_F^2 - k^2}{k \cdot k_F} \cdot \log|\frac{k + k_F}{k - k_F}| \equiv \frac{k_F}{2\pi} F(\frac{k}{k_F}) \quad (6)$$

The GS energy per electron is easily calculated in the two cases. For the free-electron approximation, it is purely kinetic:

$$\varepsilon_{kin}(\rho) = (3/10)/(\alpha^2 r_s^2) = (3/10)[3\pi^2]^{2/3} \quad (7)$$

In the HF case, it contains, in addition, an exchange term: $\varepsilon_{HF}(\rho) = \varepsilon_{kin}(\rho) + \varepsilon_x(\rho)$, with:

$$\varepsilon_x(\rho) = -3/(4\pi\alpha r_s) = -(3/4)/(3\rho/\pi)^{1/3}. \quad (8)$$

The results just presented have had great historical importance in creating a bad reputation for the HF approximation among solid state physicists. Consider first the density of states (DOS), that is the number of one-electron states per unit energy interval and per unit volume. It is tempting to use the information above to calculate the distribution of electronic excitation energies, an excited state being created by letting one or more electrons pass from an occupied level, ε, to an empty one, ε', with a corresponding energy cost: $\Delta\varepsilon = \varepsilon' - \varepsilon$. This assumption is strictly valid within the free-electron approximation, and finds a loose justification in Koopmans' theorem for the HF model (see section 2.3). Using the free-electron levels $[\varepsilon(k) = k^2/2]$ gives satisfactory agreement with experiment for many electronic properties (low temperature specific heat, electrical conductivity, spin susceptibility among others) of alkali metals such as lithium or sodium, for which the electron gas model is best justified. The addition of the exchange correction, $v_x(k)$ (equation 6), instead of improving the agreement, introduces serious discrepancies with respect to experiment. First of all, the energy spread of occupied levels is too large. More fundamentally, it is easily seen that the derivative with respect to energy of the $F(k/k_F)$ factor in equation 6 exhibits a logarithmic divergence when $k = k_F$: this corresponds to zero density of states at the Fermi level, obviously contrary to experimental evidence. This paradoxical result is not inherent in the homogeneous electron gas model, but has been shown to be a necessary outcome of the HF equations when applied to metallic systems [8, 9]. It is the result of neglecting correlation effects in the HF approximation, which are essential for screening the r^{-1} dependence of the Coulomb interactions in a metal.

Correlation effects are of decisive importance also for the cohesive energy of the electron gas. In the HF expression for the specific energy: $\varepsilon_{HF}(\rho) = \varepsilon_{kin}(\rho) + \varepsilon_x(\rho)$, the exchange term makes the electron gas stable for $r_s > 2.41$, but in all cases, the calculated cohesion energy is low with respect to the experimental cohesion energy of alkali metals. A great deal of theoretical and computational work has been devoted to obtaining a good estimate of the correlation energy for the electron gas. The many-body study by Gell-Mann and Bruckner for high densities [10], and the more recent quantum-Montecarlo results of Ceperley and Alder for low densities (see section 5) [11] are particularly important. The calculated data have been recently parameterized by Perdew and Zunger [12], in order to obtain an analytic expression, which is useful for DF calculations (see section 2.2):

$$\begin{aligned} \varepsilon_c(\rho) &= 0.0311 \log r_s - 0.048 + 0.0020 r_s \log r_s - 0.0116 r_s \quad [\text{for } r_s < 1] \\ \varepsilon_c(\rho) &= -0.1423/(1 + 1.0529\sqrt{r_s} + 0.3334 r_s) \quad [\text{for } r_s > 1] \quad (9) \end{aligned}$$

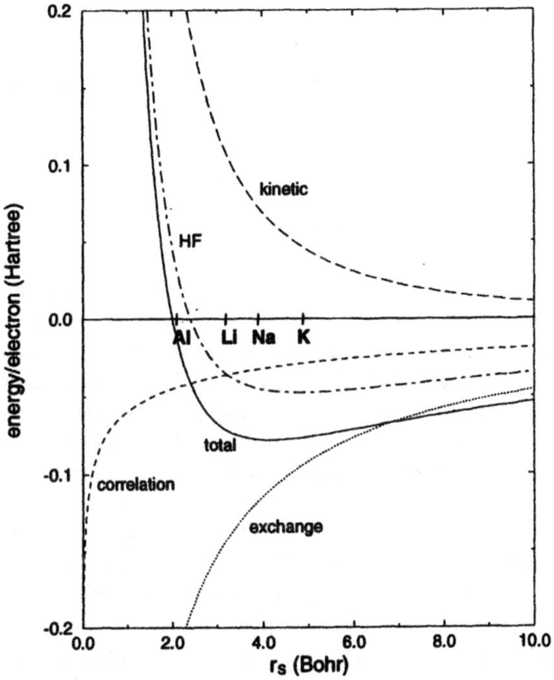

Figure 1. Contributions to the energy per electron for the homogeneous electron gas, as a function of the Wigner-Seitz radius, r_s. The r_s values of valence electrons of a few simple metals are reported.

Figure 1 shows the importance of correlation corrections in the region of electron densities of chemical importance. The effect of external fields on the properties of the electron gas is a subject of great interest, still intensively studied, especially for its applications in DF work (see section 2.2).

2.2. Density functional theory

DF techniques are based on two theorems, proved by Hohenberg and Kohn in 1964 [13], and on a computational scheme proposed by Kohn and Sham (KS) the following year [14].

The first theorem states that a given ground state (GS) electron density, $\rho_0(\mathbf{r})$, cannot arise from two different external potentials, unless the two differ by a constant. Hence, the GSES (Ψ_0 and all derived GS observables, in particular E_0) is uniquely determined by $\rho_0(\mathbf{r})$.

The second theorem establishes a variational criterion for determining $\rho_0(\mathbf{r})$ and E_0 together for a given external potential created, for example, by a set of nuclear charges: $V(\mathbf{r}) = -\sum_A Z_A/|\mathbf{r}-\mathbf{r}_A|$. E_0 is found by minimizing an expression of the form:

$$\begin{aligned} E[\rho(\mathbf{r})] &= \int \rho(\mathbf{r})V(\mathbf{r})d\mathbf{r} + \frac{1}{2}\int [\rho(\mathbf{r})\rho(\mathbf{r}')]/|\mathbf{r}-\mathbf{r}'|\, d\mathbf{r}\, d\mathbf{r}' + \\ &\quad + \int \rho(\mathbf{r})g(\mathbf{r};[\rho])\, d\mathbf{r} + \sum_{A>B} Z_A Z_B/r_{AB} \\ &= E_{Ext} + E_J + E_{kxc} + E_{RepNuc} \end{aligned} \tag{10}$$

with respect to an arbitrary function, $\rho(\mathbf{r})$, which represents an n-electron density. The minimum is obtained precisely in correspondence to $\rho_0(\mathbf{r})$. In this expression, the first term gives the interaction energy with the external potential, the second term is the classical self-interaction of the charge density $\rho(\mathbf{r})$, while the last term is the usual internuclear repulsion. The third term, E_{kxc}, contains the $g(\mathbf{r})$ function, which may be interpreted as a density of kinetic, exchange and correlation energy per unit electron density at $\mathbf{r}$. $g(\mathbf{r})$ is known to be a universal *functional* of the density ρ (that is, it uniquely depends on the *whole* charge distribution). Its analytic form is not known, but different approximations are used. A local density approximation (LDA) consists of making $g(\mathbf{r})$ depend only on the value of ρ at $\mathbf{r}$. Among LDAs, the historically important Thomas-Fermi approximation [15] is obtained by giving $g(\mathbf{r})$ the value of the specific kinetic energy of an electron gas with uniform density, $\rho(\mathbf{r})$: $g(\mathbf{r}) \approx (3/10)[3\pi^2\rho(\mathbf{r})]^{2/3}$.

The most critical aspect of $g(\mathbf{r};[\rho])$, however, is precisely that it contains the kinetic energy term: the Thomas-Fermi approximation is not very accurate but it is difficult to improve on it. The KS method is based on the two theorems above but permits us to calculate explicitly the most important part of the kinetic energy. In fact, it consists of solving a parallel problem for a system of n pseudo-independent electrons, characterized by the same GS density as the actual system. For the pseudo-system, the pseudo-kinetic energy E_{kps} can be explicitly calculated and the problem is solved according to the following self-consistent field (SCF) procedure:

a) Find the $n/2$ lowest eigenvalues and the corresponding eigenfunctions (*orbitals*) of a one-electron effective Hamiltonian:

$$\hat{h}_{eff}\, \psi_i(\mathbf{r}) \equiv [-\frac{1}{2}\nabla^2 + V_{eff}(\mathbf{r})]\psi_i(\mathbf{r}) = \varepsilon_i\psi_i(\mathbf{r}); \tag{11}$$

b) Calculate the density:

$$\rho(\mathbf{r}) = 2\sum_i |\psi_i(\mathbf{r})|^2; \tag{12}$$

c) Re-calculate $V_{eff}(\mathbf{r})$ as a functional of $\rho(\mathbf{r})$:

$$V_{eff}(\mathbf{r}) = V(\mathbf{r}) + \int \rho(\mathbf{r}')/|\mathbf{r}-\mathbf{r}'|d\mathbf{r}' + \mu_{xc}(\mathbf{r};[\rho]); \tag{13}$$

if self-consistency is not reached, go to step a).

d) At self-consistency, $\rho(\mathbf{r}) = \rho_0(\mathbf{r})$ (GS density) and the GS energy can be calculated:

$$\begin{aligned} E_0 &= E_{kps} + \int \rho_0(\mathbf{r})V(\mathbf{r})d\mathbf{r} + \frac{1}{2}\int [\rho_0(\mathbf{r})\rho_0(\mathbf{r}')]/|\mathbf{r}-\mathbf{r}'|\, d\mathbf{r}\, d\mathbf{r}' + \\ &+ E_{xc} + E_{RepNuc} \\ E_{kps} &= \sum_i \int \psi_i(\mathbf{r})^*(-\frac{1}{2}\nabla^2)\psi_i(\mathbf{r})d\mathbf{r} \\ E_{xc} &= \int \rho_0(\mathbf{r})\varepsilon xc(\mathbf{r};[\rho_0])d\mathbf{r} \end{aligned} \tag{14}$$

The two universal functionals $\mu_{xc}(\mathbf{r};[\rho])$ in equation 13 (effective exchange-correlation potential) and $\varepsilon_{xc}(\mathbf{r};[\rho])$ in equation 14 (specific exchange-correlation energy) are related to each other by a *functional derivative* equation:

$$\mu_{xc}(\mathbf{r};[\rho]) = \delta E_{xc}([\rho])/\delta\rho(\mathbf{r}) \equiv \delta\{\int \rho(\mathbf{r}')\varepsilon_{xc}(\mathbf{r}';[\rho])d\mathbf{r}'\}/\delta\rho(\mathbf{r}) \tag{15}$$

The following can be noted:

- The beauty and the originality of the KS scheme lies in the fact that its aim is not to give an approximation to Ψ_0, but to yield the exact GS energy E_0 and the exact GS electron density, $\rho_0(\mathbf{r})$, directly. In this sense, it is important to observe that the pseudo-wave-function, defined as an antisymmetrized product of the pseudo-spin orbitals: $\Psi_{KS} = \widehat{A}\,[\prod_i(\psi_i(\mathbf{r}_{2i-1}\alpha(\sigma_{2i-1})\psi_i(\mathbf{r}_{2i}\beta(\sigma_{2i}))]$, has nothing to do, in principle at least, with the true Ψ_0, except that it defines the same GS density. This is equally true for the pseudo-eigenvalues ε_i, which should not be used as indicative of one-electron excitation energies. On the other hand, it has been customary in solid state applications to adopt such an inclusive interpretation of the results of the KS equations: this attitude has some theoretical ground (see section 6) and is justified by many remarkable successes. Part of the theoretical work in the field of DF theory aims at defining an effective potential which generates pseudo-eigenvalues ε_i more strictly related to one-particle excitations [16].
- The crucial problem in the KS method is the definition of the exchange-correlation functional (it should be noted that this definition is not fully appropriate, because E_{xc} must contain, in addition to the exchange and correlation energies, a correction to the pseudo-kinetic energy E_{kps}). For this purpose, reference is most often made to the results for the homogeneous electron gas. The LDA is by far the most frequent choice:

$$\varepsilon_{xc}(\mathbf{r};[\rho]) \approx \varepsilon_x(\rho(\mathbf{r})) + \varepsilon_c(\rho(\mathbf{r}))$$

where $\varepsilon_x(\rho)$ and $\varepsilon_c(\rho)$ are given by expressions (8) and (9) (when applied to the electron gas, LDA-DF theory is exact but tautological). In the LDA, the relationship between μ_{xc} and ε_{xc} simplifies to: $\mu_{xc}(\rho) = d[\rho\,\varepsilon_{xc}(\rho)]/d\rho$. The celebrated X-$\alpha$ method of Slater [17] can be interpreted as an anticipation of the KS-LDA equations, with only the exchange LDA potential included, multiplied by an empirical parameter $(3\alpha/2)$.

- Given a fairly regular density, any functional $\alpha(\mathbf{r}; [\rho])$ is a *function* of ρ and all its derivatives at $\mathbf{r}$. The LDA can be viewed as the zero term in a gradient expansion of the exact functionals [13]. In recent years, the use of gradient-corrected (GC) functionals has become more and more wide-spread, which corresponds to going a step farther in the gradient series. In these cases, the value of the ε_{xc} functional depends both on the value of the density at r, $\rho(\mathbf{r})$, as well as on the magnitude of the density gradient there, $|\nabla\rho(\mathbf{r})|$. For the parameterization of such functionals, reference is made to results for the inhomogeneous electron gas [18]. On the other hand, the convergence properties of the gradient series are unknown and making real progress by adding more of the neglected, higher-order terms (see equation 4) seems a hopeless task.
- Up to now, we have considered only closed-shell systems. The extension of DF theory to spin polarized systems is quite natural [19, 12]. One must define an electron density for each spin state: $\rho^{\uparrow}$, $\rho^{\downarrow}$ or equivalently, a total density, $\rho = \rho^{\uparrow} + \rho^{\downarrow}$ and a spin electron density $m = \rho^{\uparrow} - \rho^{\downarrow}$. All functionals, for instance ε_{xc}, will depend on both densities: $\varepsilon_{xc} \equiv \varepsilon_{xc}(\mathbf{r}; [\rho, m])$. The discussion of this class of important problems goes beyond the scope of this chapter.
- The electron-electron Coulomb interaction energy (see equation 10) contains a spurious contribution arising from the interaction of each electron with itself. This should be exactly cancelled by an exchange term of opposite sign but this does not happen in standard applications of DF theory, due to the approximations involved in the calculation of E_{xc}. There is a need for a self-interaction-correction (SIC) to the KS equations, especially in their application to molecular systems. Different SIC methods have been proposed, which consist essentially of subtracting, from the total energy expression, the sum of the self-interactions in the individual pseudo-orbitals [12, 20]. For a recent critical account, see references [21].
- The history of DF applications in solid state physics is too rich to be even summarily sketched here; some examples will be given in subsequent Chapters of this book.The pioneering work of M.L. Cohen and collaborators at Berkeley should however be mentioned, who produced in the seventies and eighties an ample and convincing documentation of the power, generality and reliability of the *ab initio* DF treatment of crystalline systems [22].

2.3. Hartree-Fock theory and the electron correlation problem

The HF equations are homologous to the KS ones, in the sense that we are again looking for a set of $n/2$ orbitals as eigenfunctions of a one-electron Hamiltonian, the Fock Hamiltonian $\hat{f}$ (here, reference is made to a closed-shell system; the extension to spin-polarized systems is considered in Chapter 6):

$$\begin{aligned}\hat{f}\,\psi_i(\mathbf{r}) \;&\equiv\; [-(1/2)\nabla^2 + v(\mathbf{r}) + \int d\mathbf{r}'\rho(\mathbf{r}')/|\mathbf{r}-\mathbf{r}'|]\;\psi_i(\mathbf{r}) + \\ &\quad -\sum_j \int d\mathbf{r}'[\psi_j(\mathbf{r}')^*\,\psi_i(\mathbf{r}')]/|\mathbf{r}-\mathbf{r}'|]\;\psi_j(\mathbf{r}) = \\ &= \;\varepsilon_i\,\psi_i(\mathbf{r}) \end{aligned} \tag{16}$$

The first three terms in the Fock Hamiltonian coincide with the corresponding KS ones (the expression for the electron density is the same). However, instead of the effective exchange-correlation potential, $\mu_{xc}(\mathbf{r};[\rho])$, here we have the "exact" non-local exchange operator (the fourth term in equation 16). Again, the HF equations must be solved through an SCF procedure, corresponding to steps a-d of section 2.2, since both the Coulomb and the exchange operator depend on the set of functions, $\psi_i(\mathbf{r})$. The essential difference, with respect to the KS approach, is that the antisymmetrized product of the Fock spin orbitals: $\Psi_{HF} = \hat{A}[\prod_i(\psi_i(\mathbf{r}_{2i-1}\alpha(\sigma_{2i-1})\psi_i(\mathbf{r}_{2i}\beta(\sigma_{2i})]$ is an approximation of the GS wave function: more precisely, it is the best single-determinant approximation (in a variational sense) of Ψ_0. Similar comments apply concerning the one-matrix (see Chapter 14). The HF energy, E_{HF}, is defined as the expectation value of the non-relativistic Hamiltonian for Ψ_{HF}:

$$\begin{aligned} E_{HF} &= \langle\Psi_{HF}|H^{nr}|\Psi_{HF}\rangle \\ &= 2\sum_i \int d\mathbf{r}\, \{\psi_i(\mathbf{r})^* \,[-\nabla^2/2 + V(\mathbf{r})]\psi_i(\mathbf{r})\} + \\ &\quad +\sum_{ij} \int d\mathbf{r}\, d\mathbf{r}'\, [\psi_i(\mathbf{r})^*\psi_i(\mathbf{r})]\,[\psi_j(\mathbf{r}')^*\psi_j(\mathbf{r}')^*]\,/\,|\mathbf{r}-\mathbf{r}'| + \\ &\quad -1/2\sum_{ij} \int d\mathbf{r}\, d\mathbf{r}'\, [\psi_i(\mathbf{r})^*\psi_j(\mathbf{r})]\,[\psi_j(\mathbf{r}')^*\psi_i(\mathbf{r}')^*]\,/\,|\mathbf{r}-\mathbf{r}'| \end{aligned} \quad (17)$$

Here are a few comments on the HF method:

- The non-local exchange term makes the HF equations more difficult to solve than the KS ones, where $\mu_{xc}(\mathbf{r};[\rho])$ is simply a multiplicative operator, no matter how complicated its determination. This is specially true for crystalline systems, as we shall see in section 4. If one adds the fact that HF performs poorly for the electron gas, the simplest of all periodic systems (section 2.1), the lack of popularity of this approximation among solid state physicists would appear well justified. This contrasts with its almost universal acceptance in molecular studies. Early attempts to formulate and implement HF computational schemes for periodic systems have had scant success, in spite of the high quality of those studies [23, 24, 25, 26]. Only in recent years has it become possible to formulate a fair judgment about the usefulness of the HF approach in solid state physics, since the advent of powerful computers and of general purpose computational schemes [27, 28], which have allowed us to assess its performance for a variety of systems (see section 7).
- At variance with DF theory, one-electron HF eigenvalues *can* be attributed a physical meaning. If one creates an $(n-1)$-electron state, represented by a single-determinant Ψ' which may be obtained from Ψ_{HF} by eliminating a spin-orbital with eigenvalue ε_i, it is easily shown that $\varepsilon_i = \Delta E \equiv \langle\Psi_{HF}|H^{nr}|\Psi_{HF}\rangle - \langle\Psi'|H^{nr}|\Psi'\rangle$ (Koopmans' theorem) [1]. ΔE (or ε_i) can be considered as an ionization potential of the system, if one assumes that the removal of the electron does not affect the other occupied orbitals (the frozen-orbital approximation). The neglect of relaxation tends to produce ionization potentials that are too positive; correlation effects, whose importance grows with the number of electrons, correct partly for this error. For molecules,

Koopmans' ionization potentials are usually in reasonable agreement with experiment [1].

- As a by-product of the solution of the HF equations, virtual orbitals and energy levels are obtained, corresponding to solutions of equation 11 with eigenvalues higher than those of occupied orbitals. Using Koopmans' theorem, one can estimate electron affinities, by adding an electron in a virtual orbital. In this case, however, the correlation correction adds to the relaxation error, and electron affinities are far too negative. Similarly, it is incorrect to use one-electron HF eigenvalues to calculate the excitation energy resulting from the promotion of an electron from an occupied (ψ_i) to a virtual (ψ_j') energy level as $\Delta E = \varepsilon_j' - \varepsilon_i$. Therefore, HF band structures are not suitable for the estimation of optical excitations in crystals: the HF main gap for insulators and semi-conductors is usually about twice as large as that given by experiment [27]. On the other hand, it is relatively easy to correct HF band-structures to reproduce the spectrum of optical excitations, by using, for instance, the GW approximation [5, 29, 30].
- The main limitation of the HF solution lies in the neglect of correlation effects. Consideration of the interaction of each electron with the average field of the others leads to an over-estimation of electron-electron repulsions: in fact, electrons move so as to keep apart from each other (otherwise stated, each electron carries along a *correlation hole*). The recent book by Fulde discusses the correlation problem in crystalline studies and can be used as a fundamental reference [6].
- If we are interested only in an estimate of the correlation energy, $E_c = E_0 - E_{HF}$, simple and powerful techniques are available, based on the precepts of DF theory. As was shown in the preceding sub-section, it is possible to define an LDA or a GC expression for *correlation-only* functionals. In Chapter 11, indications are given of the efficiency of these techniques.
- While in molecular quantum chemistry the electron correlation problem has been studied in detail for several decades now and many different powerful techniques have been devised to cope with it, the state of the art in solid state theory is far less advanced, not only because of the difficulty of obtaining good crystalline HF wave-functions. Most standard molecular schemes use the set of occupied and virtual orbitals as basic ingredients. In crystals, these are de-localized and correspond to a continuum of levels, which makes their treatment awkward. Furthermore, a correlation scheme must be *size-consistent* in order to be transferable to solid state applications. The size-consistency requirement can be stated as follows. The application of the scheme to a system composed of N identical and independent subsystems, must give exactly N times the correlation energy obtained for the individual system. The most standard of all correlation schemes, the variational Configuration-Interaction technique, restrained to single and double excitations (SD-CI) does not satisfy this requirement. The reason is easily understood: the product of the wave-functions for the individual sub-systems, each containing SD excitations, would indeed give N times the individual energy, but contains excitations involving up to 2N electrons. Among size-consistent schemes which are beginning to be transferred from molecular to solid state applications, we note the coupled-cluster (CC) technique [31], the Moeller-Plesset perturbative series and the Green-function many-body approach, which is particularly

appropriate for correcting the HF band-structure [30]. The *local* approach, which is a variant of the CC method specially designed for crystals, is amply illustrated in Fulde's book [6]. Finally, if one does not impose symmetry restrictions on the spin component of the wave-function and looks for unrestricted HF solutions, one often obtains a lower energy by separating spin-up from spin-down electrons in space; interesting results have been obtained for simple metals [32].

- Fulde's comment about the value of HF computations for solids can be reported in conclusion [6]:

  "Provided that the correlations are not too strong, HF calculations are a good starting point, and allow solids and molecules to be treated the same way and with the same accuracy. The development of quantum chemistry has proven that ab initio calculations based on controlled approximations capable of systematic improvement have made simpler computational schemes based on uncontrolled simplifications obsolete. Whether or not the same will eventually hold true for solid-state theory remains to be seen."

2.4. Pseudo-potentials

In many respects, core electrons are unimportant for determining the stability, structure and low-energy response properties of molecules and of crystals. It is a well-established practice to modify the one-electron part of the Hamiltonian by replacing the bare nuclear attraction with a PP operator, $\widehat{V}_{ps}$, which permits us to restrict the calculation to valence electrons, for both types of system. This choice is necessary in practice when the system includes heavy atoms ($Z > 20$). Consider a system comprising a set of nuclei A and n electrons, of which n' are valence electrons. We can write equations of the form:

$$[-\nabla^2/2 + \int \rho'(\mathbf{r}')/|\mathbf{r}-\mathbf{r}'|d\mathbf{r}' + \mu_{xc}(\mathbf{r};[\rho']) + \sum_A \widehat{V}_{psA}]\,\psi_i'(\mathbf{r}) = \epsilon_i'\,\psi_i'(\mathbf{r}) \quad (18)$$

which coincide with the DF-KS equations 11-13, except for the substitution of V(**r**) by the PP operator: $\widehat{V}_{ps} = \sum_A \widehat{V}_{psA}$ (a parallel procedure can be followed in the HF case). Primed symbols have been introduced to indicate that the SCF solution of these equations is limited to valence electrons; that is, we will consider only the $n'/2$ eigenfunctions corresponding to the lowest eigenvalues, and use them to calculate the valence density, $\rho'(\mathbf{r})$, which in turn is used to define the electrostatic, exchange and correlation interactions between valence electrons. The PP operator must reproduce screened nuclear attractions, but must also account somehow for the Pauli exclusion principle, which requires that valence orbitals are orthogonal to core ones. Consider an atom A at $\mathbf{A}$, with n_{cA} core electrons, whose highest angular quantum number is L. Suppose $L = 1$, that is, the core contains only s and p electrons. At long range, we must have: $\widehat{V}_{psA} = -(Z_A - n_{cA})/r_A$, with $r_A = |\mathbf{r} - \mathbf{A}|$. At short range, $\widehat{V}_{psA}$ must act differently on functions of s and p symmetry, while it operates equivalently on functions of higher angular quantum number, that is:

$$\widehat{V}_{ps} = -(Z_A - n_{cA})/r_A + \sum_{\ell=0}^{L} U_\ell^{s-r}(r_A)\,\{\sum_{m=-\ell}^{\ell} |\ell m\rangle\langle\ell m|\}_A + W^{s-r}(r_A) \quad (19)$$

$W^{s-r}(r_A)$ and $U_\ell^{s-r}(r_A)$ are short-range functions; the term in braces is a projector, which makes U_ℓ^{s-r} act only on functions which have ℓ-symmetry with respect to $\mathbf{r}_A$. In establishing the explicit form of $\widehat{V}_{psA}$, a number of characteristics are sought (not all of them can be optimally satisfied):

i) Pseudo-valence eigenvalues, ε_i', should coincide with the true ones, ε_i ;
ii) Pseudo-orbitals, $\psi_i'(\mathbf{r})$, should resemble as closely as possible the true ones, $\psi_i(\mathbf{r})$, in an external region as well as being smooth and node-less in the core region;
iii) Pseudo-orbitals, $\psi_i'(\mathbf{r})$, should be properly normalized;
iv) The functional form of the PPs, though preserving non-local character, should be designed so as to simplify as far as possible their use in computations;
v) Explicit expressions for the functions $W^{s-r}(r_A)$ and $U_\ell^{s-r}(r_A)$ should be provided for the different atomic species, to be used independently of the environment;
vi) For heavy atoms, relativistic effects should be taken into account (see Introduction): that is, the form of the PP should be derived from relativistic KS or HF atomic calculations.

The performance of PP techniques in solid state physics is usually very good, except for some critical cases (for instance, if core relaxation effects are important, which may occur when simulating very high pressures, or when the electron configuration in the crystalline environment is very different from that of the isolated atom) [22, 33]. Among PPs designed for KS-LDA calculations, the norm-conserving ones tabulated for all atoms by Bachelet et al. [34] are perhaps the most popular; *ultrasoft* PPs, which ensure the very smooth behaviour of the pseudo-valence orbitals in the core region are useful in applications where plane waves are used as a basis set (see section 3.2 and Chapter 10) [35]. With regard to PPs for HF calculations, we may mention those introduced by the Toulouse group [36] twenty years ago, which are still widely used, and the more recent ones designed by Hay and Wadt [37]. In all these cases, relativistic corrections are added for heavy atoms.

3. Analytic representation of Wave-function and Operators

3.1. Introductory remarks

There is a huge amount of information contained in the GSES and associated operators. An efficient computational scheme must be able to manipulate this information economically with regard to CPU time, storage requirements and the number of input/output operations. If possible, the algorithms should be simple and general enough to make accumulation and transfer of know-how from one program to another easier. Linear representations of one-electron functions are usually adopted:

$$f(\mathbf{r}) = \sum_{\mu=1}^{p} c_\mu \phi_\mu(\mathbf{r}) \tag{20}$$

With few exceptions, the basic ingredients in solid state applications for the construction of the basis-set functions, $\phi_\mu(\mathbf{r})$, are plane-waves (PW) and/or Gaussian type orbitals (GTO), and they are the only ones that are considered in the following. The basic advantage obtained from the use of PWs and GTOs is related to the fact that they make the computation of integrals in direct and/or reciprocal space very easy. If *numerical techniques* are used to perform integrals throughout the calculation, much greater freedom is possible in the choice of the representative functions: these may include Slater type orbitals (see below), PWs, numerically tabulated orbitals or any combination of these. A KS computational scheme based on a numerical approach has been prepared recently by Baerends and collaborators [38]. All types of symmetry and periodicity in one, two and three dimensions can be handled. The efficiency of the scheme in handling complicated structures is still to be assessed, however.

When using linear representations, the problem of the errors related to the use of an incomplete expansion set arises. There is the need of making such errors as small as possible, while using manageable basis sets. Each application requires a careful analysis: the art of devising good basis sets is a very important one, and is based on experience and competence (see Chapter 8). In this respect, even *ab initio* methods can profit from *empirical knowledge*. One of the advantages that are obtained from the use of Baerends' numerical approach is that the problem of basis set incompleteness, which often plagues schemes based on PWs and GTOs, becomes almost irrelevant.

3.2. Plane waves

PWs are, in a sense, the ideal basis functions for a periodic system. If one denotes the general wave-vector as $\mathbf{k}$ and the translationally equivalent vector in the first Brillouin Zone (BZ) as $\boldsymbol{\kappa}$, we can write:

$$\phi_{\mathbf{k}}(\mathbf{r}) = \Omega^{-1/2} \exp\,(\imath \mathbf{k}\cdot\mathbf{r}) = \Omega^{-1/2} \exp\,[\imath(\boldsymbol{\kappa}+\mathbf{K})\cdot\mathbf{r})] \equiv \phi_{\mathbf{K}}(\mathbf{r};\boldsymbol{\kappa}) \tag{21}$$

The last symbol indicates that the general PW is a Bloch function (BF) associated with the point $\boldsymbol{\kappa}$ within the BZ and labelled with a discrete index $\mathbf{K}$, corresponding to a vector of the reciprocal lattice (see Chapter 2, equation 41). PWs are an orthonormal, complete set: any function belonging to the class of continuous, normalizable functions (which are those of interest in QM) can be expanded with arbitrary precision in the PW set. The set is universal, in the sense that it depends neither on where atoms are located within the cell nor on their nature. As we shall see in section 5, this characteristic is particularly valuable in Car-Parrinello calculations, where nuclear positions are continually changing.

In practice, one must use a finite number of PWs: how many are necessary? Normally one chooses a *cutoff energy*, T, that is, for each $\boldsymbol{\kappa}$ in the BZ, one includes all PWs with $(\boldsymbol{\kappa}+\mathbf{K})^2 < T = X^2$ (X is the radius of the sphere centered at the origin in reciprocal space which includes all selected PWs) in the expansion set. In typical calculations, T is of the order of 20 Ry (for historical reasons, cutoffs are usually measured in Rydbergs: 1 Ry = 0.5 a.u.). If the volume, V, of the unit cell in direct space is 1000, say, then the volume of the Brillouin zone is $V_{BZ} = 8\pi^3/V = 0.248$ and the number of **K** vectors, hence of PWs, $\phi_{\mathbf{K}}(\mathbf{r};\boldsymbol{\kappa})$, contained in the sphere is $(4\pi X^3/3)/V_{BZ} = 534$. While this is not a small number of basis functions, the finest detail that can be revealed in direct space with such a set is of the order of $1/X = 0.58$, about half the radius of the hydrogen atom. This may be a reasonable figure if we are interested only in valence electrons, but is totally unacceptable for the description of core electrons. For a factor ten in detail, which would still be insufficient except for very light atoms, the number of PWs per $\boldsymbol{\kappa}$ point should increase by a factor 1000! Unless one *augments* the PW set with some additional function (see section 5 and Chapter 9) the use of PPs is mandatory. In practice, ultrasoft PPs are preferred for obvious reasons (section 2.3).

The use of PWs makes most algebraic manipulations very simple and it is convenient to perform all calculations in the momentum representation. For example, consider the potential $v_H(\mathbf{r}) = \int \rho'(\mathbf{r}')/|\mathbf{r}-\mathbf{r}'|d\mathbf{r}'$ generated at the general point $\mathbf{r}$ by the distribution of pseudo-valence electrons, which is one of the components of the KS Hamiltonian (equation 18). Since $\rho'(\mathbf{r})$ is a periodic function in direct space, so is $v_H(\mathbf{r})$ and we can write, after setting to zero the average value of $v_H(\mathbf{r})$:

$$\rho'(\mathbf{r}) = \sum_{\mathbf{K}} \rho_{\mathbf{K}} \exp(\imath\mathbf{K}\cdot\mathbf{r}) \quad \text{and} \quad v_H(\mathbf{r}) = \sum_{\mathbf{K}\neq 0} v_{\mathbf{K}} \exp(\imath\mathbf{K}\cdot\mathbf{r}) \tag{22}$$

The coefficients, $\rho_{\mathbf{K}}$, are obtained immediately from the knowledge of the occupied orbitals:

$$\psi_i(\mathbf{r};\boldsymbol{\kappa}) = \sum_{\mathbf{K}} c_{i,\mathbf{K},\boldsymbol{\kappa}}\phi_{\mathbf{K}}(\mathbf{r};\boldsymbol{\kappa}) \rightarrow \rho_{\mathbf{K}} = 2\sum_{\boldsymbol{\kappa}}\sum_{i} |c_{i,\mathbf{K},\boldsymbol{\kappa}}|^2 \tag{23}$$

and as a result of Poisson's equation:

$$\nabla^2 v_H(\mathbf{r}) = 4\pi\rho'(\mathbf{r}) \tag{24}$$

we have :

$$v_{\mathbf{K}} = -4\pi\rho_{\mathbf{K}}/K^2 \tag{25}$$

3.3. Gaussian functions

Two kinds of basis functions have been traditionally used in molecular calculations: Slater-type orbitals (STOs) and GTOs. They are both the product of spherical harmonics $Y_{\ell m}(\theta,\phi)$ by a radial function $R(r)$. For STOs, $R(r) = r^n \exp(-\zeta r)$, while for the basic form of GTOs (Gaussian *primitives*), $R(r) = r^\ell \exp(-\alpha r^2)$. In spite of the fact that STOs reproduce much better the cusps of the wave-function in the proximity of the nuclei [it may be proven that in the vicinity of a point charge Z, the radial part of the one-electron ground state wave-function has the form: $A(1 - Zr)$], their use has become less and less frequent in favour of GTOs, for which the calculation of multi-center two-electron integrals is essentially simpler. A few important characteristics of GTOs can be mentioned:

- In molecular applications, GTOs are usually constructed as a linear combination of Gaussian primitives $g_j(\mathbf{r})$, characterized by the same center (usually coinciding with a nucleus), the same angular numbers but different exponents:

$$\varphi_\mu(\mathbf{r}) = \sum_{j=1}^{p_\mu} d_j g_j(\mathbf{r}) \; ; \; g_j(\mathbf{r}) \equiv g(\mathbf{r}; \alpha, \ell, m) = r^\ell Y_{\ell m}(\theta, \phi) \exp(-\alpha_j r^2) \quad (26)$$

  These combinations are often called *contracted Gaussians*, where α_j represents *exponents* and d_j, *contraction coefficients*. By a suitable choice of d_j and α_j, one can prepare basis functions which have, from the start, good characteristics (for example, they exhibit an almost correct cusp behavior at the core). Different conventions are used to denote the basis set. A common one can be exemplified with reference to a calculation by Rauk *et al.* concerning NH_3 [39]: $(13s8p2d/8s2p) \rightarrow [8s5p2d/4s1p]$. The information in the round parentheses concerns primitives: for nitrogen, 13 of them are of type s ($\ell = 0$), 8 of type p ($\ell = 1, m = -1, 0, 1$), 2 of type d ($\ell = 2, m = -2, -1, 0, 1, 2$); for hydrogen, 8 of type s, 2 of type p. The contracted set is reported in square parentheses; for hydrogen, the two p primitives have been contracted to one GTO, the eight s primitives participate in four GTOs; note that the two d primitives on nitrogen have remained uncontracted.
- In solid state applications, a finite number (p) of GTOs are attributed to the various atoms in the reference zero cell ($\mathbf{A}_\mu$ will denote the coordinate of the nucleus on which φ_μ is centered); the same GTOs are then associated with all translationally equivalent atoms in the crystal. In total, we have Np GTOs, from which we can construct Np Gaussian-type Bloch orbitals:

$$\phi_\mu(\mathbf{r}; \boldsymbol{\kappa}) = \sum_{\mathbf{T}} \varphi_\mu(\mathbf{r} - \mathbf{A}_\mu - \mathbf{T}) \exp(\imath \boldsymbol{\kappa} \cdot T) \qquad (\mu = 1, \cdots, p; \boldsymbol{\kappa} = 1, \cdots, N) \quad (27)$$

- From experience with molecular calculations, a large number of proposals have emerged to help computational chemists in the preparation of efficient basis sets; excellent reviews exist of this subject [40, 41]. Huzinaga's sets may be mentioned [42] as an example: for each atom, several options are proposed according to the required accuracy, differing in the number of GTOs, contraction coefficients and exponents. Molecular basis sets are not directly transferable to crystalline applications. In particular, there are at least three reasons for avoiding the use of very diffuse primitives (low exponents) in solid state studies: first, the number of integrals to be explicitly calculated increases explosively; secondly, the accuracy of the calculation must be particularly high in order to avoid pseudo-linear dependence catastrophes; thirdly, diffuse functions are not of much use in densely packed crystals, because their tails are found in regions where there is large variational freedom associated with functions on other atoms. These problems are re-considered in Chapters 8 and 11.
- With respect to PWs, the use of suitably contracted GTOs permits us to describe accurately electronic distributions both in the valence and in the core region with a limited number of basis functions. The price is the loss of orthogonality, of universality and the need for more sophisticated algorithms for the calculation of the required integrals. The latter are expressed in terms

of the primitives; their evaluation is made simpler by the fundamental property that the product of two Gaussians is a Gaussian (Boys' theorem). In particular for s-type primitives:

$$g(\mathbf{r}-\mathbf{A};\alpha,0,0)\,g'(\mathbf{r}-\mathbf{A}';\alpha',0,0)=Kg''(\mathbf{r}-\mathbf{A}'';\alpha'',0,0) \qquad (28)$$

with: $\alpha''=\alpha+\alpha'$; $\mathbf{A}''=(\alpha\mathbf{A}+\alpha'\mathbf{A}')/\alpha''$; and $K=\exp(-\alpha\alpha'|\mathbf{A}-\mathbf{A}'|^2/\alpha'')$. This result is easily generalized to higher quantum numbers.

- The most complicated integrals that enter the calculations are the so-called *two-electron, four-center integrals* which result from calculating the Coulomb or exchange term in the Fock Hamiltonian (equation 16) or the corresponding energy term (equation 17), after having expressed the spin orbitals $\psi_i(\mathbf{r})$ in the GTO basis set:

$$\begin{aligned}(\mu\nu|\kappa\lambda) \;=\; &\int d\mathbf{r}\, d\mathbf{r}'\, \{[\varphi_\mu(\mathbf{r}-\mathbf{A}_\mu-\mathbf{T}_\mu)\,\varphi_\nu(\mathbf{r}-\mathbf{A}_\nu-\mathbf{T}_\nu))]\times \\ &\times[\varphi_\kappa(\mathbf{r}'-\mathbf{A}_\kappa-\mathbf{T}_\kappa)\,\varphi_\lambda(\mathbf{r}'-\mathbf{A}_\lambda-\mathbf{T}_\lambda)]/|\mathbf{r}-\mathbf{r}'|\} \qquad (29)\end{aligned}$$

This kind of integral, as well as the simpler nuclear attraction ones, are obtained by starting from an auxiliary function of one variable:

$$F_0(w)=\int_0^1 ds\,\exp(-ws^2) \qquad (30)$$

with the help of powerful recurrence relations [43]. A large part of the success of GTOs in molecular and solid state QM applications is due precisely to the efficiency of the algorithms developed for integral calculation.

- An additional advantage of GTOs is due to the fact that their Fourier transform is another Gaussian [$F_p\{\exp(-\alpha r^2)\}\propto\exp(-p^2/4\alpha)$], and their use in combination with PW techniques is therefore easy [29].
- A given set of GTOs, associated with atomic positions, usually performs better when the atoms are close to each other than if they are far apart. This is a consequence of basis set incompleteness; the region between two atoms is better described if use is made of functions centered on both atoms. There is then an over-estimation of binding energies, which is called basis-set superposition error (BSSE). BSSE can be very important with poor basis sets; the counterpoise technique for estimating this error gives usually reasonable results [41, 44].

4. Standard Methods of Solution

4.1. κ-factorization of the one-electron equations for crystals

Although DF and HF schemes correspond to totally different approaches from a theoretical point of view, in practical applications there are profound similarities between them. They are both specially designed for the GSES, and have little significance for excited states; in both cases, the one-electron hamiltonian depends functionally on the GSES, and the equations must be solved through an SCF procedure; the basis set problems are similar. We describe in this section standard methods of solution for periodic systems in very general terms, without specifying which one-electron Hamiltonian $\hat{h}$ we are referring to, except where needed. A more detailed description is provided in Chapters 8, 9 and 10.

As shown in Chapter 2, BFs associated with different $\boldsymbol{\kappa}$ points within the first BZ belong to different irreducible representations of the group of one-electron Hamiltonians, $\hat{h}$. It is then possible to factorize the problem into separate parts for each $\boldsymbol{\kappa}$:

a) Consider the p BFs, $\phi_\mu(\mathbf{r};\boldsymbol{\kappa})$, associated with $\boldsymbol{\kappa}$ (equation 21 or 23);
b) Calculate the matrix elements: $H_{\mu\nu}^{\boldsymbol{\kappa}} = \langle\phi_\mu^{\boldsymbol{\kappa}}|\hat{h}|\phi_\nu^{\boldsymbol{\kappa}}\rangle$; $S_{\mu\nu}^{\boldsymbol{\kappa}} = \langle\phi_\mu^{\boldsymbol{\kappa}}|\phi_\nu^{\boldsymbol{\kappa}}\rangle$ (the latter equals $\delta_{\mu,\nu}$ in the PW set);
c) Solve the $p \times p$ matrix equation:

$$H^{\boldsymbol{\kappa}}\, C^{\boldsymbol{\kappa}} = S^{\boldsymbol{\kappa}}\, C^{\boldsymbol{\kappa}}\, E^{\boldsymbol{\kappa}}, \tag{31}$$

where the diagonal matrix $E^{\boldsymbol{\kappa}}$ contains the eigenvalues $\varepsilon_i^{\boldsymbol{\kappa}}$, and the matrix $C^{\boldsymbol{\kappa}}$ contains, columnwise, the coefficients of the crystalline orbitals (COs):

$$\psi_i(\mathbf{r};\boldsymbol{\kappa}) = \sum_{\mu=1}^{p} c_{\mu i}^{\boldsymbol{\kappa}}\, \phi_\mu(\mathbf{r};\boldsymbol{\kappa}). \tag{32}$$

The above procedure should be carried on for the complete set of $\boldsymbol{\kappa}$ points in the first BZ, so as to determine the complete set of COs (that is, the KS or HF spin-orbitals: see equations 11 and 16) with the precision granted by the basis set adopted and by the accuracy of the algorithms. There are, however, some points which need closer analysis:

i) The problem of how many and which $\boldsymbol{\kappa}$ points should be considered is an extremely important one. In order to reconstruct the $\hat{h}$ operator and to calculate the crystalline properties from the solution, we need all occupied spin-orbitals, in principle. Due to the continuity of eigenvalues and eigenfunctions with respect to $\boldsymbol{\kappa}$ (see Chapter 2), it is, however, possible to derive the required information from the results obtained at a few suitably sampled $\boldsymbol{\kappa}$ points, as discussed in the next subsection.
ii) In order to calculate $H_{\mu\nu}^{\boldsymbol{\kappa}}$ matrix elements, we must consider functions and operators which are extended to the whole crystalline structure. Except in particularly favorable cases, we are faced with a problem of summations over one or more indices associated with the different crystalline cells. The accurate and efficient handling of such series (*lattice sums*) often determines the final quality of a computational scheme (section 4.3).
iii) It has been mentioned already that the process must be repeated until self-consistency is achieved, that is, until eigenvalues and eigenvectors coincide, within a given tolerance, with those used for the reconstruction of $\hat{h}$. This may be very difficult to achieve at times. All techniques that accelerate convergence are valuable (section 4.4).
iv) To solve the matrix equation 31, the diagonalization of $p \times p$ complex matrices is needed. The time required is proportional to p^3. With very complex systems, p can become very large, and the diagonalization step will be the slowest one, perhaps to the point where the system is untreatable. Different, non-standard procedures, which by-pass this step, may then become preferable (section 5).

4.2. Reciprocal space integration and interpolation

From the knowledge of the eigenvalues, $\varepsilon_i^{\boldsymbol{\kappa}}$ and the eigenfunctions, $\psi_i(\mathbf{r};\boldsymbol{\kappa})$, at a few, sample $\boldsymbol{\kappa}$ points (to be indicated in the following by $\underline{\boldsymbol{\kappa}}$), we want to obtain accurate estimates of quantities such as the number of states below a certain energy, $i(e)$ and the matrix of integrated densities of states (IDOS), $I(e)$, which imply a sum over *all* $\boldsymbol{\kappa}$ points:

$$\begin{aligned} i(e) &= 2V_{BZ}^{-1}\sum_i \int_{BZ} d\boldsymbol{\kappa}\, \theta(e-\varepsilon_i^{\boldsymbol{\kappa}}) \\ I_{\mu\nu}^{\mathbf{T}}(e) &= 2V_{BZ}^{-1}\sum_i \int_{BZ} d\boldsymbol{\kappa}\, c_{\mu i}^{\boldsymbol{\kappa}*}\, c_{\nu i}^{\boldsymbol{\kappa}}\, \exp(\imath\boldsymbol{\kappa}\cdot\mathbf{T})\, \theta(e-\varepsilon_i^{\boldsymbol{\kappa}}) \end{aligned} \tag{33}$$

The sum has been replaced with an integral over the BZ, due to the fact that $\boldsymbol{\kappa}$ points are uniformly distributed in reciprocal space (see Chapter 2); the condition that only the orbitals of energy less than e are included in the sum is expressed by the presence in the integrand of the step function $\theta(e-\varepsilon_i^{\boldsymbol{\kappa}})$, whose value is 1 if $\varepsilon_i^{\boldsymbol{\kappa}}$ is less than e, and is zero otherwise. The quantities $i(e)$ and $I(e)$ are very important. The Fermi energy, e_F, is determined by imposing the condition: $i(e_F) = 2n$, that is, by requiring that there are exactly $2n$ spin-orbitals per cell with energy less than e_F. As shown in Chapter 2, the total density of states (DOS): $n(e) = di(e)/de$ and the IDOS derivative, $N(e) = dI(e)/de$, which is called the projected density of states (PDOS), give rich information on the chemical structure of the system and allow all one-electron properties to be obtained within the independent-electron approximation [27]. The value of $I(e)$ at the Fermi energy, e_F, is the P matrix [$P = I(e_F)$]. When using a basis set of localized functions, the one-matrix, $\gamma_0(\mathbf{r},\mathbf{r}')$, (equation 3) corresponding to a detor constructed on the occupied HF or KS spin-orbitals, is conveniently expressed in terms of the P matrix:

$$\gamma_0(\mathbf{r},\mathbf{r}') = \sum_{\mu,\nu=1}^{p}\sum_{\mathbf{T},\mathbf{T}'} P_{\mu\nu}^{T'-T}\, \varphi_\mu(\mathbf{r}-\mathbf{A}_\mu-\mathbf{T})\, \varphi_\nu(\mathbf{r}'-\mathbf{A}_\nu-\mathbf{T}')$$

The determination of the Fermi energy, e_F, is a delicate problem only in the case of metals (see Chapter 4 for details). For insulators, n bands are fully occupied, the others are void. The $\theta(e-\varepsilon_i^{\boldsymbol{\kappa}})$ function can then be dropped from the integrand and the sum over i is limited to the n lowest eigenvalues at each $\boldsymbol{\kappa}$ point.

In summary, we must estimate integrals of the form:

$$\Xi(e) = V_{BZ}^{-1}\int_{BZ} d\boldsymbol{\kappa}\, [\xi(\boldsymbol{\kappa})\, \theta(e-\varepsilon(\boldsymbol{\kappa}))] = V_{BZ}^{-1}\int_{BZ'(e)} d\boldsymbol{\kappa}\, \xi(\boldsymbol{\kappa}) \tag{34}$$

where $\xi(\boldsymbol{\kappa})$ and $\varepsilon(\boldsymbol{\kappa})$ are well-behaved, periodic functions in reciprocal space, and $BZ'(e)$ is the portion of BZ where $\varepsilon(\boldsymbol{\kappa}) < e$: it coincides with the whole BZ, if $\varepsilon(\boldsymbol{\kappa})$ is less than e everywhere and vanishes if $\varepsilon(\boldsymbol{\kappa})$ is always greater than e.

If $BZ'(e) = BZ$, *special points* techniques may be used [45]. This means that one may select a special set of points $\{\, \underline{\boldsymbol{\kappa}}\, \}$ within the BZ, with a weight, $w(\underline{\boldsymbol{\kappa}})$, associated with each of them, evaluate $\xi(\underline{\kappa})$ at each of them and substitute the integral with a weighted sum: $\Xi \approx \sum_{\underline{\kappa}} w(\underline{\kappa})\, \xi(\underline{\kappa})$. A discussion of these techniques and a documentation of how good their performance can be is provided in Chapter 4.

In the more general case ($BZ'(e) \neq BZ$), the linear [46] or quadratic [47] tetrahedron techniques are usually adopted. The BZ is subdivided into tetrahedral mini-cells: the integral will be the sum of sub-integrals over each of them. After evaluating ξ and ε at the vertices of the tetrahedra, $\underline{\kappa}$, a linear or quadratic approximation is obtained for both $\xi(\kappa)$ and $\varepsilon(\kappa)$ inside the tetrahedron: the sub-integral is finally evaluated analytically using these approximate expressions. An alternative method can be used for evaluating directly Ξ integrals, starting from the value of $\xi(\kappa)$ and $\varepsilon(\kappa)$ at a few special points, $\underline{\kappa}$: this is called the Fourier-Legendre expansion technique, because it is based first, on a Fourier representation of bands, second, on the expansion of the DOS associated with each band into an orthonormal set of Legendre functions [48, 27]. It is not as accurate as the tetrahedron technique, but has the advantage of providing analytic expressions for the Ξ dependence on e and of being much faster.

4.3. Lattice sums

The problem of Coulomb lattice sums is present in both DF and HF schemes and must be treated with extreme care, because the Coulomb series is only conditionally convergent. In HF theory, we have the additional problem of the exchange series, which is involved in the construction of the Fock hamiltonian, $\hat{f}$ (see equation 16), and gives a contribution, E_{ex}, to the HF energy (see the last term in equation 17). Since both these problems are discussed explicitly in Chapter 8, we will not examine them here.

4.4. The SCF problem

In variational approaches like KS or HF, the SCF procedure can be viewed as a path in the space of the variational coefficients $\{c_{\mu i}\} \equiv \mathbf{c}$ (for simplicity, the κ index is dropped). This path begins at some initial good (?) guess ($\mathbf{c} = \mathbf{c}_{start}$) and ends, hopefully, at $\mathbf{c} = \mathbf{c}_{scf}$, where the minimum of the functional $E_0[\mathbf{c}]$ occurs, subject to the orthonormality condition of the spin-orbitals:

$$\sum_{\mu,\nu=1}^{p} c^{*}_{\mu i} c_{\nu j} S_{\mu\nu} \quad = \quad \delta_{ij}$$

At $\mathbf{c}_{scf}$, the self-consistency condition is satisfied. On the other hand, self-consistency does not necessarily imply that a minimum has been reached, although this is always the case, in practice. The standard movement of the representative point $\mathbf{c}$ in the coefficient space is as described in section 4.1. At the general n-th cycle of the SCF procedure, the current point $\mathbf{c}_n$ is used to generate the new hamiltonian, $\hat{h}_n$; this in turn (equation 25) generates $\mathbf{c}_{n+1}$, satisfying the orthonormality constraint. There is, however, no guarantee that the new point moves in the direction of $\mathbf{c}_{scf}$. In fact, critical cases are frequent: oscillations in energy occur, requiring a large number of steps in the procedure, sometimes of the order of hundreds or even ending in a numerical catastrophe. Many procedures have been devised to accelerate the SCF procedure. Some of them simply damp oscillations by mixing quantities belonging to the current and preceding cycles before starting the new one (typically, the Hamiltonian matrix in HF calculation and the density in DF calculations): $\hat{h}_n \leftarrow \alpha\hat{h}_{n-1}+(1-\alpha)\hat{h}_n$, where the mixing coefficient a may be larger than 0.5. More sophisticated procedures have been proposed in the field of molecular calculations, which try to obtain good guesses for **c** for use in subsequent cycles by extrapolation [49]. These techniques are just now beginning to be used in solid state applications. In the next section, we present briefly the Car-Parrinello technique, which combines the SCF problem with the problem of the optimization of nuclear coordinates in an original way.

5. Non-standard Methods of Solution

To give an idea of the variety of approaches that can be used for the *ab initio* treatment of periodic systems, a few of those in current practice are sketched here.

KKR (Korringa-Kohn-Rostoker)
There are some aspects of this powerful technique [50] which make it awkward to use for those accustomed to the ordinary tools and concepts of molecular quantum chemistry (for simplicity, we consider its application to elementary crystals with one atom per cell, in the framework of DF theory). First, space is partitioned into ***atomic spheres*** centered on atoms, and a complementary ***interstitial region***; the atomic sphere is the largest that can be inscribed in the Wigner-Seitz cell. Secondly, a simplified form is assumed for $V_{eff}(\mathbf{r})$ (equation 13), that is, $V_{eff}(\mathbf{r})$ is zero in the interstitial region and is a radial function $[V_{eff}(\mathbf{r}) = V(r_A)]$ inside the atomic spheres: this is called the muffin-tin (MT) approximation. Thirdly, for the solution of the KS equations, Green-function techniques are adopted, that is, the eigenfunction at a given $\boldsymbol{\kappa}$ point, corresponding to an eigenvalue $\varepsilon < 0$, $\psi(\boldsymbol{\kappa}, \mathbf{r}; \varepsilon)$, is the solution of the Lippmann-Schwinger integral equation:

$$\psi(\boldsymbol{\kappa},\mathbf{r};\varepsilon) = \int d\mathbf{r}' \{[-\exp(-|\mathbf{r}-\mathbf{r}'|\sqrt{-\varepsilon}) / (4\pi|\mathbf{r}-\mathbf{r}'|)] V_{eff}(\mathbf{r}') \psi(\boldsymbol{\kappa},\mathbf{r}';\varepsilon)\} \quad (35)$$

The expression in square brackets is the Green function for the free electron. The MT form of the potential and the fact that:

$$\psi(\boldsymbol{\kappa},\mathbf{r}+\mathbf{T};\varepsilon) = \exp(\imath\boldsymbol{\kappa}\cdot\mathbf{T})\psi(\boldsymbol{\kappa},\mathbf{r};\varepsilon)$$

for all lattice vectors, **T**, as a result of Bloch's theorem (see Chapter 2, equation 41), are now taken into account. The Lippmann-Schwinger equation can then be recast in a very convenient form, where the integral is restricted to the interior of a single atomic sphere, in the reference zero cell. After expressing $\psi(\boldsymbol{\kappa},\mathbf{r};\varepsilon)$ as a linear combination of "partial waves", the eigenvalues, ε, are obtained by looking for non-trivial solutions, which correspond to the zeros of a determinant ($|I - G(\varepsilon)V| = 0$: see Chapter 13, Appendix 1). A remarkable example of the power of this technique is provided by the detailed study of the electronic properties of 32 elementary metals from hydrogen to indium, by Moruzzi, Janak and Williams [51].

APW(augmented plane waves)
The partition of space between atomic spheres and the interstitial region is the characteristic feature of a whole family of techniques, derived from Slater's original proposal [52] through continuous improvements and transformations: APW, LAPW (linearized plane waves) [53], FLAPW (Full potential LAPW) [54]. Essentially, a plane-wave expansion is adopted in the interstitial region, while spherical harmonics multiplied by radial functions are used within the spheres; the continuity of the wave-function and its derivative across the boundary is the principal condition to be obeyed. All restrictions concerning the form of the potential have been progressively removed, complicated crystalline structures can be treated without problems and linearized forms of the equations have been adopted which allow all eigenvalues at a given κ to be found at the same time through matrix diagonalization. These techniques are specially suited to the study of properties related to the distribution of inner electrons, such as the electric field gradient at the nuclear positions [55]. The FLAPW method is discussed in detail in Chapter 9.

Recursion
This method shares with KKR the use of Green function techniques. Consider a system described by one-electron equations, and a representative basis set of localized functions $\phi_\mu(\mathbf{r}\}$. The originality of the recursion technique [56] lies in the fact that its aim is not the determination of eigenvalues and eigenfunctions, but rather, to determine the PDOS directly (section 4.2):

$$N_{\mu\nu}(e) \quad = \quad \sum_i c^*_{\mu i} c_{\nu i} \delta(e - \varepsilon_i)$$

For each e, the sparse nature of the H and S matrices is exploited. Starting from a given "centre", ϕ_μ, one considers its neighbours (those that have non-negligible matrix elements with ϕ_μ), then the next nearest neighbours, and so on. Through a continued fraction expansion, the PDOS is determined with progressively better precision and the procedure is truncated after considering a certain number of steps. An appealing aspect of this method is that it does not exploit translational symmetry, except for the determination of H and S matrices, and is largely insensitive to boundary conditions: for these reasons, it can be applied quite naturally to the study of defects in solids. A drawback lies in the requirement of matrix sparseness, which has made difficult its use with *ab initio* Hamiltonians, until very recently [57].

CP(Car-Parrinello)
The CP technique [58] differs from those described up to now because it treats the problem of the nuclear motion and that of the electron response to nuclear displacements contemporaneously. It is expedient to give a short account of this method which is gaining enormous importance in solid state physics; for more details, reference can be made to the review article by Remler and Madden [59]. As anticipated in section 4.4, the SCF process, which consists of finding the optimal

c_i coefficients of the basis functions defining the occupied COs, can be viewed as a motion of the system in the coefficient space towards the energy minimum, subject to an orthonormality constraint. Such problems of constrained motion can be treated in a classical Lagrangian formalism, by introducing a fictitious time, t, a fictitious mass, μ and a fictitious kinetic energy, $\mu \cdot \sum_i (\dot{c}_i)^2/2$. At each time step, dt, the electronic variables evolve in a direction dictated by the acting forces ($F_i = -\partial E[\mathbf{c}]/\partial c_i$) which drive the system towards lower total energy, subject to constraints and to ***inertial forces. Attrition*** terms are included in the Lagrangian, so that the system evolves towards an energy minimum, hopefully the absolute one. With fixed nuclei, this procedure is usually much less efficient than the standard one, described in section 4.4.

The brilliant idea in the CP method was to include in the Lagrangian both the real nuclear motion in R-space and the fictitious electron motion in c-space, on the same footing. The system (nuclei + electrons) starts from a given initial configuration $(R, \dot{R}; \mathbf{c}, \dot{\mathbf{c}})$ and evolves in the phase space: the forces acting on nuclei are calculated using the Hellman-Feynman theorem [59]. This process is also called ***simulated annealing***, because the initial configuration can be made to correspond to a given temperature, and the attrition terms are, therefore, "cooling" the system by subtracting real and fictitious kinetic energy. With a proper choice of the time step dt, of the fictitious mass μ, and of the attrition terms, the nuclei are found to move close to the ever-changing Born-Oppenheimer energy surface, $E_0(R)$ (see equation 2) and eventually to reach a local energy minimum. At a stroke, both the electronic SCF problem has been solved and the equilibrium configuration has been reached, an achievement that could require painstaking calculations with standard techniques. A fringe benefit is that one obtains the electron response to nuclear motion automatically.

The large majority of applications of the CP method are concerned with this energy minimization problem. However, the same technique can be used for studying real problems of molecular dynamics, such as the distribution of phonon frequencies (see for instance reference [60] concerning highly compressed water ice). In solid state applications, the standard ingredients of the CP method are:

1) the DF-LDA Hamiltonian (derivatives are easily calculated);
2) the super-cell technique (in the different super-cells, nuclei and electrons are moving identically; only the $\boldsymbol{\kappa} = \mathbf{0}$ point is considered in integrations);
3) the soft-core pseudo-potentials for core electrons and plane waves for valence electrons (these choices permit the use of a simple, orthonormal, universal basis set, and of highly efficient fast-Fourier-transform algorithms).

GFMC and VMC (Green-Function Montecarlo and Variational Montecarlo)
The impact of Montecarlo techniques in solid state physics is mostly due to the calculations by Ceperley and Alder, which have provided information of unprecedented accuracy concerning the homogeneous and inhomogeneous electron gas at low densities [11] (see section 2.1). These techniques allow the $2n$-electron closed-shell wavefunction, Ψ_0, to be treated without the need for separation into the products of one-electron wave-functions. This is possible because integrals of the type $\langle \Psi_0 | \widehat{H}^{nr} | \Psi_0 \rangle$ in $3n$ spatial co-ordinates are evaluated accurately and economically by a statistical sampling procedure, the so-called Metropolis algorithm. There are two fundamental types of Montecarlo methods. According to GFMC, the evolution of the wave-function in imaginary time, $\tau = \imath t$, is followed using statistical techniques, starting from a trial function, Ψ_T. For $\tau \to \infty$, all

components in the wave- function, except Ψ_0, are damped because of the time evolution operator: $\exp(-\imath Ht) = \exp(-H\tau)$.

This procedure has been adopted, for instance, in the electron-gas calculations just mentioned. In VMC, the trial wave-function is made dependent on some variational parameter, q $[\Psi_T \equiv \Psi(q)]$. The optimal value of the parameters is determined by minimizing the expectation value of the energy:

$$E(q) = \langle\Psi(q)|\hat{H}^{nr}|\Psi(q)\rangle/\langle\Psi(q)|\Psi(q)\rangle$$

The trial function is usually written so as to include explicitly the most important part of the correlation. The form most often adopted is the product of a Jastrow factor, $J(q)$ [61], by a single determinantal approximate wave-function, usually the HF solution, perhaps with the introduction of adjustable scaling factors, q': $\Psi_T \equiv J(q)\Psi_{HF}(q')$. The Jastrow factor is totally symmetric in the electron coordinates (therefore, Ψ_T is antisymmetric) and has the form $\sum_{i>j}[ar_{ij}/(1+br_{ij})]$, where a and b are variational parameters. Thanks to this factor, the trial wave-function automatically contains electron-electron ***correlation cusps*** of the correct form, that is, each electron moves accompanied by a correlation hole. The main difficulty in the application of Montecarlo techniques to solid state problems, is that they are very inefficient in the case of heavy atoms where a large majority of the sampling points in the statistical integration would end in core regions, the most unintersting ones: computational requirements increase very rapidly with Z ($\propto Z^{5.5}$). The use of pseudo-potentials would allow us to get rid of core-electrons, but their non-local form (equation 19) makes their use with standard Montecarlo techniques very difficult. Nevertheless, interesting attempts have been made in this direction: pseudo-potential calculations of crystalline silicon have been performed both using VMC [62], and GFMC techniques [63].

N-scaling

When one considers such a bewildering variety of techniques, the question can be raised as to which one has best performance or if there is a theoretical limit to the efficiency of new techniques that can be devised. In a system containing N non-equivalent atoms, all physically relevant information is proportional to N (in the rest of this lecture, N designates the number of crystalline cells; here we use it in a different sense, to conform with standard nomenclature). However, standard schemes depend on N^3, due to the necessity of determining a number of orbitals proportional to N, which involves diagonalization or inversion of $(kN) \times (kN)$ matrices (equation 31: the pre-factor k depends on the basis set size). Much work has taken place in recent years in the framework of DF-LDA approaches, aimed at devising algorithms which vary linearly with N, at least for very large N values. There is a variety of proposals, still at a rough stage of implementation [64]. When such tools become available, it will be possible to study very complex systems which are beyond the accessibility of ***ab initio*** schemes at present.

6. From the Electronic Structure to the Crystalline Properties

A large part of this book is devoted to the calculation of ground state crystalline properties: here, we make a few general remarks. Consider an observable quantity

O, depending on the electronic variables. Its average ground state value, O^{av}, is given by the expression:

$$O^{av} = \langle \Psi_0 | \hat{O} | \Psi_0 \rangle \tag{36}$$

where the operator $\hat{O}$ is obtained from the classical expression of the observable O, via the correspondence principle: $\mathbf{r} \rightarrow \mathbf{r}\cdot$; $\mathbf{p} \rightarrow \imath\nabla$ (for instance: $T = \sum_i |\mathbf{p}|_i^2/2 \rightarrow -\sum_i \nabla|_i^2/2$). If O is a one-electron observable, that is, $O \equiv O^1 = \sum_i o_i$, which is often the case, then expression 36 can be simplified by making reference to the one-electron density matrix, $\gamma_0(\mathbf{r}, \mathbf{r}')$ (equation 3):

$$O^{1av} = \int d\mathbf{r}[\hat{o}(\mathbf{r}')\gamma_0(\mathbf{r}, \mathbf{r}')]_{\mathbf{r}' \rightarrow \mathbf{r}} \tag{37}$$

In HF theory, Ψ_{HF} is an approximation of Ψ_0 in the form of an antisymmetrized product of spin-orbitals. It is easily seen that:

$$\gamma_{HF}(\mathbf{r}, \mathbf{r}') = 2\sum_i \psi^*_{HF,i}(\mathbf{r})\psi_{HF,i}(\mathbf{r}')$$

for closed-shell systems, the sum being restricted to the $n/2$ occupied orbitals and we can write, for one-electron operators:

$$O^{1av}_{HF} = 2\sum_i \int d\mathbf{r}\psi^*_{HF,i}(\mathbf{r})\hat{o}(\mathbf{r})\psi_{HF,i}(\mathbf{r}) \tag{38}$$

A similar expression is obtained for two-electron observables. O^{1av}_{HF} will be affected by a correlation error which can be estimated according to the usual techniques.

When considering DF theory, Hohenberg and Kohn's first theorem (section 2.2) insures that for each observable, a functional $O([\rho_0]; \mathbf{r})$ must exist such that:

$$O^{av} = \int d\mathbf{r}O([\rho_0]; \mathbf{r})\rho_0(\mathbf{r}) \tag{39}$$

However, there is no general and simple prescription for generating such functionals. By considering the formal similarity of HF and KS procedures, it would be tempting to use equation 38 again, this time making reference to the KS pseudo-orbitals $\psi_{KS,i}(\mathbf{r})$. This seems to be impossible because Ψ_{KS} is *not* an approximation of Ψ_0. However, it turns out that it is, in fact, convenient to isolate such a contribution, and to write [65]:

$$O^{1av} = 2\sum_i \int d\mathbf{r}\psi^*_{KS,i}(\mathbf{r})\hat{o}(\mathbf{r})\psi_{KSi}(\mathbf{r}) + \Delta O[\rho] \tag{40}$$

Different prescriptions are proposed to estimate the correction ΔO; in LDA:

$$\Delta O_{LDA}[\rho] = \int d\mathbf{r}[o_h(\rho(\mathbf{r})) - o_f(\rho(\mathbf{r}))] \tag{41}$$

where $o_h(\rho)$ and $o_f(\rho)$ are the average value (per electron) of the observable O, calculated for the homogeneous gas of interacting electrons and of free electrons, respectively. Similar expressions, with the addition of a SIC term (see section

2.2), have been used, for example, to calculate Compton profiles from KS pseudo-wave-functions (see Chapter 13).

There are important quantities, notably the macroscopic dielectric polarization, which cannot be expressed in the form of equation 36: in these cases, the availability of an approximate form of the wave-function, as provided by HF theory, allows a more satisfactory formulation of the theory (see Chapter 14).

7. Conclusions

There has been, in recent years, an explosive expansion in the field of *ab initio* techniques for the treatment of solid state problems. Very rapidly, new proposals emerge, existing schemes become obsolete, or conversely, receive new impetus from the efficient solution of a certain critical computational step. The solution of problems which appeared untreatable just a few years ago is now commonplace. For users not directly involved in the development of the methodologies and in the implementation of computer codes, such a rapid evolution may be disturbing. Getting accustomed to the use of one of these complex programs, knowing how to obtain answers to one's problems and how to interpret results involves an important human investment in terms of study, time and practice. The simplicity of input, the transparency of output and the quality of the documentation can be more important than a factor two, say, in speed.

It is fortunate that both in the field of DF and of HF approaches, powerful, general-purpose programs are now available, which tend to incorporate in, consecutive releases, the natural evolution of techniques in the field. The "life-time" of these projects is becoming much longer (of the order of ten years at least) and justifies the effort required to become familiar with the programs. Another advantage of these public codes is that one can compare their performance over a variety of systems. Due to their *ab initio* nature, they should give comparable results when applied to the same system. There is now the trend to test these powerful codes with whole families of crystalline compounds. We may cite a few examples from recent literature: seventeen III-V and IV-IV semiconductors (LCAO-HF) [66]; nine oxides (LCAO-HF) [67]; ten wurtzite crystals (LCAO-LDA) [68]; carbides and nitrides of eight transition metals [69]; all hcp metals from Be to Cd [55]. This kind of documentation may help final users to make an unbiased, sensible choice. Having at least two codes based on different schemes available would be preferable, of course, because the optimal area of application is usually different for the different approaches.

Acknowledgments

This work was supported in part by the Italian MURST (fondi 40%), and by the Italian CNR.

References

1. A. Szabó and N.S. Ostlund, *Modern Quantum Chemistry*, McGraw-Hill, New York (1993).
2. R.E. Moss, *Advanced Molecular Quantum Mechanics*, Chapman and Hall, London (1973).

3. P.P. Singh, Phys. Rev. B **49**, 4954 (1994).
4. J.C. Inkson, *Many-body Theory of Solids*, Plenum, New York (1984).
5. L. Hedin, Phys. Rev. **139**, A796 (1965).
6. P. Fulde, *Electron Correlations in Molecules and Solids*, Solid-State Sciences, Vol. 100, Springer, Berlin (1993).
7. S. Raimes, *The Wave Mechanics of Electrons in Metals*, North Holland, Amsterdam (1961).
8. H.J. Monkhorst, Phys. Rev. B **20**, 1504 (1979).
9. J. Delhalle and J.L. Calais, J. Chem. Phys. **85**, 5286 (1986).
10. M. Gell-Mann and K.A. Bruckner, Phys. Rev. **106**, 364 (1957).
11. D.M. Ceperley, Phys. Rev. B **18**, 3126 (1978); D.M. Ceperley and B.J. Alder, Phys. Rev. Lett. **45**, 566 (1980).
12. J.P. Perdew and A. Zunger, Phys. Rev. B **23**, 5048 (1981).
13. P. Hohenberg and W. Kohn, Phys. Rev. **136**, B864 (1964).
14. W. Kohn and L.J. Sham, Phys. Rev. **140**, A1133 (1965).
15. L.H. Thomas, Proc. Cambr. Phil. Soc. **23**, 542 (1927); E. Fermi, Z. f. Phys. **49**, 550 (1928).
16. L. Hedin and B.I. Lundqvist, J. Phys. C: Solid St. Phys. **4**, 2064 (1971).
17. J. C. Slater, Phys. Rev. **81**, 385 (1951).
18. C. Lee, W. Yang and R.G. Parr, Phys. Rev. B **37**, 329 (1975); A.D. Becke, J. Chem. Phys. **88**, 1053 (1988); L.C. Wilson and M. Levy, Phys. Rev. B **41**, 12930 (1990); J.P. Perdew and Y. Wang, Phys. Rev. B **45**, 13244 (1991)
19. O. Gunnarsson and B.I. Lundqvist, Phys. Rev. B **13**, 4274 (1976).
20. M.R. Pederson and C.C. Lin, J. Chem. Phys. **88**, 1807 (1988).E.S. Fois, J.I. Penman and P.A. Madden, J. Chem. Phys. **98**, 6352 (1993).
21. B.G. Johnson, C.A. Gonzales, P.M.W. Gill and J.A. Pople, Chem. Phys. Lett. **221**, 100 (1994); M. Biagini, Phys. Rev. B **49**, 2156 (1994).
22. M.L. Cohen, in *Structure and Bonding in Crystals*, Vol I (M. O'Keeffe and A. Navrotsky, eds), Academic, New York, p.25 (1981).
23. J.M. André, L. Gouverneur and G. Leroy, Intern. J. Quantum Chem. **1**, 451 (1967); J.L. Brédas, J.M. André, J.G. Fripiat and J. Delhalle, Gazz. Chim. Ital. **108**, 307 (1978).
24. G. Del Re, J. Ladik and G. Biczó, Phys. Rev. **155**, 997 (1967).
25. R.N. Euwema, D.L. Wilhite and G.T. Surratt, Phys. Rev. B **7**, 818 (1973).
26. A.B. Kunz, Phys. Rev. B **6**, 606 (1972).
27. C. Pisani, R. Dovesi and C. Roetti, *Hartree-Fock ab initio Treatment of Crystalline Systems*, Lecture Notes in Chem. Vol. 48, Springer, Berlin (1988).
28. R. Dovesi, C. Pisani, C. Roetti, M. Causà and V.R. Saunders, *CRYSTAL 88, Program No. 577*, QCPE, Indiana University, Bloomington, IN (1989); R. Dovesi, C. Roetti, and V.R. Saunders, *CRYSTAL 92 User Documentation*, University of Torino, Torino (1992)
29. S. Baroni, G. Pastori Parravicini and G. Pezzica, Phys. Rev. B **32**, 4077 (1985).
30. R. Hott, Phys. Rev. B **44**, 1057 (1991).
31. P. Piecuch, S. Zarrabian, J. Paldus and J. Cizek. Phys. Rev. B **42**, 3351 (1990).
32. M.B. Lepetit, E. Aprà, J.P. Malrieu and R. Dovesi, Phys. Rev. B **46**, 12974 (1992).
33. M. Causà, R. Dovesi and C. Roetti, Phys. Rev. B **43**, 11937 (1991).
34. G.B. Bachelet, D.R. Hamann and M. Schlüter, Phys. Rev. B **26**, 4199 (1982); D.R. Hamann, Phys. Rev. B **40**, 2980 (1989).
35. D. Vanderbilt, Phys. Rev. B **41**, 7892 (1990).
36. P. Durand and J.C. Barthelat, Theor. Chim. Acta **38**, 283 (1975); J.C. Barthelat, P. Durand and A. Serafini, Molec. Phys. **33**, 159 (1977).
37. P.J. Hay and W.R. Wadt, J. Chem. Phys. **82**, 270; 284; 299 (1985).
38. G. te Velde and E.J. Baerends, Phys. Rev. B **44**, 7888 (1991).
39. A. Rauk, L.C. Allen and E. Clementi, J. Chem. Phys. **52**, 4193 (1970).
40. S.F. Boys and F. Bernardi, Molec. Phys. **19**, 553 (1970).
41. E.R. Davidson and D. Feller, Chem. Rev. **86**, 681 (1986).
42. S. Huzinaga, *Gaussian Basis Sets for Molecular Calculations*, Elsevier, Amsterdam (1984).
43. L.E. McMurchie and E.R. Davidson, J. Comp. Phys. **26**, 218 (1978); **44**, 289 (1981); V.R. Saunders, in *Methods in Computational Molecular Physics* (G.H.F. Diercksen and S. Wilson, Eds.) Reidel, Dordrecht (1983).

44. J. Tomasi, in *Basi Teoriche ed Applicazioni Numeriche* (R. Moccia and C. Pisani, Eds.) Polo Editoriale Chimico, Milano, p. 129 (1990).
45. A. Baldereschi, Phys. Rev. B **7**, 5212 (1973); D.J. Chadi and M.L. Cohen, Phys. Rev. B **16**, 1746 (1977); H.J. Monkhorst and J.D. Pack, Phys. Rev. B **13**, 5188 (1976); C. Pisani and P. Ugliengo, Phys. Stat. Sol. (b) **96**, 537 (1982).
46. O. Jepsen and O.K. Andersen, Solid St. Comm. **9**, 1763 (1971); G. Lehmann and M. Taut; Phis. Status Sol. (b) **54**, 469 (1972).
47. M.H. Boon, M.S. Methfessel and F.M. Müller, J. Phys. C **19**, 5337 (1986); G. Wiesenekker, G. te Velde and E.J. Baerends, J. Phys. C **21**, 4263 (1988).
48. G. Angonoa, R. Dovesi, C. Pisani and C. Roetti, Phil. Mag. **44**, 413 (1981).
49. I.H. Hillier and V.R. Saunders, Int. J. Quantum Chem. **4**, 503 (1970); R. Seeger and J.A. Pople, J. Chem. Phys. **65**, 265 (1976); P. Pulay, J. Comp. Chem. **3**, 556 (1982); J. Cioslowski, J. Chem. Phys. **89**, 2126 (1988); H.B. Schlegel and J.J.W. Douall, *Computational Advances in Organic Chemistry*, Kluwer, New York (1991).
50. J. Korringa, Physica **13**, 392 (1947); W. Kohn and N. Rostoker, Phys. Rev. **94**, 1111 (1954).
51. V.L. Moruzzi, J.F. Janak, A.R. Williams, *Calculated Electronic Properties of Metals*, Pergamon, New York (1978).
52. J.C. Slater, Phys. Rev. **51**, 846 (1937).
53. O.K. Andersen, Phys. Rev. B **12**, 3060 (1975).
54. E. Wimmer, H. Krakauer, M. Weinert and A.J. Freeman, Phys. Rev. B **24**, 864 (1981); P. Blaha, K. Schwarz , Intern. J. Quantum Chem. **23**, 1535 (1983).
55. P. Blaha, K. Schwarz and P.H. Dederichs, Phys. Rev. B **37**, 2792 (1988).
56. V. Heine, R. Haydock and M. Kelly, in *Solid State Physics, Vol. 35* (H. Ehrenreich, F. Seitz and D. Turnbull, Eds.), Academic, New York (1980).
57. A. Gibson, R. Haydock and J.P. LaFemina, Phys. Rev. B **47**, 9229 (1993).
58. R. Car and M. Parrinello, Phys. Rev. Lett. **55**, 2471 (1985).
59. D. Remler and P.A. Madden, Molec. Phys. **70**, 921 (1990).
60. C. Lee, D. Vanderbilt, K. Laasonen, R. Car and M. Parrinello, Phys. Rev. B **47**, 4863 (1993).
61. R. Jastrow, Phys. Rev. **98**, 1479 (1955).
62. S. Fahy, X.W. Wang, G. Louie, Phys. Rev. B **42**, 3503 (1990).
63. X.-P. Li, D.M. Ceperley and R.M. Martin, Phys. Rev. B **44**, 10929 (1991).
64. W. Yang, Phys. Rev. Lett. **66**, 438 (1991); S. Baroni and P. Giannozzi, Europhys. Lett. **17**, 547 (1992); G. Galli and M. Parrinello, Phys. Rev. Lett. **69**, 3547 (1992); E.B. Stechel, A.R. Williams and P.J. Feibelman, Phys. Rev. B **49**, 10088 (1994); W.Yang and T.S. Lee, J. Chem. Phys. **103**, 5674 (1995).
65. G.E.W. Bauer, Phys. Rev. B **27**, 5912 (1983).
66. M. Causà, R. Dovesi and C. Roetti, Phys. Rev. B **43**, 11937 (1991).
67. R. Dovesi, C. Roetti, C. Freyria-Fava, E. Aprà, V.R. Saunders and N.M. Harrison Phil. Trans. R. Soc. London **A341**, 203 (1992).
68. Y-N Xu and W.Y. Ching, Phys. Rev. B **48**, 4335 (1993).
69. J. Häglund, G. Grimvall, T. Jarlborg and A. Fernández Guillermet, Phys. Rev. B **43**, 14400 (1991).

Reciprocal Space Integration and Special-Point Techniques

Andrea Dal Corso

Institut Romand de Recherche Numérique en Physique des Matériaux (IRRMA), IN-Ecublens, 1015 Lausanne, Switzerland.

May 17, 1996

Summary. We describe the special-point technique to integrate a periodic function over the BZ. The method of Monkhorst and Pack for special-point generation is explained and an example of its application to the fcc lattice is discussed. Several convergence tests performed with the PWSCF code are presented. The problem of sampling the Fermi surface in metallic systems is briefly discussed.

Key words: Reciprocal Space Techniques – Special Points – Fermi Surface Sampling

1. Introduction

Many calculations in crystalline materials involve the averaging over the BZ of a periodic function of the wave vector: the electronic charge density is the sum over the BZ of the square modulus of the Bloch wavefunctions and the band contribution to the total energy is the average of the eigenvalues of the KS [1] equations over the wave vectors [2]. In a finite solid, with N unit cells and periodic boundary conditions, the $\boldsymbol{\kappa}$ points are a useful way of classifying the energetic levels of the electrons. The sum over N $\boldsymbol{\kappa}$ points is actually an average over the electronic states. To compute the bulk properties of a crystalline material, it is necessary to perform the thermodynamic limit where $N \rightarrow \infty$. In this limit, the sum over N becomes an integral over a finite region of the reciprocal space: the BZ. In principle, it is possible to compute this integral by a standard numerical technique [3], but in practice, this approach requires the evaluation of the integrand function on a very large number of wave vectors. Since the cost of evaluating this function at any $\boldsymbol{\kappa}$ point is usually high, it is

Correspondence to: A. Dal Corso - e-mail dalcorso@eldpa.epfl.ch important to perform the integration with a method which minimizes the number of such evaluations.

For sufficiently smooth functions, it is possible to perform the integration by computing the function in a carefully selected set of a few points in reciprocal space. These points are called ***special points*** [4]. Several special-point sets exist for each Bravais lattice and for each point group symmetry. Historically, the use of the special points in electronic structure calculations was introduced in 1973 by A. Baldereschi [5] who showed that the computation of the charge density of semiconductors could be performed by using a single point. Since then the method has been generalized in various ways. Nowadays, several points are used and many recipes are known to compute the special points of a lattice with a given point group. The technique is now a tool of paramount importance in many electronic structure calculations. Since the method is based on a Fourier expansion of the integrand function, it is particularly useful to integrate smooth functions in reciprocal space whose Fourier transform decays rapidly in real space. This fact does not create any particular problem when the special points are used in the study of semiconductors or insulators, but in the study of metals it is often necessary to integrate the Fermi distribution function. This function has a discontinuity at the Fermi energy and its Fourier transform is not localized in real space. Therefore, the application of the special-point method to metals requires particular care.

The plan of this chapter is as follows. In Section 2, we present the theory necessary to understand the special-point technique. In Section 3, we describe the technique of Monkhorst and Pack [6] which is a simple and powerful method of finding sets of special points in a Bravais lattice which can then be enlarged in a systematic way. In Section 4, we illustrate the method by applying it to an fcc Bravais lattice. For this lattice, we give several series of special points which will be used in Chapter 4 to perform some calculations on Si, NaCl and Al. In Section 5, we discuss the application of the special-point technique to the computation of the charge density and of the band contribution to the total energy. Tests of convergence of the special-point sets are given in Section 6, using the Bloch orbitals of the PWSCF code [2]. One method to deal with metallic systems is briefly discussed in Section 7. Section 8 contains conclusions drawn from this work.

2. The special-point theorem

Let us start by recalling some well known facts about the symmetry properties of a solid (see Chapter 1 and reference [7]). The position of the atoms in a crystalline solid are determined by the Bravais lattice vectors, $\mathbf{T}_\mu$ and by a finite set of vectors, $\mathbf{d}_s$, which define the atomic positions inside the unit cell. The vectors, $\mathbf{T}_\mu$, are defined by introducing three principal lattice vectors $\mathbf{a}_i$:

$$\mathbf{T}_\mu = T_{\mu,1}\mathbf{a}_1 + T_{\mu,2}\mathbf{a}_2 + T_{\mu,3}\mathbf{a}_3 \tag{1}$$

where $T_{\mu,j}$ are integer numbers. The most general symmetry operation which maps the crystal onto itself is a combination of a rotation (or pseudo-rotation) S^m plus a translation $\mathbf{f}^m$. The set of the symmetry operations is called the space group of the crystal. This group contains all the operations $(S^m|\mathbf{f}^m)$ $(m =$

$1 \cdots N_S$) such that, for each $\mathbf{T}_\mu$ and $\mathbf{d}_s$, it is possible to find a direct lattice vector $\mathbf{T}_\nu$ and a vector $\mathbf{d}_{S^m(s)}$ such that:

$$S^m(\mathbf{T}_\mu + \mathbf{d}_s) + \mathbf{f}^m = \mathbf{T}_\nu + \mathbf{d}_{S^m(s)}, \tag{2}$$

and atom $S^m(s)$ is of the same type as atom s. The set of operations $(S^m|0)$ is itself a group, called the point group of the crystal. We may define similarly the point group of the Bravais lattice. In general, the point group of the crystal is a subgroup of the point group of the Bravais lattice. In fact, a basis of N_{at} atoms per unit cell might lower the symmetry of the Bravais lattice. The special-point technique can be formulated without any hypothesis being made about the point group associated with the reciprocal lattice which may be the point group of the Bravais lattice or a subgroup of it.

We are interested in the integral:

$$\bar{f} = \frac{\Omega}{(2\pi)^3} \int_{BZ} d\boldsymbol{\kappa}\, f(\boldsymbol{\kappa}) \tag{3}$$

where $f(\boldsymbol{\kappa})$ is a periodic function of the wave vector $\boldsymbol{\kappa}$ with period given by the reciprocal lattice vectors and Ω is the unit cell volume. If $f(\boldsymbol{\kappa})$ does not have the same symmetry as the crystal, we can consider the symmetrized function:

$$f_\Gamma(\boldsymbol{\kappa}) = \frac{1}{N_S} \sum_{m=1}^{N_S} f(S^m \boldsymbol{\kappa}) \tag{4}$$

The integral over the BZ of $f_\Gamma(\boldsymbol{\kappa})$ is equal to $\bar{f}$ and f_Γ has the symmetry of the point group of the crystal. Therefore, it is possible to assume that $f(\boldsymbol{\kappa})$ has the same symmetry as the crystal. In Section 5, we show how to apply equation (4) in a practical case where this hypothesis is not valid.

Let us expand $f(\boldsymbol{\kappa})$ in Fourier components. Since $f(\boldsymbol{\kappa})$ is periodic, with period given by the vectors of the reciprocal lattice, the expansion is over the vectors of the direct lattice $\mathbf{T}_\mu$:

$$f(\boldsymbol{\kappa}) = \sum_\mu f(\mathbf{T}_\mu) e^{\imath \boldsymbol{\kappa} \mathbf{T}_\mu} \tag{5}$$

Furthermore, if $\mathbf{T}_\mu$ and $\mathbf{T}_\nu$ are related by a symmetry operation, $f(\mathbf{T}_\mu) = f(\mathbf{T}_\nu)$, because $f(\boldsymbol{\kappa})$ is invariant with respect to the operations of the point group. Therefore, we can divide all the direct lattice vectors into shells of symmetry related vectors:

$$C_n = \{\mathbf{T}_m = S^m \mathbf{T}_0 \mid m = 1, \cdots, N_S\} \tag{6}$$

where the index n is used to order the shells according to increasing modulus of their $\mathbf{T}$ vectors. In all the elements of a given shell, C_n, the Fourier transform, $f(\mathbf{T}_m)$, has the same value: $f(\mathbf{T}_m) = f(\mathbf{T}_0) = f_n$ Therefore, it is possible to write the Fourier expansion of $f(\boldsymbol{\kappa})$ as

$$f(\boldsymbol{\kappa}) = \sum_{n=0}^{\infty} f_n \sum_{|\mathbf{T}| \in C_n} e^{\imath \boldsymbol{\kappa} \mathbf{T}} = \sum_{n=0}^{\infty} \tilde{f}_n A_n(\boldsymbol{\kappa}) \tag{7}$$

where:

$$A_n(\boldsymbol{\kappa}) = \frac{1}{\sqrt{N_n}} \sum_{|\mathbf{T}| \in C_n} e^{\imath \boldsymbol{\kappa} \mathbf{T}} \tag{8}$$

N_n is the number of distinct vectors of the shell and $\tilde{f}_n = \sqrt{N_n} f_n$. From the definition of $A_n(\boldsymbol{\kappa})$, it is easy to see that:

$$\int_{BZ} A_n(\boldsymbol{\kappa}) = 0 \quad \text{if} \quad |\mathbf{T}| \neq 0 \tag{9}$$

This result shows that the integral over the BZ of $f(\boldsymbol{\kappa})$ is equal to f_0. Let us now consider what happens when we evaluate $f(\boldsymbol{\kappa})$ in a point $\boldsymbol{\kappa}_i$. We have:

$$f(\boldsymbol{\kappa}_i) = f_0 + \sum_{n=1}^{\infty} \tilde{f}_n A_n(\boldsymbol{\kappa}_i) \tag{10}$$

The value $f(\boldsymbol{\kappa}_i)$ differs from f_0 by a quantity which depends on the values of $A_n(\boldsymbol{\kappa}_i)$. If we could find a single point, $\boldsymbol{\kappa}^*$ such that $A_n(\boldsymbol{\kappa}^*) = 0$ for each $n \neq 0$, we could have $f_0 = f(\boldsymbol{\kappa}^*)$. If we choose a $\boldsymbol{\kappa}^*$ which gives $A_n(\boldsymbol{\kappa}^*) = 0$ for $n = 1, 2, \ldots N_p$ with the largest possible N_p, we obtain the best approximation of f_0 with a single point. Note that, by the mean-value-point theorem, it is always possible to find a point $\boldsymbol{\kappa}'$ inside the integration region such that $f_0 = f(\boldsymbol{\kappa}')$ but, in this case, the point $\boldsymbol{\kappa}'$ depends on the integrand function. In the special-point method, $\boldsymbol{\kappa}^*$ does not depend on the function f but only on the lattice and on the point group which is used to build $A_n(\boldsymbol{\kappa})$. The number of shells, N_p for which $A_n(\boldsymbol{\kappa}) = 0$, is limited by the compatibility between these equations. In general, we expect $N_p = 3$ because there are three independent components of $\boldsymbol{\kappa}^*$. It might happen that this set of equations is incompatible. In this case, $N_p = 2$ and the point $\boldsymbol{\kappa}^*$ can be defined as the one which minimizes $A_3(\boldsymbol{\kappa})$ [5].

In order to integrate a generic function a single point does not suffice and it is useful to use larger and larger sets of points. Let us suppose that we could find a set of $N_{\boldsymbol{\kappa}}$ points and relative weights $\{\boldsymbol{\kappa}_i, w_i\}$ such that:

$$\sum_{i=1}^{N_{\boldsymbol{\kappa}}} w_i = 1 \tag{11}$$

$$\sum_{i=1}^{N_{\boldsymbol{\kappa}}} w_i A_n(\boldsymbol{\kappa}_i) = 0 \quad n = 1, \cdots, N_p \tag{12}$$

then we use these equations in the Fourier expansion of $f(\boldsymbol{\kappa})$, equation (5) and we obtain:

$$\sum_{i=1}^{N_{\boldsymbol{\kappa}}} w_i f(\boldsymbol{\kappa}_i) = f_0 + \sum_{n=N_p+1}^{\infty} \tilde{f}_n \sum_{i=1}^{N_{\boldsymbol{\kappa}}} w_i A_n(\boldsymbol{\kappa}_i) \tag{13}$$

In this case, we can approximate f_0 with:

$$f_0 \simeq \sum_{i=1}^{N_{\boldsymbol{\kappa}}} w_i f(\boldsymbol{\kappa}_i) \tag{14}$$

the remaining error being:

$$\epsilon = - \sum_{n=N_p+1}^{\infty} \tilde{f}_n \sum_{i=1}^{N_{\kappa}} w_i A_n(\kappa_i) \tag{15}$$

The magnitude of ϵ depends on the values of the Fourier transform f_n for $n > N_p$. For sufficiently smooth functions, the values of f_n tend rapidly to zero as n becomes large; the precision of the calculation can be increased simply by increasing the number of sampling points and, therefore, of shells N_p for which equation (12) is true.

3. The special-point generation

In principle, in order to determine the coordinates of a special-point set, it is sufficient to solve equation (12) with the constraints given by equation (11). Although this appears to be a formidable task because equation (12) is a non-linear system, the periodicity and the point group symmetry of $A_n(\kappa)$ lead to useful simplifications. Chadi and Cohen [8] have given a procedure to generate special-point sets and have presented sets for the cubic and hexagonal lattice. Monkhorst and Pack [6] have developed a more systematic prescription based on equally spaced points. At present, it is not clear which is the "best" method for κ-point generation but the Monkhorst and Pack scheme appears to be reliable and simple to use. In this Section, we describe their method. First, the number q which determines the size of the mesh is chosen. For this given q, the method is as follows:

a) Define the sequence of q numbers:

$$u_r = \frac{2r - q - 1}{2q} \quad \text{and } r = 1, 2, \cdots q \tag{16}$$

b) Choose the following q^3 κ points:

$$\kappa_{prs} = u_p \mathbf{b}_1 + u_r \mathbf{b}_2 + u_s \mathbf{b}_3 \tag{17}$$

where $\mathbf{b}_i$ represents the principal reciprocal lattice vectors and u_p and u_s are defined in the same way as u_r.

c) Choose an irreducible zone of the reciprocal lattice. If a point κ_{prs} is inside this irreducible zone, it is considered to be a special point and its weight is set to 1. If it is outside, we find the equivalent point inside the irreducible BZ (IBZ). If this point coincides with one of the special points already found, the weight of that point is increased by one, otherwise the point is considered as a new special point and its weight is set to one.

d) Renormalize all the weights by dividing by q^3. This procedure assures that equation (11) is verified.

If the point group has a large number of symmetry operations, this procedure can significantly reduce the total number of points.

We now show that we actually obtain a special-point set which fulfills equation (12) using this procedure. In order to do this, it is sufficient to show that the original mesh, κ_{prs}, fulfills equation (12) because then the points obtained by a rotation of the point group symmetry or translated by a reciprocal lattice vector will also fulfill the same set of equations. In fact, from the definition of

$A_n(\boldsymbol{\kappa})$ we have $A_n(\boldsymbol{\kappa}+\mathbf{G}) = A_n(\boldsymbol{\kappa})$ and $A_n(S^m\boldsymbol{\kappa}) = A_n(\boldsymbol{\kappa})$. We have to show that:

$$\sum_{p,r,s=1}^{q} A_n(\boldsymbol{\kappa}_{prs}) = 0 \tag{18}$$

for each $n \le N_p$ and we have to compute N_p. Actually, it is possible to show a more general relationship. Let us define:

$$S_{mn}(q) = \frac{1}{q^3} \sum_{p,r,s=1}^{q} A_m^*(\boldsymbol{\kappa}_{prs}) A_n(\boldsymbol{\kappa}_{prs}) \tag{19}$$

We show that $S_{mn}(q) = \delta_{mn}$ if $m < N_p$ and $n < N_p$. Equation (18) then follows as a special case with $m = 0$ and $n \neq 0$. The demonstration is straightforward. We insert the definition of $A_n(\boldsymbol{\kappa})$ into equation (19) and we note that:

$$\boldsymbol{\kappa}_{prs}\mathbf{T}_\mu = \frac{\pi}{q}\Big[(2p-q-1)T_{\mu,1} + (2r-q-1)T_{\mu,2} + (2s-q-1)T_{\mu,3}\Big] \tag{20}$$

Then we can write:

$$S_{mn}(q) = \frac{1}{\sqrt{N_m N_n}} \sum_{a=1}^{N_m}\sum_{b=1}^{N_n}\prod_{j=1}^{3} W_j^{ab}(q) \tag{21}$$

where:

$$W_j^{ab}(q) = \frac{1}{q}\sum_{r=1}^{q} e^{\left(\frac{i\pi}{q}\right)(2r-q-1)(T_{b,j}-T_{a,j})} \tag{22}$$

and where $T_{a,j}$ is an integer number which expresses the component of $\mathbf{T}_a$ over the j-th principal vector. In equation (21), the sums over $a = 1,\ldots,N_m$ and $b = 1,\ldots,N_n$ are performed over the members of the stars C_m and C_n, respectively. The computation of $W_j^{ab}(q)$ requires the sum of a finite geometric series. We have the following possibilities: $W_j^{ab}(q) = 1$ if $|T_{b,j} - T_{a,j}| = 0, 2q, 4q\ldots$; $W_j^{ab}(q) = (-1)^{q+1}$ if $|T_{b,j} - T_{a,j}| = q, 3q, 5q\ldots$; $W_j^{ab}(q) = 0$ in all the other cases. It should be remembered that q, $T_{a,j}$ and $T_{b,j}$ are integers. If we consider only $|T_j| < q/2$, we have the result that $|T_{a,j} - T_{b,j}|$ is always lower than q. In this case, $S_{mn}(q)$ is non-zero only when $T_{a,j} = T_{b,j}$. Therefore, $S_{m,n} \neq 0$ only if $m = n$ and $S_{mm}(q) = 1$. The value of N_p depends upon the lattice. We discuss the case of the fcc lattice in the next section.

4. Application to the fcc lattice

In this section, we show an example of the application of the above technique to an fcc lattice. As the symmetry group, we choose the O_h [7] point group. This point group has 48 symmetry operations. There are six rotations of $\pm\pi/2$ about a coordinate axis, three rotations of π about the same axis, six rotations of π about a cube face diagonal and eight rotations of $\pi/3$ about a cube diagonal. Together with the identity operation, these represent all of the 24 proper group operations. Multiplying each operation by the inversion, we obtain the other 24 operations. For a cubic system, it is convenient to use even values of q. Using $q = 2$, we obtain 8 points in the mesh, κ_{prs}, with $u_r = -1/4, 1/4$.

Table 1. Special-point coordinates of the fcc lattice obtained with the method of Monkhorst and Pack.

q=2	κ_x	κ_y	κ_z	w_κ
1	1/4	1/4	1/4	2
2	1/4	1/4	3/4	6
6 points set				
1	1/6	1/6	1/6	2
2	1/6	1/6	1/2	6
3	1/6	1/6	5/6	6
4	1/6	1/2	1/2	6
5	1/6	1/2	5/6	6
6	1/2	1/2	1/2	1
q=4				
1	1/8	1/8	1/8	2
2	1/8	1/8	3/8	6
3	1/8	1/8	5/8	6
4	1/8	1/8	7/8	6
5	1/8	3/8	3/8	6
6	1/8	3/8	5/8	12
7	1/8	3/8	7/8	12
8	1/8	5/8	5/8	6
9	3/8	3/8	3/8	2
10	3/8	3/8	5/8	6

We choose an IBZ which is 1/48 of the BZ. One possible choice is given by the conditions:

$$0 \leq \kappa_x \leq \kappa_y \leq \kappa_z \leq \frac{2\pi}{a_0} \tag{23}$$

$$\kappa_x + \kappa_y + \kappa_z \leq \frac{3\pi}{a_0} \tag{24}$$

Reducing the points using the method described in the previous section, we obtain two distinct points which are given in Table 1 with the correct weight.

These points fulfill equation (12) up to $N_p = 7$ and the first failure is at $\mathbf{T} = a_0(2,0,0)$.

The choice of $q = 4$ in equation (16) yields 64 points which are determined by the coordinates $u_r = -3/8, -1/8, 1/8, 3/8$. They correspond to 10 distinct points in the irreducible zone. These points fulfill equation (12) up to $N_p = 39$ and the first failure is at $\mathbf{T} = a_0(4,0,0)$. For simple semiconductors and insulators, they generally give converged values for many quantities, with the exception of the quantities which depend on an electric field such as the dielectric constant. Where necessary, it is possible to increase the precision by using $q = 6$ which corresponds to 28 points in the IBZ. These points satisfy equation (12) up to $N_p = 114$ and the first failure is at $\mathbf{T} = a_0(6,0,0)$.

In Table 1, we report all the special-point coordinates in units $2\pi/a_0$. The weights are given before the normalization. Each weight corresponds to the number of points which are equivalent to the given point in the full BZ. For completeness, we report in the same table also an intermediate set of 6 points. These points are used in Chapter 10 to perform some calculations on silicon, NaCl and aluminum. They fulfill equation (12) up to $N_p = 20$. The first failure is for $\mathbf{T} = a_0(3,0,0)$.

5. Symmetry of the charge density

In some cases, the function $f(\boldsymbol{\kappa})$ does not have the point group symmetry of the crystal. In the above discussion, we have avoided the problem by showing that only the symmetrized function, $f_\Gamma(\boldsymbol{\kappa})$, contributes to the integral (see equation (4)). However, to compute the function $f_\Gamma(\boldsymbol{\kappa})$, we must evaluate the function $f(\boldsymbol{\kappa})$ on all the symmetry related points, $S^m\boldsymbol{\kappa}$. This means that we would have to compute $f(\boldsymbol{\kappa})$ on a mesh of q^3 points and the advantages of using the special-point method would be partially lost. Clearly, there is not a general solution to this problem. The solution depends on the function and can be illustrated on real examples. In many real cases, the value of the function $f(\boldsymbol{\kappa})$ on the points $S^m\boldsymbol{\kappa}$ can be inferred without actually evaluating it. We illustrate here the case of charge density. In this instance, the density at the point $\mathbf{r}$ is given by the integral over the BZ of the square modulus of the Bloch wavefunctions computed at the point $\mathbf{r}$:

$$\rho(\mathbf{r}) = \frac{\Omega}{(2\pi)^3}\sum_v \int_{BZ} d\boldsymbol{\kappa}\ |\psi_v(\mathbf{r};\boldsymbol{\kappa})|^2 \tag{25}$$

The sum v is over the valence bands. The integrand function is $f_v(\boldsymbol{\kappa}) = |\psi_v(\mathbf{r};\boldsymbol{\kappa})|^2$ which depends parametrically on the position $\mathbf{r}$. We must evaluate the integral over the BZ for each point of the real space mesh. Although the Bloch wavefunctions are not invariant with respect to a point group rotation of the wave vector, they fulfill the following relationship [9]:

$$\psi_{S^m\boldsymbol{\kappa},v}(\mathbf{r};S^m\boldsymbol{\kappa}) = \psi_v((S^m)^{-1}\mathbf{r} - \mathbf{f}^m;\boldsymbol{\kappa}) \tag{26}$$

Furthermore, the time-reversal symmetry guarantees that $\psi_v(\mathbf{r};-\boldsymbol{\kappa}) = \psi_v(\mathbf{r};\boldsymbol{\kappa})$ so that even if the point group of the crystal does not contain the inversion symmetry, we can use a larger group, which does contain the inversion, in the computation of the charge density. These properties can be used to compute the symmetrized function, $f_\Gamma(\boldsymbol{\kappa})$, without actually computing $f(\boldsymbol{\kappa})$ at a point $\boldsymbol{\kappa}$ outside the IBZ. We have:

$$f_\Gamma(\boldsymbol{\kappa},\mathbf{r}) = \frac{1}{N_S}\sum_{m=1}^{N_S} |\psi_v(\mathbf{r};(S^m)\boldsymbol{\kappa})|^2 = \frac{1}{N_S}\sum_{m=1}^{N_S} |\psi_v((S^m)^{-1}\mathbf{r}-\mathbf{f}^m;\boldsymbol{\kappa})|^2 \quad (27)$$

Using this relationship, we write the density at a point $\mathbf{r}$ as:

$$\begin{aligned}\rho(\mathbf{r}) &= \frac{\Omega}{(2\pi)^3}\sum_v \int_{BZ} f_{\Gamma,v}(\boldsymbol{\kappa},\mathbf{r})d\boldsymbol{\kappa} \\ &= \frac{\Omega}{(2\pi)^3}\frac{1}{N_S}\sum_v\sum_{m=1}^{N_S}\int_{BZ} |\psi_v((S^m)^{-1}\mathbf{r}-\mathbf{f}^m;\boldsymbol{\kappa})|^2 d\boldsymbol{\kappa} \quad (28)\end{aligned}$$

In a practical computation, the sum over the BZ of the square modulus of the Bloch wavefunctions is first performed with special-point discretization, giving the unsymmetrized charge density:

$$\tilde{\rho}(\mathbf{r}) = \sum_v \sum_{\boldsymbol{\kappa}\in IBZ} w_{\boldsymbol{\kappa}} |\psi_v(\mathbf{r};\boldsymbol{\kappa})|^2 \quad (29)$$

and then equation (28) is used to symmetrize the result and to compute the total charge:

$$\rho(\mathbf{r}) = \frac{1}{N_S}\sum_{m=1}^{N_S} \tilde{\rho}((S^m)^{-1}\mathbf{r}-\mathbf{f}^m) \quad (30)$$

In the following test, we also studied the convergence of the sum of the eigenvalues of the KS equation over the BZ with special-point sampling. This is the band contribution to the total energy (see equation (7) of Chapter 10). In this case, $f_v(\boldsymbol{\kappa}) = \epsilon_{\boldsymbol{\kappa},v}$. These eigenvalues must be invariant with respect to the point group symmetry of the solid [9] so that no particular problem is present: $f_{\Gamma,v}(\boldsymbol{\kappa}) = f_v(\boldsymbol{\kappa})$.

6. Test of convergence

In this section, we show some examples of integration over the BZ performed with the special-point technique. We concentrate on two quantities: the band contribution to the total energy and the electronic charge density along the (111) direction. The band energy is defined as:

$$E_B = \frac{\Omega}{(2\pi)^3}\sum_v \int_{BZ} \epsilon_{\boldsymbol{\kappa},v} d\boldsymbol{\kappa} \quad (31)$$

The charge density is computed as explained in the previous section and is plotted, taking the points $\mathbf{r} = a_0(t,t,t)$ with $0 \le t \le 1$, as a function of $|\mathbf{r}|$. To give some numerical values, we also consider the charge on the three points $\mathbf{r}_1 = (0,0,0)$, $\mathbf{r}_2 = a_0(1/2,1/2,1/2)$ and $\mathbf{r}_3 = a_0(1/8,1/8,1/8)$. As an example, we studied the silicon crystal, which is considered in greater detail in Chapter 10. Here, we simply note that it has an fcc Bravais lattice with two atoms per cell. The point $\mathbf{r}_1$ is on one silicon, the point $\mathbf{r}_2$ is at the center of the cell, while $\mathbf{r}_3$ is at the center of the bond which connects the two atoms. We computed these quantities using several special-point sets. We first performed a self-consistent calculation with 10 special points in the IBZ and a 20 Ry cut-off (see Chapter 10) to obtain a completely converged Hartree, exchange and correlation potential and then we performed the computations of the charge and of the energy with this fixed potential. In Table 2, we report the band energy and the charge density on $\mathbf{r}_1$, $\mathbf{r}_2$ and $\mathbf{r}_3$ obtained by using the points of Table 1.

Table 2. Values of band energy and of the charge density of the silicon crystal. The energy is expressed in Ry while the charge is in 10^{-3}electrons per $(\text{a.u.})^3$.

	E_B	$\mathbf{r}_1$	$\mathbf{r}_2$	$\mathbf{r}_3$
2 pt.	0.57533	2.491	3.123	85.395
6 pt.	0.56445	2.543	3.366	84.318
10 pt.	0.56325	2.556	3.407	84.146
28 pt.	0.56304	2.558	3.416	84.096

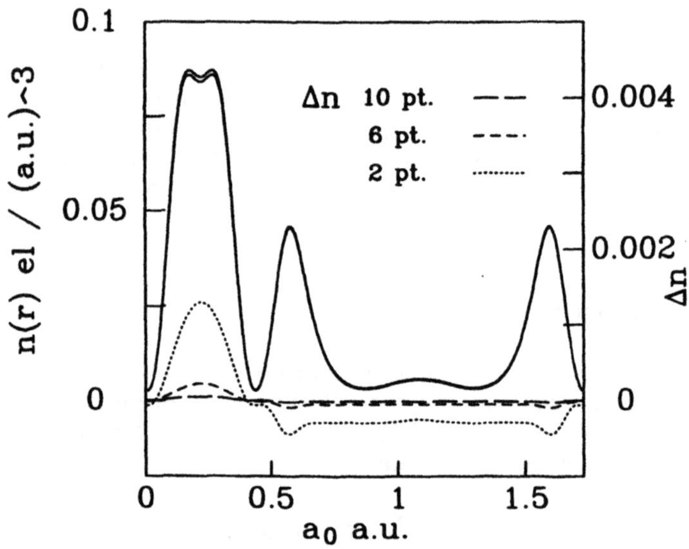

Figure 1. Valence pseudo-charge density of silicon along the (111) direction and errors due to the finite sampling of the BZ. The total charge is indicated with a continuous line, the errors are indicated with dotted, dashed and long dashed lines.

We also report in Table 2, the results obtained with 28 points, which we consider to be completely converged. It is interesting to note that the value of the band energy obtained using 2 special points has already converged to within 2% of the limit. In a total energy computation, a precision of the order of 1 mRy is required and in this case, 6 points are sufficient. The charge density, both on the maximum ($\mathbf{r}_3$) and on the minimum ($\mathbf{r}_1$) has converged within a few percent using only two κ points.

In Figure 1, we report with a continuous line the charge density along the (111) direction, computed both with 28 and with 2 special points. On the scale of the total charge, the two charges almost coincide. In the same plot, we show, on an enlarged scale to the right, the differences between the charge obtained with 2, 6 or 10 special points and the converged charge obtained with 28 points. For all practical purposes, 6 κ points give a converged charge density.

7. The case of metals

In the study of metals, we wish to evaluate:

$$\bar{f} = \frac{\Omega}{(2\pi)^3} \int_{BZ} d\kappa\ \theta(e_F - \epsilon(\kappa)) f(\kappa) \tag{32}$$

where e_F is the Fermi energy and $\theta(x)$ is the step function $\theta(x) = 0$ if $x < 0$ and $\theta(x) = 1$ if $x > 0$. Since the integrand function is discontinuous in reciprocal space, its Fourier transform does not decay rapidly in real space. To reach sufficient precision with the special-point technique a large number of points must be used. Furthermore, there is a second problem which may also cause instabilities in an electronic structure calculation. If we approximate the integral as in equation (14):

$$\bar{f} = \sum_{i=1}^{N_\kappa} w_i \theta(e_F - \epsilon(\kappa_i)) f(\kappa_i) \tag{33}$$

it may happen that, with a small change in the value of the Fermi energy, a κ point enters or exits from the Fermi surface, producing a discontinuous change of the values of $\bar{f}$. This pathology can create a numerical instability in the self-consistent cycle of an electronic structure code.

One method to avoid these problems is to use the linear tetrahedron method [14] without special points. This procedure allows accurate values of the BZ integrals to be obtained, but requires many more κ points. There are also simple methods to avoid these effects using the special-point technique. One possibility is to introduce a sequence of continuous functions $\bar{\theta}(\Delta, x)$, parametrized by a parameter Δ such that in the limit $\Delta \to 0$, $\bar{\theta}(\Delta, x)$ becomes the step function. In this hypothesis, the integral over the BZ can be performed by substituting the discontinuous step function with $\bar{\theta}$ which is continuous. The convergence of the result can be checked by decreasing the value of Δ. If Δ is large the BZ integral converges with a small number of κ points, but—in general—to an inacurate result. As Δ is decreased, the integral tends to the correct result, but a larger and larger set of κ points is required to perform the integral. For a given Δ, after reaching convergence with the κ points, the sampling error depends on the functional form of $\bar{\theta}(\Delta, x)$. One of the most popular forms of $\bar{\theta}(\Delta, x)$ is due to Fu and Ho [10]: it consists in "smearing" the Fermi surface with a Gaussian distribution so that the step function is approximated by:

$$\bar{\theta}(\Delta, x) = \frac{2}{\sqrt{\pi}} \int_{-\infty}^{x/\Delta} \exp(-y^2)dy \tag{34}$$

The Fermi energy is determined by imposing the constraint of conservation on the number of electrons with the same Δ:

$$\frac{\Omega}{(2\pi)^3} \int_{BZ} d\boldsymbol{\kappa}\, \bar{\theta}(\Delta, e_F - \epsilon(\boldsymbol{\kappa})) = N_{el} \tag{35}$$

This approximation creates smooth integrand functions different from the true function by an error which depends on Δ. When the technique is applied in a code which computes the total energy of a solid within DFT (see Chapter 10) it is possible to estimate this error and to add, to the total energy functional, a Δ dependent correction [11, 12]. This procedure allows a correct value of the total energy to be obtained with a large Δ and a few $\boldsymbol{\kappa}$ points. The smearing error can also be corrected in a few other cases but a correction valid for a general function $f(\boldsymbol{\kappa})$ is not yet available. Furthermore, the introduction of an additional term to the total energy functional creates some problems for the computation of the derivatives of the energy.

A second possibility is to use a form of the functions $\bar{\theta}(\Delta, x)$ which systematically corrects for the smearing error. A set of functions has been introduced by Methfessel and Paxton [13] and is briefly described here. The first step is to transform the BZ integral into an energy integral:

$$\bar{f} = \frac{\Omega}{(2\pi)^3} \int_{BZ} d\boldsymbol{\kappa}\, \theta(e_F - \epsilon(\boldsymbol{\kappa}))f(\boldsymbol{\kappa}) = \int_{-\infty}^{+\infty} \theta(e_F - \epsilon)F(\epsilon)d\epsilon \tag{36}$$

where:

$$F(\epsilon) = \frac{\Omega}{(2\pi)^3} \int_{BZ} d\boldsymbol{\kappa}\, \delta(\epsilon - \epsilon(\boldsymbol{\kappa}))f(\boldsymbol{\kappa}) \tag{37}$$

The error due to the use of $\bar{\theta}(\Delta, x)$ is:

$$\delta\bar{f} = \int_{-\infty}^{+\infty} \Big[\bar{\theta}(\Delta, x) - \theta(x)\Big] F(\epsilon)d\epsilon \tag{38}$$

This integral is actually over the small energy interval around the discontinuity where $\bar{\theta}(\Delta, x)$ differs from $\theta(x)$. If the function $F(\epsilon)$ is sufficiently smooth in this interval, it can be approximated by a polynomial of degree P. In this case, a function $\bar{\theta}_P(\Delta, x)$ can be found for which $\delta\bar{f}$ is exactly zero. In particular, if $F(\epsilon)$ is a constant in the interval, the choice of Fu and Ho is sufficient to give $\delta\bar{f} = 0$. A better approximation is to choose $\bar{\theta}$ so that $\delta\bar{f} = 0$ even if $F(\epsilon)$ is linear in ϵ. In this case, we have:

$$\bar{\theta}_1(\Delta, x) = \frac{1}{2\sqrt{\pi}} \int_{-\infty}^{x/\Delta} e^{-y^2} dy + \frac{1}{4\sqrt{\pi}} \frac{2x}{\Delta} \exp\left[-\left(\frac{x}{\Delta}\right)^2\right] \tag{39}$$

Higher order functions are also available [13], but it is not possible to use a very high value of P, otherwise the functions θ_P coincide with the step function. In simple metals, the functions $\bar{\theta}_1(\Delta, x)$ are sufficient to give the correct value of the integral even with a large value of Δ and a few $\boldsymbol{\kappa}$ points. One possible difficulty with this scheme is the fact that some states have a weight larger than two or lower than zero. In systems having regions with low charge density (*e.g.* nearby surfaces), a negative charge density could be obtained with this method. Due to the fact that $\theta_1(\Delta, x)$ is no longer monotonic, there is a non-unique definition of the Fermi energy. However, these problems do not occur in bulk calculations where the charge density is fairly uniform throughout the whole system. One example of the application of this technique is given in Chapter 10, where we compute the ground state of aluminum.

8. Conclusions

In this chapter, we discussed the special-point method for the integration of a periodic function over the BZ. We presented the special-point theorem and we illustrated the method of Monkhorst and Pack for the special-point generation. The method has been applied to an fcc lattice for which several special-point sets have been given. In insulators and semiconductors, the application of the method allows one to perform the BZ integral by using a very small number of sampling points. In the case of silicon, we have shown that 2 points are sufficient to give a charge density and a band energy which differs only by a few percent from the converged value. The treatment of metals with this technique is slightly more difficult due to the presence of the Fermi surface. Some possible solutions of this problem have been briefly discussed.

Acknowledgements

I gratefully acknowledge many useful discussions with A. Devita, F. Mauri, S. Massida and R. Resta. This work was supported by the Swiss National Science Foundation under Grant No. 20-30272.90.

References

1. W. Kohn and L.J. Sham, Phys. Rev. **140**, 1133 (1965).
2. See the chapter "The PWSCF code" this issue.
3. W.H. Press, B.P. Flannery, S.A. Teukolsky, and W.T. Vetterling, "Numerical Recipies" (Cambridge University Press, Cambridge, 1986).
4. For a comprehensive review see R.A. Evarestov and V.P. Smirnov, Phys. Stat. Sol. (b) **119**, 9 (1993).
5. A. Baldereschi, Phys. Rev. B **7**, 5212 (1973).
6. H.J. Monkhorst and J.D. Pack, Phys. Rev. B **13**, 5188 (1976).
7. G.F. Koster, in *Solid State Physics*, edited by H. Ehrenreich, F. Seitz and D. Turnbull, vol. 5 (Academic, New York, 1957).
8. D.J. Chadi and M. L. Cohen, Phys. Rev. B **8**, 5747 (1973).
9. F. Bassani and G. Pastori Parravicini, *Electronic states and optical transitions in solids* (Pergamon Press, Oxford, 1975).
10. C.-L. Fu and K.M. Ho, Phys. Rev. B **28**, 6687 (1983).
11. A. Devita, Ph.D. Thesis, Keele University (1992).
12. C. Elsässer, M. Fähnle, C.T. Chan, and K.M. Ho, Phys. Rev. B **49**, 13 975 (1994).
13. M. Methfessel and A.T. Paxton, Phys. Rev. B **40**, 3616 (1989).
14. G. Lehmann and M. Taut, Phys. Status Solid **54**, 469 (1962).

Numerical Integration in Density Functional Methods with Linear Combination of Atomic Orbitals

Mauro Causà

Department of Inorganic, Physical and Materials Chemistry, University of Torino, via Giuria 5, I-10125 Torino, Italy

May 17, 1996

Summary. The generalization of Becke's numerical integration scheme to periodic functions is presented, which allows the LCAO-Kohn-Sham equations for crystals to be solved efficiently. The computational implementation of the scheme and its calibration are briefly discussed.

Key words: Numerical Integration of Periodic Functions – Density Functional Theory (DFT) – LCAO

1. Introduction

The application of the density functional (DFT) method (see references[1]-[6]) has become a powerful technique to study the electronic structure of atoms, molecules and solids. Due to the analytical complexity of the correlation and exchange energy formulas, its application requires numerical integrations to be performed. The integrands, usually analytic functions of the electron density and its derivatives, have a very inhomogeneous spatial distribution, with high cusps corresponding to the positions of the nuclei: this is true not only for All Electron (AE) calculations, but also for Effective Core Potential (ECP) calculations, in the small-core approximation [7].

Becke [8] introduced a very powerful integration method, which subdivides the numerical integral into a set of integrals, each centered on a nucleus, recognizing the essentially spherical nature of the cusps: this scheme has been successfully applied to many molecular systems, including those using LCAO [9] as well as fully numerical schemes [10]. The HF-LCAO method for polymers, surfaces and bulk solids [11], developed in our laboratories and coded in the computer program CRYSTAL [12], has been recently generalized to include DFT techniques, both in the *hybrid* HF/DFT [9, 13] and in the Kohn-Sham [14] approximations. The numerical integration of DFT energy functionals for periodic densities presents the same difficulties as for molecules but is further complicated by the infinite number of contributions to the integrand. This work aims to show the formal generalization of Becke's method to solids and its computational implementation and calibration, in order to find a reasonable compromise between numerical accuracy and computer time.

2. Description of the method

In the periodic DFT-LCAO method [15], the electron-electron interaction's contribution to the energy must be calculated accurately by the following integration:

$$E^{DFT} = \int_{\text{UNIT CELL}} \varepsilon^{DFT}(\mathbf{r})d\mathbf{r} \qquad (1)$$

where $\varepsilon^{DFT}(\mathbf{r})$ is the DFT energy density, representing exchange-correlation or correlation-only interactions. In AE calculations, the integrand has very high cusps, corresponding to the nuclear positions: the exchange energy density is proportional to $\rho^{4/3}$, while the correlation energy density is roughly proportional to ρ (where ρ is the electron density). The numerical integration method must take into account the cusp-shaped nature of the integrand functions. The homogeneous sampling methods that are applied successfully in LCAO crystalline calculations to integrate numerically the smoother reciprocal space quantities, cannot be used to integrate the highly inhomogeneous quantities defined in direct space.

The atomic partition method, applied by Becke [10] and Savin [17] to molecular systems, defines atomic domains and performs spherical integrations that concentrate many sampling points in the core cusp regions. The atomic domains are overlapping and therefore, any border discontinuity, which can make the numerical quadrature processes difficult, is avoided. In the following, the formulas for a molecular case are derived, so the integration domain is extended to all space.

The atomic domains are obtained after formally multiplying the integrand by a unit weight function:

$$\begin{aligned} E^{DFT} &= \int \varepsilon^{DFT}(\mathbf{r})d\mathbf{r} = \int \varepsilon^{DFT}(\mathbf{r}).1.d\mathbf{r} = \\ &= \int \varepsilon^{DFT}(\mathbf{r}) \frac{\sum_{A}^{n\ atoms} w_A(\mathbf{r})}{\sum_{B}^{n\ atoms} w_B(\mathbf{r})} d\mathbf{r} \end{aligned} \qquad (2)$$

The introduction of the sum over n atoms in the unit weight function, allows the splitting of the integral into a sum of integrals:

$$E^{DFT} = \sum_{A}^{n\ atoms} \int \varepsilon^{DFT}(\mathbf{r}) \frac{w_A(\mathbf{r})}{\sum_{B}^{n\ atoms} w_B(\mathbf{r})} d\mathbf{r} \equiv$$

$$\equiv \sum_{A}^{n\ atoms} \int \varepsilon^{DFT}(\mathbf{r}) W_A(\mathbf{r}) d\mathbf{r} \tag{3}$$

The atomic weight functions, w, are chosen so that the weights, W_A, have the value of 1 near atom A and go to zero close to the other atoms. In this way, the integration space is partitioned into overlapping domains (each of the atomic integrals extends to all space) but the integrand function, $\varepsilon^{DFT}(\mathbf{r})W_A(\mathbf{r})$, has a cusp at atom A and goes rapidly to zero in the direction of the other atoms. Following Becke, the weight functions, w, are defined as follows:

$$\begin{aligned} w_A(\mathbf{r}) &= \prod_{C \neq A} p_{A,C}(\mathbf{r}) \quad \text{where} \\ p_{A,C}(\mathbf{r}) &= f(\mu_{AC}), \quad \text{and} \\ \mu_{AC} &= \frac{r_A - r_C}{R_{AC}} \end{aligned} \tag{4}$$

The elliptical coordinate, μ_{AC}, defines a projection of the point $\mathbf{r}$ along the segment, which connects the atoms A and C. The function, $f(\mu_{AC})$, is a step smoothed function that has the value of 1 up to the middle of the segment $\mathbf{R}_{AC}$ and goes to zero approaching the atom C. The product of functions, $f(\mu_{AC})$, for each of the pairs $AC_1, AC_2, \ldots$ defines a polyhedron around atom A: the weight function, $w_A(\mathbf{r})$, is 1 inside the polyhedron. These Voronoi polyhedra fill all space. If smooth step functions, $f(\mu_{AC})$, are used, the weight function goes smoothly to zero outside the polyhedra: this avoids numerical integration problems but atom pairs beyond nearest neighbours must be considered if a high accuracy is required. In addition, when calculating the weight functions, $W_A(\mathbf{r})$, following Becke's method, we need to compute sums of products of pair functions and this requires a high computational cost for large molecules and condensed systems. Simpler weights can be defined, following Savin:

$$\begin{aligned} w_A(\mathbf{r}) &= C_A e^{-\zeta_A r_A} \\ W_A(\mathbf{r}) &= \frac{C_A e^{-\zeta_A r_A}}{\sum_B C_B e^{-\zeta_B r_B}} \end{aligned} \tag{5}$$

The weight functions are simply exponential functions that roughly mimic the electron density around atom A: the exponential is normalized to the number of electrons that can be attributed to atom A following a Mulliken analysis and the exponential coefficient, ζ_A, is defined to reproduce the atomic shape, taking into account the atomic or ionic dimensions. The evaluation of the weight, W_A, requires only the summing of the exponential functions centered on the neighbours of atom A. This method can be applied to very complex molecules and crystals, and the loss of accuracy is less than 1 order of magnitude with respect to the weights as defined by Becke.

These methods can be applied to periodic systems after some modification and careful calibration. The integration is then confined to the unit cell, which is a finite solid for 3D periodic systems (crystals), a polyhedron infinite in 1 dimension for 2D systems (slabs) and a cylinder with infinite radius for 1D systems (polymers). The same transformation from a single integral to a sum of atomic integrals as for molecules can be performed but infinite lattice sums arise:

$$E^{DFT} = \sum_{A,\mathbf{g}}^{\infty} \int_{\text{CELL}} \varepsilon^{DFT}(\mathbf{r}) \frac{w_{A,\mathbf{g}}(\mathbf{r})}{\sum_{B,\mathbf{m}}^{\infty} w_{B,\mathbf{m}}(\mathbf{r})} \, d\mathbf{r} =$$

$$= \sum_{A,\mathbf{g}}^{\infty} \int_{\text{CELL}} \varepsilon^{DFT}(\mathbf{r}) W_{A,\mathbf{g}}(\mathbf{r}) \, d\mathbf{r} \tag{6}$$

The integrand functions are periodic for lattice translations:

$$\varepsilon^{DFT}(\mathbf{r}-\mathbf{g}) = \varepsilon^{DFT}(\mathbf{r}) \, ; \qquad W_{A,0}(\mathbf{r}-\mathbf{g}) = W_{A,\mathbf{g}}(\mathbf{r}) \tag{7}$$

Then a translation of the coordinate $\mathbf{r} \rightarrow \mathbf{r}-\mathbf{g}$ can be applied to each integral:

$$\sum_{A,\mathbf{g}}^{\infty} \int_{\text{CELL}} \varepsilon^{DFT}(\mathbf{r}) W_{A,\mathbf{g}}(\mathbf{r}) \, d\mathbf{r} = \sum_{A,\mathbf{g}}^{\infty} \int_{\text{CELL}} \varepsilon^{DFT}(\mathbf{r}-\mathbf{g}) W_{A,0}(\mathbf{r}-\mathbf{g}) \, d\mathbf{r}$$

$$= \sum_{A}^{n \text{ atoms in cell}} \int_{\text{SPACE}} \varepsilon^{DFT}(\mathbf{r}) W_{A,0}(\mathbf{r}) \, d\mathbf{r} \tag{8}$$

The infinite lattice sum of unit cell integrals has been transformed to a space integration, strictly analogous to the molecular case. The weight function, W, contains a lattice sum and can be truncated with the usual periodic LCAO techniques: good accuracy can be obtained by including neighbours within a radius of 8 Å. The lattice sums, necessary to calculate the weights, are the only difference between the periodic and molecular sums: the analysis of the numerical approximation , which follows in section 4, will not distinguish between the two cases.

3. Treatment of lattice sums in the weight function

In the periodic case, the weight functions, W, contain lattice sums in the denominator:

$$W_A(\mathbf{r}) = \frac{w_A(\mathbf{r})}{\sum_{B,\mathbf{m}} w_{B,\mathbf{m}}(\mathbf{r})} \tag{9}$$

When Savin's exponential weight functions are employed, the lattice sums are truncated for exponentials smaller than a given threshold, T_W. In 0, 1 and 2 dimensional periodicity, there are regions far from all atomic centres, where the limit of W is finite:

$$\lim_{r\to\infty} W_A(\mathbf{r}) = \frac{\lim_{r\to\infty} w_A(\mathbf{r})}{\sum_{B,\mathbf{m}} \lim_{r\to\infty} w_{B,\mathbf{m}}(\mathbf{r})} = 1 \tag{10}$$

In such regions, the electron density and its functionals decay exponentially. Very small thresholds must be used to reach an acceptable accuracy for the integrated charge. The treatment of the lattice sums is more complex when Becke's weight functions are employed.

4. Numerical Quadratures of the spherical atom-centred integration

Considering the atomic, quasi-spherical shape of the integrand, efficient numerical integration methods can be applied. Spherical coordinates can be used to separate the radial and angular integration.

$$\begin{aligned} I_A &= \int_{\text{SPACE}} \varepsilon^{DFT}(\mathbf{r}) W_{A,0}(\mathbf{r})\, d\mathbf{r} = \int_{\text{SPACE}} f(\mathbf{r})\, d\mathbf{r} = \\ &= \int_0^\infty dr_A r_A^2 \int\!\!\int_{4\pi} d\omega f(r_A, \vartheta_A, \varphi_A) \end{aligned} \tag{11}$$

Many quadrature techniques exist to evaluate 1-dimensional integrals [18]: the numerical evaluation of multiple integrals is less common and is usually performed by using direct products of 1-dimensional quadratures, that is, by using a 1-dimensional quadrature for each dimension. However, 2-dimensional quadrature on the sphere surface has been fully exploited by Lebedev [19] who derived very efficient sets of sampling points: the richest set of 435 points can integrate exactly a 35-degree polynomial with angular coordinates (ϑ_A, φ_A). This is enough to integrate exactly the density function of atoms with angular quantum number up to $l = 6$ and $m = 6$ (i function), which is enough to have, in principle, high accuracy for any atomic coordination. Lebedev's set of sampling points has octahedral symmetry. We may readily select an irreducible set of points, using the local symmetry of each atomic site; no simple sampling technique in (ϑ_A, φ_A), based on 1-dimensional quadrature, can exploit the local symmetry, which is relatively more important in condensed systems. As will be explained below, Lebedev's grids are adopted, together with products of 1-dimensional quadratures for the angular integration; in both cases, the angular integral, $\Theta(r_A)$, is approximated as follows:

$$\begin{aligned} I_A &= \int_0^\infty dr_A r_A^2 \Theta(r_A) \\ \Theta(r_A) &= \int\!\!\int_{4\pi} d\omega f(r_A, \vartheta_A, \varphi_A) \\ &\approx \sum_{i=1}^{\text{Nang}(r_A)} w_i f(r_A, \vartheta_{Ai}, \varphi_{Ai}) \end{aligned} \tag{12}$$

The computational scheme and the parameters which control several features of the numerical integration method must be chosen to be the best compromise between accuracy and computation time. In our case, the same criteria must be applied to periodic (crystals, slabs or polymers) and non-periodic systems (atoms and molecules). A set of about 20 systems has been used, including closed and open-shell atoms, molecules, polymers, slabs, covalent, semi-ionic and ionic crystals. Table 1 reports the systems and the relative computational conditions.

Table 1. Systems used in the test, basis sets and computational conditions

System	Periodicity	Basis Set	Conditions
Be	atom	Dunning (9/5)	AE
C	atom	Dunning (9/5)	AE, open shell
CO	molecule	STO-3G	AE
CH_4	molecule	STO-3G	AE
urea	molecule	6-21G*	AE
SN	polymer	STO-3G*	AE
Al_2O_3	slab	6-21G*	AE
Si	crystal	STO-2G	AE
Be	crystal	6-11G	AE
MgO	crystal	8-61G (Mg); 8-51G (O)	AE
LiF	molecule	6-1G* (Li); 7-311G (F)	AE
LiF s3	slab	6-1G* (Li); 7-311G (F)	AE
Si	crystal	6-21G*	AE
Si	crystal	ECP-21G	ECP
NaF	crystal	8-511G (Na); 7-311G (F)	AE
NaCl	crystal	8-511G (Na); 86-311G (Cl)	AE
Li	polymer	3-1G*	anti-ferromagnetic
NiO	crystal	ECP-41(d)-21G (Ni); ECP-31G (O)	ECP open shell

Some information about the accuracy of the DFT energy integration can be obtained from the electronic charge density integration. The relative error in the integrated charge with respect to the number of electrons, Δ (per unit cell for the periodic case, usually expressed in parts per million (ppm)), has been used to select the optimal computational conditions, even though the error in the energy integration is only weakly correlated to the error in the charge integration, as will be shown in the next section. The modulus of the relative error in the integrated charge, averaged over the 20 test systems ($<|\Delta|>$), is used to illustrate the influence of the computational parameters on the accuracy. The set of parameters, P, has been subdivided into homogeneous subsets: P^a are the angular parameters; P^r, the radial parameters; and P^w, the weight parameters. One set of parameters has been varied at a time, starting from a set of parameters which resulted from a previous optimization, that we define as P_0.

5. Radial integration

In the first implementation of our method, the radial integration has been subdivided into several segments and a separate numerical quadrature has been employed in each.

$$\int_0^\infty f(r)dr = \sum_{i=1}^{N} \int_{r_{i-1}}^{r_i} f(r)dr + \int_{r_N}^{\infty} f(r)dr \equiv$$

$$\equiv \sum_{i=1}^{N} \sum_{k=1}^{n_i} f(r_k^i) w_k + \sum_{l=1}^{n_{N+1}} f(r_l^i) w_l' \tag{13}$$

Gauss-Legendre quadrature has been used to approximate the N finite integrations and Gauss-Laguerre points have been used for the last semi-infinite quadrature. The upper limit of each interval was fixed by using an estimate of the atomic or ionic radius. Improved accuracy was obtained by employing three finite intervals with this method; the best choices for the upper limits of these intervals were, respectively: 0.5, 1.0 and 2.0 times the atomic radius. For each interval, 20 quadrature points were used, so that the total number of radial points used was 80, considering the semi-infinite as well as the three finite intervals. In principle, with this method, the quadratures are limited to intervals in which the integrand function spans ranges smaller than the whole function range, resulting in faster convergence. Greater difficulties arise from the quadrature of the integration in a radial interval close to the nucleus, that is, within half of the atomic radius: here the integrand functions, which are functional of the electron density, have cusps spanning several orders of magnitude. As a result, high numbers of orthogonal polynomials are needed to produce a reasonable approximation and therefore, a high number of quadrature points are required in the calculation. For this reason, the most important characteristic of the radial integration is the *degree* (number of points) of quadrature for the cusps. Where the most internal radial quadrature has a high degree, there is no further computational advantage in the subdivision into segments. The most recent implementation of our integration method contains a single interval radial quadrature.

The semi-infinite, one-dimensional quadrature of the integrals of functions can be performed using the weights and nodes of orthogonal polynomials with exponential weight functions, such as the Laguerre polynomials. However, orthogonal polynomials on a finite interval can be used by mapping to a semi-infinite interval by means of an appropriate transformation of the coordinate:

$$\int_0^\infty f(r)dr = \int_{-1}^{1} f(r(t)) \frac{dr}{dt} dt \tag{14}$$

A logarithmic transformation of the coordinates allows the Gauss-Legendre quadrature to be used for a semi-infinite integration:

$$r(t) = \log\left(\frac{2}{1-t}\right) \tag{15}$$

The Gauss-Legendre semi-infinite radial quadrature gives a better accuracy than that of Gauss-Laguerre or other methods proposed in the literature, such as those of Euler-McLaurin or Gauss-Chebyshev.

6. Angular integration

At each stage of the implementation of our method, a different level of angular accuracy has been used for the various values of r, the radius. The angular accuracy, and therefore, the number of points on the sphere, has been decreased near the nucleus, following the hypothesis that the electron density and its functionals have spherical symmetry when the radius of the sphere is vanishingly small.

A set of points for a spherical quadrature can be defined by the degree, L, of the angular polynomial that can be exactly integrated. Up to $L = 35$, Lebedev's octahedral grids can be used; for $L \geq 35$, a product formula can be applied. In this latter case, the Gauss-Legendre quadrature in ϑ is directly multiplied by a *trapezoidal* quadrature in φ, where the points are equispaced: $N\vartheta$ points by $2N\varphi$ points integrate exactly spherical polynomials of N-th degree.

Many scaling criteria have been tested for the angular degree, L: power, logarithmic and exponential functions gave poorer results, in terms of accuracy, than a simple linear scaling:

$$L(r) = \min\left(\frac{\mathrm{r}}{\mathrm{r_{atom}}}, \mathrm{L_{atom}}\right) \tag{16}$$

If $L < 35$, the next larger Lebedev grid is adopted. In this way, L is discontinuously varied in the set: $L = 9, 11, 17, 23, 27, 29, 35$. If $L \geq 35$, Legendre-trapezoidal grids are used; values up to $L = 60$ have been explored with this method.

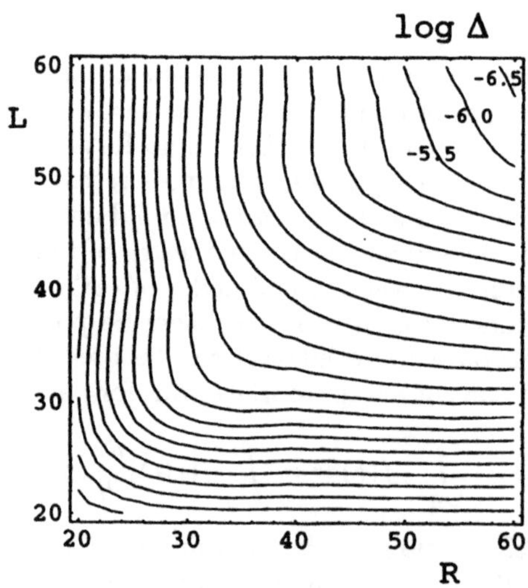

Figure 1. Isoline plot of the functions $\Delta(R, L)$ (integration error in ppm) versus the angular degree L, and the number of radial points R. The surface is a spline interpolation over a grid with ΔL and ΔR equal to five.

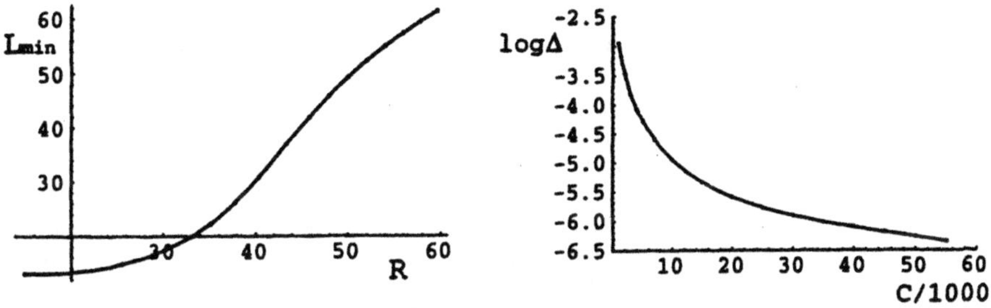

Figure 2. (Left) The polynomial degree of the angular grid, $L_{\min}$, versus the number of radial points, R. The function reported in this plot is the path of minimum error, Δ, calculated from the data reported in Fig. 1; (right) the average integration accuracy (relative to the electronic charge), $\Delta_{\min}$, versus the number of grid points per atom, C.

7. Calibration of the method

The choice of radial integration on a single, semi-infinite interval, with a number of points, R, and the introduction of linear scaling of the angular degree, L, simplifies the parameter space to be optimised. Figures 1, 2 (left) and 2 (right) show the steps of the optimisation path which was followed.

As a result of many integration tests for the systems contained in Table 1, the integration accuracy, $(< |\Delta| >)$ (the modulus of the relative error in the electronic charge averaged over the test systems) and the cost of calculation, C (the number of grid points per atom, without symmetry) were measured and the functions $\Delta(R, L)$ and $C(R, L)$ were interpolated. The function $\Delta(R, L)$ was resolved with respect to L, with the condition that $\Delta(R, L_{\min}) = \min$, in the form $L_{\min}(R)$. In this way, the angular degree, L which gives the minimum integration error Δ was found for each value of R.

Figure 2 (right) shows the most important function, $\Delta_{\min}(C)$, which gives the minimal error for a given number of points per atom: this relation makes choosing predefined sets of grid parameters very easy. The calculation of the density functional, self-consistent potential does not need a very high accuracy: integration errors of 100 ppm (10^{-4}) are acceptable, corresponding to 4000 points per atom for the systems explored. The density functional energy calculation needs an accuracy of 10 ppm (10000 points per atom) and 1 ppm is required if the elastic properties are explored (33000 points per atom).

References

1. M.L. Cohen, in *Structure and Bonding in Crystals*, Vol.I (M. O'Keeffe and A. Navrotsky, eds), Academic, New York, p.25 (1981).
2. A. Zunger, in *Structure and Bonding in Crystals*, Vol.I (M. O'Keeffe and A. Navrotsky, eds), Academic, New York, p.73 (1981).
3. O.K. Andersen, Phys. Rev. B **12**, 3060 (1975).

4. T. Ziegler, Chem. Rev, **91**, 651 (1991).
5. B.G Johnson, P.M.W. Gill, and J.A. Pople, J. Chem. Phys. **98** 5612(1992).
6. J. Andzeim and E. Wimmer, J. Chem. Phys, **96**, 1280 (1992).
7. P. Durand and J.C. Barthelat, Chem. Phys Lett. **27**, 191 (1974).
8. A.D. Becke, J. Chem. Phys. **88**, 2547 (1988).
9. P.W.M. Gill, B.G. Johnson, J.A. Pople, and M.J. Frisch, Chem. Phys. Lett. **197**, 499 (1992).
10. A.D. Becke, J. Chem. Phys **96**, 2155 (1992).
11. C. Pisani, R. Dovesi and C. Roetti, *Hartree-Fock Ab Initio Treatment of Crystalline Systems*, Lecture Notes in Chem. Vol. 48, Springer, Berlin (1988).
12. R. Dovesi, C. Pisani, C. Roetti, M. Causà, and V.R.Saunders, *CRYSTAL 88, Program No. 577*, QCPE, Indiana University, Bloomington, IN (1989); R. Dovesi, V.R. Saunders and C. Roetti, *CRYSTAL92 User Documentation*, University of Torino, Torino (1992).
13. M.Causà and A. Zupan, Chem. Phys. Lett. **220**, 145 (1994).
14. M.Causà and A. Zupan, Int. J. Quantum Chem. **S28**, 633 (1994).
15. A. Zupan and M. Causà, Int. J. Quantum Chem, **S32**, 446 (1995).
16. F.E. Harris and H.J. Monkhorst, Phys. Rev. Lett. **23**, 1026 (1969).
17. A. Savin, Int. J. Quantum Chem. **S22**, 457 (1988).
18. G. Stroud, *Approximate Calculations of Multiple Integrals*, Prentice-Hall (1971)
19. V.I. Lebedev, Zh. Vychisl. Mat. Fiz. **16**, 293 (1976).

Hartree-Fock Treatment of Spin-Polarized Crystals

Edoardo Aprà

High Performance Computational Chemistry Group,
Environmental Molecular Sciences Laboratory,* Pacific Northwest Laboratory, Richland, WA 99352, USA.

September 11, 1996

Summary. A brief introduction to the Hartree-Fock method for spin-polarized crystalline systems is presented, together with a description of a few applications.

Key words: Magnetic Systems – Hartree-Fock – Unrestricted Hartree-Fock (UHF) – Open-shell Systems – Spin-polarization – Ferromagnetism – Anti-ferromagnetism

1. Introduction

The local-spin-density (LSD) approximation [1, 2] of the density functional theory (DFT) and the Unrestricted Hartree-Fock (UHF) [3] approach of the Hartree-Fock formalism constitute two of the most widely used theoretical tools for the description of the electronic structure of magnetic systems.

In the following sections, we will focus our attention on the UHF formalism; we will first describe the standard molecular formalism and then its natural extension to the periodic approach.

The chapter concludes with the description of a few applications of the UHF method to crystalline systems, trying to compare UHF results with results obtained with other theoretical methods.

2. Hartree-Fock for Open-Shell systems

For the description of systems containing unpaired electrons (such as molecules with an odd number of electrons, radicals, ferromagnetic and anti-ferromagnetic

---

* Pacific Northwest Laboratory is operated for the U.S. Department of Energy by Battelle Memorial Institute under contract DE-AC06-76RL0 1830.

Correspondence to: E. Aprà - E-mail: e_apra@pnl.gov

solids) a single determinant is not, in the most general of cases, an appropriate wave-function; in order to get the correct spin eigenfunction of these systems, it is necessary to choose a linear combination of Slater determinants (whereas, in closed shell systems, a single determinant always gives the appropriate spin eigenfunction).

The Restricted Hartree-Fock Open Shell (ROHF) [4] wave-function is, in the most general of cases, a sum of Slater determinants; each determinant contains a ***closed-shell*** subset, with doubly occupied orbitals and an ***open-shell*** subset, formed by orbitals occupied by a single electron. In one particular configuration, the ROHF wave-function reduces to a single determinant: that is, in the so-called *half-closed shell* cases, where it is possible to define two sets of orbitals, the first occupied by paired electrons and the second, by electrons with parallel spins.

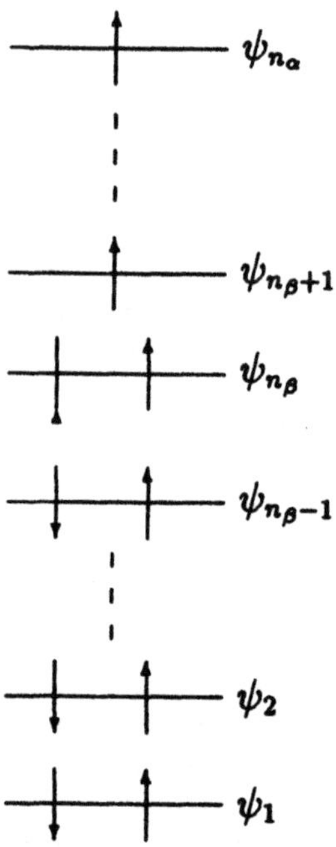

Figure 1. Molecular Orbital diagram for the Restricted Hartree-Fock Open Shell method

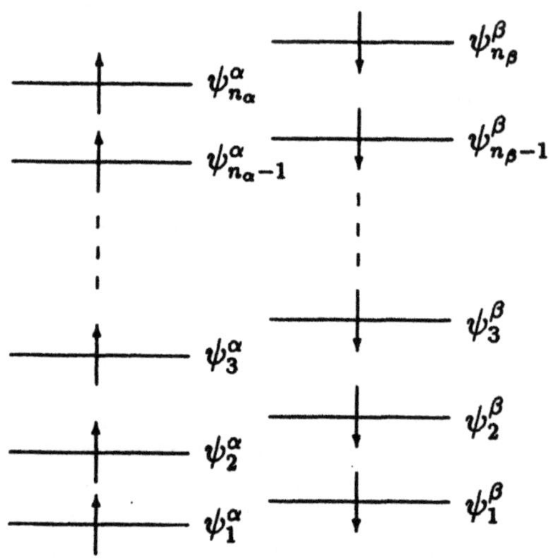

Figure 2. Molecular Orbital diagram for the Unrestricted Hartree-Fock method

Nevertheless, it is possible to keep a mono-determinantal description by using the UHF method. In this theory, the constraint of double occupancy is absent; in this way, α electrons are allowed to populate orbitals other than those occupied by the β electrons. Energy levels corresponding to a ROHF and UHF description are plotted in Fig. 1 and 2, respectively.

The double occupancy constraint allows the ROHF approach to obtain solutions that are eigenfunctions of the spin operator, $\widehat{S^2}$, whereas UHF solutions are formed by a mixture of spin states. The greater variational freedom allows the UHF method to produce wave-functions that are energetically more stable than the corresponding ROHF ones; another advantage of the UHF method is that it allows solutions with locally negative spin density (i.e. anti-ferromagnetic systems), a feature that ROHF solutions cannot possess.

In the rest of this chapter, we will focus our attention on the UHF method, since it is the most widely used approach to the study of spin-polarized systems of chemical interest.

2.1. Unrestricted Hartree-Fock

In the standard derivation of the Hartree-Fock equations relative to a closed-shell system, it is necessary to introduce the constraint that all the spin-orbitals are products of the kind: $\psi_k\alpha$ and $\psi_k\beta$; in other words, each molecular orbital is populated by two electrons (Restricted Hartree-Fock theory – RHF).

However, in some situations, the removal of this constraint allows solutions at lower energy; in the UHF approach, the mono-determinantal wave-function is computed using n_α Molecular Orbitals (MOs), ψ^α and n_β MOs, ψ^β, corresponding to the electrons with α and β spin, respectively. The wave-function that describes a system with N electrons ($N = n_\alpha + n_\beta$) has the following form:

$$\begin{aligned}\Psi &= (N!)^{-\frac{1}{2}}\det\Big\{\psi_1^\alpha(1)\alpha(1),\dots,\psi_{n_\alpha}^\alpha(n_\alpha)\alpha(n_\alpha)\cdot \\ &\qquad \psi_1^\beta(n_\alpha+1)\beta(n_\alpha+1),\dots,\psi_{n_\beta}^\beta(N)\beta(N)\Big\} \qquad (1)\end{aligned}$$

The energy expression is given by the following formula:

$$\begin{aligned}E &= \langle\Psi|\widehat{H}|\Psi\rangle = \\ &= \sum_k\langle\psi_k^\alpha|\widehat{h}+\frac{1}{2}(\widehat{J}-\widehat{K}^\alpha)|\psi_k^\alpha\rangle \; + \\ &\quad + \sum_m\langle\psi_m^\beta|\widehat{h}+\frac{1}{2}(\widehat{J}-\widehat{K}^\beta)|\psi_m^\beta\rangle \qquad (2)\end{aligned}$$

where $\widehat{J}$, $\widehat{K}^\alpha$ and $\widehat{K}^\beta$ are defined as in the standard RHF equations, that is:

$$\widehat{J}(\mathbf{r}_1)\psi(\mathbf{r}_1) = \sum_i\Big(\int d\mathbf{r}_2\psi_i^*(\mathbf{r}_2)\frac{1}{r_{12}}\psi_i(\mathbf{r}_2)\Big)\psi(\mathbf{r}_1) \qquad (3)$$

$$\widehat{K}^\alpha(\mathbf{r}_1)\psi(\mathbf{r}_1) = \sum_i\Big(\int d\mathbf{r}_2\psi_i^{\alpha*}(\mathbf{r}_2)\frac{1}{r_{12}}\psi(\mathbf{r}_2)\Big)\psi_i^\alpha(\mathbf{r}_1) \qquad (4)$$

$$\widehat{K}^\beta(\mathbf{r}_1)\psi(\mathbf{r}_1) = \sum_i\Big(\int d\mathbf{r}_2\psi_i^{\beta*}(\mathbf{r}_2)\frac{1}{r_{12}}\psi(\mathbf{r}_2)\Big)\psi_i^\beta(\mathbf{r}_1) \qquad (5)$$

2.2. LCAO approximation and Pople-Nesbet equations

If we express the molecular orbitals, ψ_i, as a linear combination of a basis set of n atomic orbitals (AO) φ_q, we can write:

$$\psi_k^\alpha = \sum_q \varphi_q c_{qk}^\alpha \tag{6}$$

$$\psi_m^\beta = \sum_q \varphi_q c_{qm}^\beta \tag{7}$$

or, in matrix form:

$$\begin{aligned} \boldsymbol{\psi}^\alpha &= \boldsymbol{\varphi}\mathbf{C}^\alpha \\ \boldsymbol{\psi}^\beta &= \boldsymbol{\varphi}\mathbf{C}^\beta \end{aligned}$$

where the matrices $\mathbf{C}^\alpha$ and $\mathbf{C}^\beta$ have dimension $n \times N$.

We now introduce the definitions of spin α and spin β *density matrices* (DM), respectively:

$$P_{rs}^\alpha = \sum_k^{n_\alpha} (c_{sk}^\alpha)^* c_{rk}^\alpha$$

$$P_{rs}^\beta = \sum_m^{n_\beta} (c_{sm}^\beta)^* c_{rm}^\beta$$

$$\mathbf{P}^\alpha = \mathbf{C}^\alpha \mathbf{C}^{\alpha\dagger} \tag{8}$$

$$\mathbf{P}^\beta = \mathbf{C}^\beta \mathbf{C}^{\beta\dagger} \tag{9}$$

and of *total density*, $\mathbf{P}^{tot}$ and *spin density*, $\mathbf{P}^{spin}$:

$$\mathbf{P}^{tot} = \mathbf{P}^\alpha + \mathbf{P}^\beta \tag{10}$$

$$\mathbf{P}^{spin} = \mathbf{P}^\alpha - \mathbf{P}^\beta \tag{11}$$

The total density matrix is used to get the expectation value of the operators that are *spin-independent*, according to the formula (where $\widehat{O}$ is a generic operator):

$$< \widehat{O} >= \sum_\mu \sum_\nu \mathbf{P}_{\mu\nu}^{tot} < \varphi_\mu \mid \widehat{O} \mid \varphi_\nu >$$

To get the energy expression as a function of the AO basis set, we introduce into equation 2 the definitions given in equations 6 and 7:

$$\begin{aligned} E &= \sum_k \left[\sum_{rs} (c_{sk}^\alpha)^* c_{rk}^\alpha \left(\langle \varphi_s | \widehat{h} + \frac{1}{2}\widehat{J} - \frac{1}{2}\widehat{K}^\alpha | \varphi_r \rangle \right) + \right. \\ &\quad \left. + \sum_{rs} (c_{sm}^\beta)^* c_{rm}^\beta \left(\langle \varphi_s | \widehat{h} + \frac{1}{2}\widehat{J} - \frac{1}{2}\widehat{K}^\beta | \varphi_r \rangle \right) \right] = \\ &= \sum_{rs} \left[\sum_k (c_{sk}^\alpha)^* c_{rk}^\alpha \right] \left[h_{rs} + \frac{1}{2} J_{rs} - K_{rs}^\alpha \right] + \\ &\quad + \sum_{rs} \left[\sum_m (c_{sm}^\beta)^* c_{rm}^\beta \right] \left[h_{rs} + \frac{1}{2} J_{rs} - K_{rs}^\beta \right] \end{aligned}$$

In more compact, matrix notation, we can write:

$$
\begin{aligned}
E &= \text{Tr } \mathbf{P}^\alpha \mathbf{h} + \frac{1}{2}\text{Tr } \mathbf{P}^\alpha \mathbf{J}[\mathbf{P}^{tot}] - \frac{1}{2}\mathbf{P}^\alpha \mathbf{K}[\mathbf{P}^\alpha] + \\
&\quad + \text{Tr } \mathbf{P}^\beta \mathbf{h} + \frac{1}{2}\text{Tr } \mathbf{P}^\beta \mathbf{J}[\mathbf{P}^{tot}] - \frac{1}{2}\mathbf{P}^\beta \mathbf{K}[\mathbf{P}^\beta] = \\
&= \text{Tr } \mathbf{P}^{tot} \mathbf{h} + \frac{1}{2}\text{Tr } \mathbf{P}^{tot} \mathbf{J}[\mathbf{P}^{tot}] + \\
&\quad - \frac{1}{2}\text{Tr } \mathbf{P}^\alpha \mathbf{K}[\mathbf{P}^\alpha] - \frac{1}{2}\text{Tr } \mathbf{P}^\beta \mathbf{K}[\mathbf{P}^\beta] \qquad (12)
\end{aligned}
$$

To get the expression of the optimal UHF orbitals, we proceed as in the standard RHF case: we compute the first-order energy variation, then we introduce the orthonormality constraint of the basis set by means of the Lagrange multipliers method. What we get at the end are the *Pople-Nesbet equations* [3]:

$$
\mathbf{F}^\alpha \, \mathbf{C}_i^\alpha = \epsilon_i^\alpha \, \mathbf{S} \, \mathbf{C}_i^\alpha \qquad (13)
$$

$$
\mathbf{F}^\beta \, \mathbf{C}_i^\beta = \epsilon_i^\beta \, \mathbf{S} \, \mathbf{C}_i^\beta \qquad (14)
$$

where $\mathbf{S}$ is the *overlap* matrix and $\mathbf{F}$ is *the Fock operator*, which assumes the form:

$$
\mathbf{F}^\alpha = \mathbf{F} - \mathbf{Z} \qquad (15)
$$

$$
\mathbf{F}^\beta = \mathbf{F} + \mathbf{Z} \qquad (16)
$$

where the matrices $\mathbf{F}$ and $\mathbf{Z}$ are defined as:

$$
\mathbf{F} = \mathbf{h} + \mathbf{G} \qquad (17)
$$

$$
\mathbf{Z} = \frac{1}{2}\, \mathbf{K}[\mathbf{P}^{spin}] \qquad (18)
$$

$$
\mathbf{G} = \mathbf{J}[\mathbf{P}^{tot}] - \frac{1}{2}\, \mathbf{K}[\mathbf{P}^{tot}] \qquad (19)
$$

We can express the energy (equation 12) as a function of the Fock matrix:

$$
E = \frac{1}{2}\text{ Tr } \mathbf{P}^{tot}\mathbf{h} + \frac{1}{2}\text{ Tr } \mathbf{P}^\alpha \mathbf{F}^\alpha + \frac{1}{2}\text{ Tr } \mathbf{P}^\beta \mathbf{F}^\beta \qquad (20)
$$

If we use the matrices $\mathbf{G}$ and $\mathbf{Z}$, the energy expression has a more compact form:

$$
E = \text{Tr } \mathbf{P}^{tot}\mathbf{h} + \frac{1}{2}\text{ Tr } \mathbf{P}^{tot}\mathbf{G} - \frac{1}{2}\text{Tr } \mathbf{P}^{spin}\mathbf{Z} \qquad (21)
$$

2.3. Solution of the Pople-Nesbet equations

In order to solve the *Pople-Nesbet* equations, it is necessary to adopt an iterative scheme. The reason is the following: the Fock matrix is computed using the density matrix as an input (see equations 17, 18 and 19), defined in terms of the $\mathbf{C}$ coefficients, which is what we want to determine through the Pople-Nesbet equations.

A possible scheme for the Self-Consistent Field (SCF) process for the solution of the Pople-Nesbet equations is based on the following steps:

1. An initial "guess" at the density matrices $\mathbf{P}^\alpha$ and $\mathbf{P}^\beta$ must be made.

2. The values of $\mathbf{J}$, $\mathbf{K}$ and, therefore, $\mathbf{F}^\alpha$ and $\mathbf{F}^\beta$ must be computed, according to the equation 15 and 16. Having obtained all of these quantities, it is possible to calculate the energy using expression 20.
3. A set of coefficients must be determined by diagonalization of the Fock matrix (see equations 13 and 14) and subsequently, the density matrix must be built.
4. The whole process must be repeated to self-consistency; the convergence criteria can consider either the density matrix or the total energy.

In order to get *spin-polarized* solutions from a UHF calculation, it is necessary that, during the SCF process, the density matrices α and β are different (see points 1 and 3 of the previous scheme).

If $\mathbf{P}^\alpha = \mathbf{P}^\beta$ for the whole SCF process, the solution produced will not necessarily have unpaired electrons, since, in the scheme just described, no element can break spin symmetry. In this scenario, we can see the RHF closed-shell solution as a special case of the UHF method, where the number of α electrons coincide with the number of β electrons ($n_\alpha = n_\beta$).

On the other hand, if we are able to get $\mathbf{P}^\alpha \neq \mathbf{P}^\beta$ (during step 1 or 3 of the process), we could have either spin-polarized solutions at the end of the SCF cycle or the spin polarization will vanish in the course of the SCF process. It is obvious that reaching a closed-shell solution from a spin-polarized starting point is favoured when the latter solution has lower energy.

3. Extension of the UHF method to periodic systems

In this section, we will briefly describe how the various contributions to the Pople-Nesbet equations change when applying the UHF method to the study of crystalline systems. In most cases, the only difference is that the matrix elements are no longer characterised by two indices, but by three; the third index identifies a particular vector, $\mathbf{T}$, of the crystalline lattice. For further references to the Hartree-Fock treatment of crystalline systems, see ref [5],[6] and other chapters of this book.

In a crystalline context, the Fock operator assumes the following form:

$$\left(\mathrm{F}^{T}_{\mu\nu}\right)^{\alpha} = \mathrm{F}^{T}_{\mu\nu} + \mathrm{Z}^{T}_{\mu\nu} \tag{22}$$

$$\left(\mathrm{F}^{T}_{\mu\nu}\right)^{\beta} = \mathrm{F}^{T}_{\mu\nu} - \mathrm{Z}^{T}_{\mu\nu} \tag{23}$$

where the F and Z matrices are defined as:

$$\mathrm{Z}^{T}_{\mu\nu} = \frac{1}{2}\,\mathrm{K}^{T}_{\mu\nu}\left[\mathrm{P}^{spin}\right] \tag{24}$$

$$\mathrm{G}^{T}_{\mu\nu} = \mathrm{J}^{T}_{\mu\nu}\left[\mathrm{P}^{tot}\right] - \frac{1}{2}\mathrm{K}^{T}_{\mu\nu}\left[\mathrm{P}^{tot}\right] \tag{25}$$

$$\mathrm{F}^{T}_{\mu\nu} = \mathrm{h}^{T}_{\mu\nu} + \mathrm{G}^{T}_{\mu\nu} \tag{26}$$

In the expression of the matrix elements J[P] and K[P], we have to introduce lattice sums:

$$\mathrm{J}^{T}_{\mu\nu}[\mathrm{P}] = \sum_{S\lambda\rho} \mathrm{P}^{S}_{\lambda\rho}\left[\sum_{\mathbf{h}}(\mu 0\ \lambda \mathbf{S} \mid \nu \mathbf{T}\ \rho\mathbf{S}+\mathbf{Q})\right]$$

$$\mathrm{K}^{T}_{\mu\nu}[\mathrm{P}] = \sum_{S\lambda\rho} \mathrm{P}^{S}_{\lambda\rho}\left[\sum_{\mathbf{h}}(\mu 0\ \lambda \mathbf{S} \mid \rho\mathbf{S}+\mathbf{Q}\ \nu \mathbf{T})\right]$$

where the notation used for the bi-electronic integral is the following:

$$(\mu 0 \lambda \mathbf{S} \mid \nu \mathbf{T} \rho \mathbf{S}+\mathbf{Q}) = \int\int d\mathbf{r}_1 d\mathbf{r}_2 \varphi_\mu^0(\mathbf{r}_1)^* \, \varphi_\nu^{\mathbf{T}}(\mathbf{r}_2)^* \frac{1}{r_{12}} \, \varphi_\lambda^{\mathbf{S}}(\mathbf{r}_1) \varphi_\rho^{\mathbf{S}+\mathbf{Q}}(\mathbf{r}_2) \tag{27}$$

We can now introduce the expression for the total energy:

$$\begin{aligned} E &= \sum_{T\mu\nu} (\mathrm{P}_{\mu\nu}^T)^{tot} \, \mathrm{h}_{\mu\nu}^T + \frac{1}{2} \sum_{T\mu\nu} (\mathrm{P}_{\mu\nu}^T)^{tot} \, \mathrm{J}_{\mu\nu}^T[\mathrm{P}^{tot}] + \\ & - \frac{1}{2} \sum_{T\mu\nu} (\mathrm{P}_{\mu\nu}^T)^\alpha \, \mathrm{K}_{\mu\nu}^T[\mathrm{P}^\alpha] - \frac{1}{2} \sum_{T\mu\nu} (\mathrm{P}_{\mu\nu}^T)^\beta \, \mathrm{K}_{\mu\nu}^T[\mathrm{P}^\beta] \end{aligned}$$

or, as a function of the Fock matrix:

$$\begin{aligned} E &= \frac{1}{2} \sum_{T\mu\nu} (\mathrm{P}_{\mu\nu}^T)^{tot} \, \mathrm{h}_{\mu\nu}^T + \frac{1}{2} \sum_{T\mu\nu} (\mathrm{P}_{\mu\nu}^T)^\alpha (\mathrm{F}_{\mu\nu}^T)^\alpha + \\ & + \frac{1}{2} \sum_{T\mu\nu} (\mathrm{P}_{\mu\nu}^T)^\beta \, (\mathrm{F}_{\mu\nu}^T)^\beta \end{aligned} \tag{28}$$

If we use the matrix **Z** and **G**, we get the following energy expression:

$$\begin{aligned} E &= \sum_{T\mu\nu} (\mathbf{P}_{\mu\nu}^T)^{tot} \, \mathbf{h}_{\mu\nu}^T + \frac{1}{2} \sum_{T\mu\nu} (\mathbf{P}_{\mu\nu}^T)^{tot} \, \mathbf{G}_{\mu\nu}^T + \\ & - \frac{1}{2} \sum_{T\mu\nu} (\mathbf{P}_{\mu\nu}^T)^{spin} \, \mathbf{Z}_{\mu\nu}^T \end{aligned} \tag{29}$$

Comparing this last equation with the equivalent one for molecular systems (eq 21), it is evident that the trace operation extended to the AOs, which we find in equation 20, is now transformed, in eq. 28, into a sum over AOs (μ and ν indices) and also over lattice vectors, **T**.

3.1. Scheme of a periodic UHF program

The scheme in Fig. 3 illustrates how the periodic UHF method is implemented in the computer code CRYSTAL [6]. If we compare this scheme with the corresponding one for the closed-shell case (see page 120 of ref [5]), we see that the most relevant quantities (Fock and density matrix, eigenvalues and eigenvectors) are duplicated, so that is possible to differentiate α orbitals from β orbitals. The two sections of the code, corresponding to α and β electrons, do not communicate for most of the program and the only interactions happen during the Fermi energy calculation and the Fock build.

It is necessary to consider some practical problems that arise when running periodic UHF calculations. In section 2.3, we have already hinted that, in an open-shell calculation, it is very important that the *alpha* DM differs from the *beta* DM. In order to get this spin unbalance, you can act on the *starting guess* needed at the beginning of the SCF cycle (step 1 in the scheme of section 2.3):

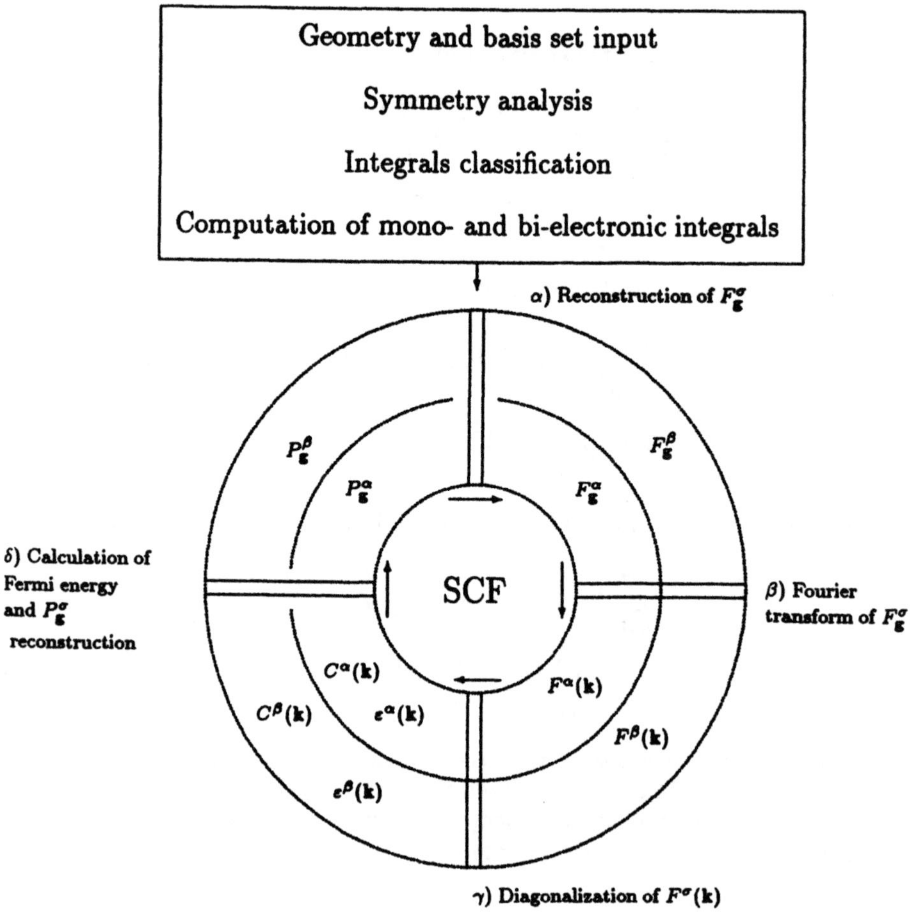

Figure 3. Scheme of the implementation of the UHF method in *CRYSTAL*

in CRYSTAL there exists an option to construct a DM formed by the superposition of atomic solutions, where it is possible to assign a specific polarization (α, β or no polarization) to each atom of the system. Once the SCF process is started, it often useful to introduce some constraints on the populations of the α and β orbitals (point 3 of the SCF scheme of section 2.3). This is performed in CRYSTAL by *ad hoc* translations of the eigenvalues spectrum, in such a way that the difference between the number of α and β electrons is equal to the value introduced in the input.

It is well known from Hartree-Fock studies of molecular systems, that it is very common to have problems of SCF convergence when studying open-shell systems; similarly, convergence problems are not rare in the Hartree-Fock treatment of spin-polarized crystals. A well-known technique for the solution of convergence problems, in the case of open-shell molecules, is the so-called *level shifting* method [7]; this approach has showed its effectiveness in the periodic HF context also, especially in the case of crystals containing transition elements.

4. Applications

In this section, we will illustrate the result of a UHF study of transition metal oxides.

Many theoretical studies have examined the description of the electronic structure of these spin-polarized solids: we can find, in the literature, a considerable number of LSDA studies ([8]-[16]), whereas periodic UHF studies appeared only quite recently ([17]-[20]).

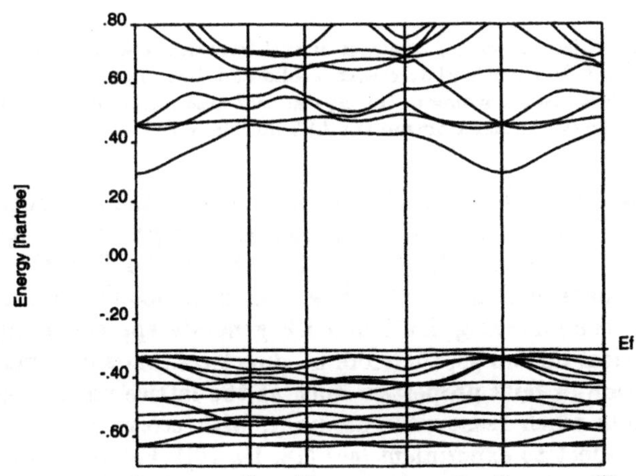

Figure 4. Band structure of anti-ferromagnetic NiO

From an experimental point of view, these materials are classified as magnetic insulators (where the magnetic character comes from being anti-ferromagnetic);

both LSDA and UHF are able to predict the insulating nature of these systems: in the band structure reported in Fig. 4, the large energy difference between the highest occupied energy level and the lowest unoccupied level (approximately 15 eV) shows the evident insulating state of the UHF solution (note: this value does not correspond to the band gap, since the UHF is a ground-state method which describes only occupied levels, whereas unoccupied levels have no physical meaning, see Chapter 2).

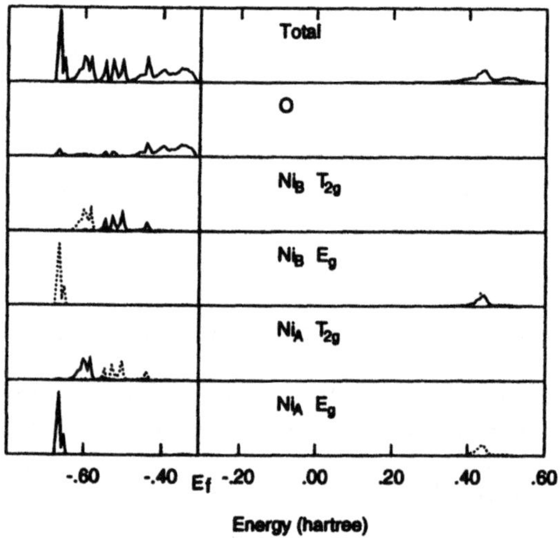

Figure 5. Total and projected density of states of anti-ferromagnetic nickel oxide. Density corresponding to α and β electrons is plotted with a solid and dashed line, respectively. The d orbitals of nickel are classified according to the crystal field description of an octahedral site. Ni_A and Ni_B identify the two nickel atoms in the AF2 double cell.

There is less agreement between the two methods for the attribution of the last occupied energy levels: UHF wave-functions show a predominant oxygen p nature for the top valence band (see Figure 5), whereas DFT methods appear to give a stronger metal d character to these bands. A possible explanation of this discrepancy is the following: DFT and HF generate spectra of eigenvalues with different physical meanings, therefore no direct comparison is possible.

For the prediction of bulk properties, such as the lattice parameter or bulk constant, HF results reach reasonable accuracy, with errors of the order of a few percent with respect to experiment (see [18, 19, 20]). Following the trend of studies of other metal oxides, HF tends to overestimate the volume of transition metal oxides; a *post-SCF* inclusion of correlation effects (by means of correlation functionals) allows the reduction of the calculated volume [21], therefore reducing the disagreement with experiment.

The UHF method also seems able to describe some of the features of the magnetism of the transition metal oxides: it correctly predicts the anti-ferromagnetic solution to be the ground state for systems such as MnO and NiO, gives magnetic moments (extracted from the Mulliken population analysis of the spin density)

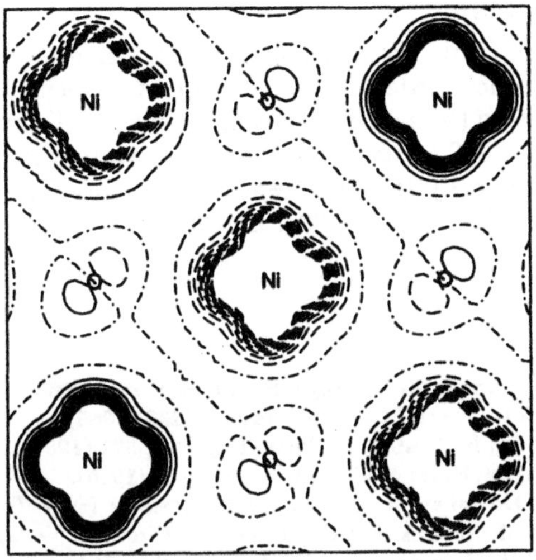

Figure 6. Spin density distribution in anti-ferromagnetic NiO. The contribution of the *d* orbitals to the spin polarization is evident. Iso-density lines are drawn at intervals of 0.01 $(\text{bohr})^{-3}$;(—) positive values, (– – –) negative and (– · · –) zero.

close to the measured ones and the relative stability of the ferromagnetic and anti-ferromagnetic solutions can be successfully related to the Curie temperature of these materials (see Fig. 7 and ref. [21, 19]). Another aspect of the magnetic

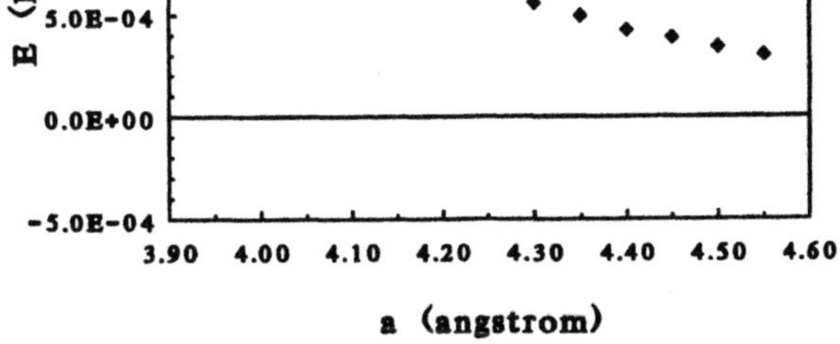

Figure 7. Hartree-Fock Energy difference between the anti-ferromagnetic and ferromagnetic solutions of NiO as a function of lattice parameter.

character of these compounds that HF accurately describes is the predominant metallic character of the spin polarization: Fig. 6 clearly shows that the spin density is localized on the *d* orbitals of the transition metal.

Acknowledgements
I wish to thank V.R. Saunders and R. Dovesi for many helpful discussions. The author also acknowledges the support of Associated Western Universities, Northwest Division (on behalf of Washington State University) under Grant No. DE-FG06-89ER-75522. Pacific Northwest Laboratory is a multiprogram national laboratory operated by Battelle Memorial Institute for the U. S. Department of Energy under contract DE-AC06-76RLO 1830.

References

1. P. Hohenberg and W. Kohn, Phys. Rev. **136**, B 864 (1964).
2. W. Kohn and L. Sham, Phys. Rev. **140**, A 1133 (1965).
3. J.A Pople and R.K. Nesbet, J. Chem. Phys. **22**, 571 (1954).
4. C.C.J. Roothaan, Revs. Modern. Phys. **32**, 179 (1960).
5. C. Pisani, R. Dovesi and C. Roetti, *Hartree-Fock ab initio Treatment of Crystalline Systems*, Vol. 48 of *Lectures Notes in Chemistry* (Springer Verlag, Heidelberg, 1988).
6. R. Dovesi, V.R. Saunders and C. Roetti, *CRYSTAL92 User's Manual*, Università di Torino, 1992.
7. M.F Guest and V.R. Saunders, Molec. Phys. **28**, 819 (1974).
8. T.M. Wilson, Int. J. Quantum. Chem. Symp. **2**, 269 (1968).
9. T.M. Wilson, Int. J. Quantum. Chem. Symp. **3**, 757 (1970).
10. J.C. Slater, Phys. Rev. **82**, 538 (1951).
11. T. Oguchi, K. Terakura and A.R. Williams, Phys. Rev. B **28**, 6443 (1983).
12. K. Terakura, T. Oguchi, A.R. Williams and J. Kübler, Phys. Rev. B **30**, 4734 (1984).
13. K. Terakura, A. Williams, T. Oguchi and J. Kübler, Phys. Rev. Lett. **52**, 1830 (1984).
14. A. Svane and O. Gunnarson, Phys. Rev. Lett. **65**, 1148 (1990).
15. M.R. Norman, Phys. Rev. B **44**, 1364 (1991).
16. V. S. P. Dufek, P. Blaha and K. Schwarz, Phys. Rev. B **49**, 10170 (1994).
17. S. Massida, M. Posternak and A. Baldereschi, Phys. Rev. B **46**, 11705 (1992).
18. W.C. Mackrodt, N.M. Harrison, V.R. Saunders, N.L.Allan, M.D. Towler, E. Aprà and R. Dovesi, Phil. Magaz. **68**, 653 (1993).
19. M.D. Towler, N.L. Allan, N.M. Harrison, V.R. Saunders, W.C. Mackrodt and E. Aprà, Phys. Rev. B **50**, 5041 (1994).
20. M. Catti, G. Valerio and R. Dovesi, Phys. Rev. B **51**, 7441 (1995).
21. E.Aprà, *Ph.D. Thesis*, University of Turin (1993).

The Quantum Theory of Periodic Systems on Modern Computers

Nicholas M. Harrison

CCLRC, The Daresbury Laboratory, Warrington, WA4 4AD, UK

September 6, 1996

Summary. We provide an overview of the current state of the art in the computational realisation of quantum theory.

Key words: Parallelization – Visualization

1. Introduction

In this Chapter, we provide a brief introduction to the problems involved in achieving an efficient computational implementation of quantum theory in periodic systems. Particular emphasis will be placed on the use of parallel computers since it is with this architecture that high performance is obtained currently. This is a large and rapidly developing field so a complete survey is not possible. We hope to provide an insight into the current state of computer hardware and the potential benefits which will arise from using it optimally.

In principle, quantum theory may be used to predict the properties of materials, given information about their composition and likely structure. As is apparent from other Chapters in this book, the application of the theory is computationally expensive and may be applied only after certain approximations have been made. In this Chapter, we will examine the implementation of two radically different schemes for solving the electronic problem in periodic materials. One scheme is based on the DFT Hamiltonian and the other, on the HF approximation (see Chapters 2, 8, 9 and 10). In each of the techniques we consider here, the wavefunctions are first expanded in a set of *basis functions*. The computational task is to choose the set of expansion coefficients which solve the equations. Before discussing the properties of these techniques in detail, we will examine the computers on which they must be implemented.

Correspondence to: N. Harrison : E-mail N.M.Harrison@dl.ac.uk

2. Computer Hardware

There is a limit to the number of computations that a single computer chip can perform in one second. This limit is imposed by two physical constraints. Information cannot move faster than the speed of light and device length scales are limited by quantum interference effects. From these facts, one may estimate that a single computer processor has a maximum speed of about 10^{11} operations per second. That is 100 Gflops. [1] This is a (very) rough estimate of peak single-CPU performance.

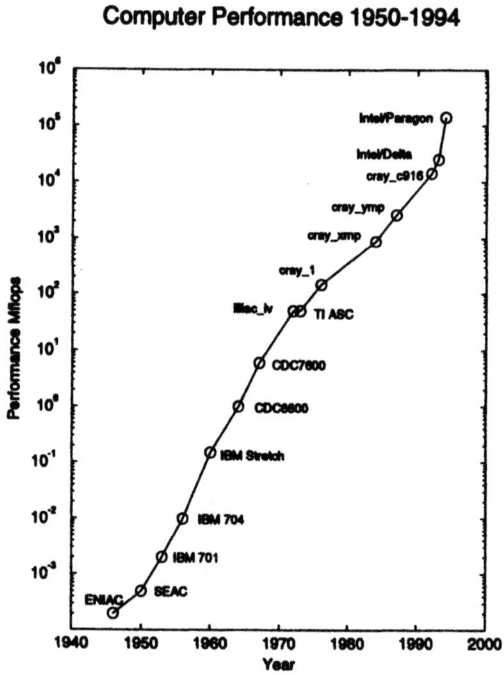

Figure 1. Representative peak computer performance for the period 1945-94

In Figure 1, the actual performance realised on some of the best computers available over the past 40 years is displayed. The first thing to notice is that the scale is almost logarithmic and that roughly every 5 years or so computer performance increases by a factor of 10. Up to the mid 1980's, these dramatic improvements were obtained using single processor machines (the Illiac IV in 1972 being a rare exception, 20 years ahead of its time). In the 1980's, machines with several processors accessing the same memory were successfully introduced. These *shared memory* parallel machines (Cray XMP, YMP, and C916 in Figure

[1] One flop is one floating point operation per second.

1) allowed performance figures to continue rising rapidly. The sudden leap in power over the last few years has been due to the introduction of massively parallel machines with thousands of processors, each of which addresses a local memory (*distributed memory* machines, the Intel Delta and Paragon in Figure 1). The fastest available computers in 1994 are running at almost 1000 Gflops, a performance which seems to be inaccessible to a single processor machine. The Cray C916 is a shared memory machine with 16 processors each yielding about 1 Gflop: 16 Gflops in all. The Intel Paragon machine has 2048 processors, each with its own memory and each performing at about 0.07 Gflops: 140 Gflops in all. It is much easier to write a computer programme to make efficient use of the shared memory machine but the potential benefit of using the massively parallel machines is enormous.

Undoubtedly computer performance will continue to rise exponentially (at least in the short term) and these improvements will be achieved by using *parallel processing*. It is difficult to use these machines optimally. On the sequential (*vector processing*) machines available in the 1980's, a reasonable fraction of the peak performance (say 20-30%) is achieved using straightforward coding in a high level language (FORTRAN, C). For parallel computers, algorithms must be redesigned to make use of the many processors available, for as much of the time as possible. A great deal of effort has been expended on this problem over the past 10 years with some notable success, but many applications have not yet been redesigned for use on parallel machines. In the next two sections, we describe the parallel implementation of two widely used techniques in electronic structure theory.

3. A Replicated Data Approach

In this section, the implementation of the CRYSTAL code [1] on distributed memory hardware is described. The strategy adopted was designed for use on machines with up to (about) 64 processors, each of which has fast access to memory and disk storage. High inter-processor communication rates are not required.

The *replicated data approach* is probably the first strategy one would think of for using a parallel computer. A complete copy of the code is run on each processor. One then imagines the code to consist of many independent computational *tasks* and each processor becomes responsible for performing a certain number of tasks. Of course, in practice, a code rarely consists of completely independent tasks and thus the individual processors must, at some stage, *communicate* results to each other. The art of designing an efficient algorithm is to minimise communications while having enough tasks to keep as many of the processors working as possible (*load balancing*). A useful method for distributing tasks to processors is the *synchronous global index*. This is a counter which may be interrogated or updated by any processor and is used to count the number of tasks which have been completed. Using such an index, it is a simple matter to allocate an initial set of tasks to the processors and then when a task is finished the processor uses the index to identify the next task which must be performed. This strategy has the benefit of being comparatively simple to implement and will balance the computational load as long as the number of tasks is much larger than the number of processors. The main disadvantage is that memory is inefficiently used since all major data structures are stored on each node.

The computational details of the CRYSTAL package are given in Chapter 8. The basic features are common to any technique which uses a basis set consisting of *local functions* and which solves the many-electron problem in a one-electron approximation using *self consistent field* (SCF) iterations.

3.1. The CRYSTAL code

In a periodic system, the single particle orbitals may be labelled with a band index n, and a crystal momentum, $\mathbf{k}$. The Hamiltonian is represented as matrix elements between these orbitals. If each $\psi_{n,\mathbf{k}}$ is expanded in local atom-centred functions, $\phi_{i,\mathbf{R}}$ (where i labels the local function within a unit cell and $\mathbf{R}$ labels the cell), then the direct space representation of the single particle Hamiltonian in the Hartree-Fock approximation will require the computation of many integrals of the form:

$$I(i\mathbf{0}, j\mathbf{R}, k\mathbf{S}, l\mathbf{T}) = \int d\mathbf{r} \int d\mathbf{r}' \frac{\phi_{i,\mathbf{0}}(\mathbf{r})\phi_{j,\mathbf{R}}(\mathbf{r})\phi_{k,\mathbf{S}}(\mathbf{r}')\phi_{l,\mathbf{S}+\mathbf{T}}(\mathbf{r}')}{|\mathbf{r}-\mathbf{r}'|} \tag{1}$$

As discussed in Chapter 8, the first stage (part 1) is to calculate and store all of the integrals required, while the second stage (part 2) consists of SCF iterations. The integrals are used to produce a Hamiltonian matrix, $H^{i,j}_{\mathbf{k}}$, which must be diagonalised at a number of $\mathbf{k}$-points in order to compute the new orbitals, $\psi_{n,\mathbf{k}}$. These orbitals may then be used to construct a new $H^{i,j}_{\mathbf{k}}$. This process must be iterated (from an initial guess) until the wavefunctions are self-consistent: a process which usually takes a few tens of cycles.

For parallelisation, the independent tasks in part 1 are taken to be blocks of integrals labelled with a common $(i\mathbf{0}, j\mathbf{R})$. In a typical computation, one might expect there to be a hundred basis functions per cell and the range of interactions to be such that about ten $\mathbf{R}$ values must be considered. There will thus be 10^5 independent tasks. Each processor will then be able to store locally the subset of integrals which it computed. There is no need for any communication between processors once the tasks have been defined and therefore, one would expect this to be a highly efficient operation.

In part 2, the parallelisation is achieved in two steps. Firstly, in the generation of the Hamiltonian matrix, each processor can only generate a partial copy from the integrals which it has stored locally. As each node may act independently, this process is naturally performed in parallel and only suffers from the possibility that the load balancing may be poor. All nodes must complete this task before the calculation can proceed, and thus, some nodes may lie idle for a period. A complete copy of the Hamiltonian must then be assembled on each processor by passing messages containing partial copies of the matrix: a process best represented as a *global summation* of the data.

The final step is to diagonalise the matrix. The diagonalisation at each $\mathbf{k}$-point is independent, so this is a natural choice for the basic computational task. There are usually a few tens of $\mathbf{k}$-points and therefore, this choice limits the number of processors on which the algorithm will run efficiently. On machines with many hundreds of processors, most will lie idle. After the diagonalisations are complete, the eigenvectors ($\psi_{n,\mathbf{k}}$) must be communicated to all processors in order to generate the next version of H. This simple strategy has the advantage that it requires a limited number of modifications of the sequential code but successfully distributes the major computational and disk accessing tasks. A limitation of this strategy is that copies of all main memory-resident data structures are stored on each processor, so that the maximum problem size is limited by the node with the smallest memory. These restrictions are lifted if a distributed data strategy is employed at the expense of more communication between processors. A strategy of this form will be examined in the next section.

3.2. The performance of CRYSTAL

As this strategy requires each processor to have fast access to disk storage and does not require fast communication between processors, it is able to take advantage of an environment in which the computers are rather weakly coupled. A cluster of workstations linked by a standard network is adequate.

In Figure 2, the performance of CRYSTAL on a workstation cluster is presented. The case considered is the computation of the electronic structure at

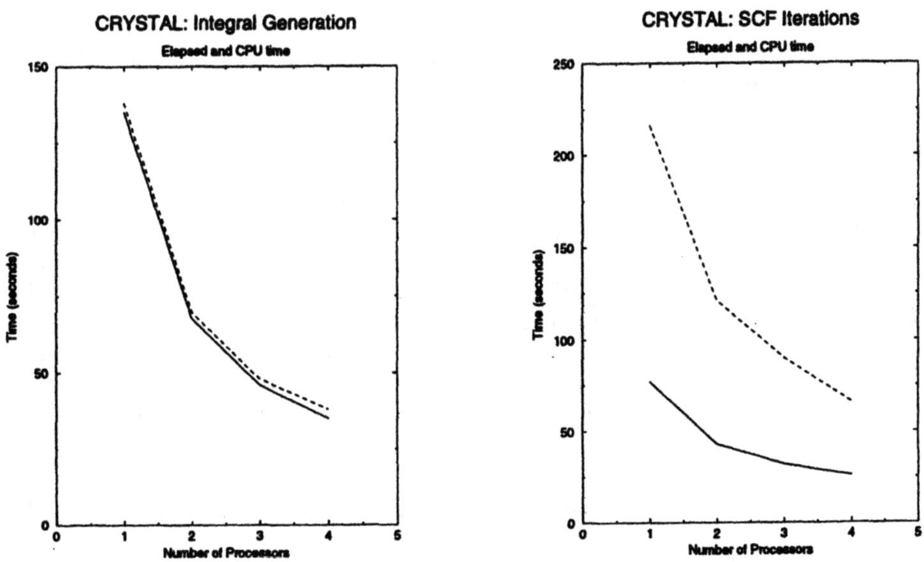

Figure 2. The performance of the CRYSTAL package on a cluster of 4 HP/735 workstations linked on Ethernet

the surface of corundum (Al_2O_3) represented as a 2D periodic slab. There are 10 atoms in each unit cell, with 100 electrons, described by 86 local basis functions. Eight irreducible **k**-points were used. This is a small calculation by modern standards and takes only a few minutes on a single workstation.

During integral generation, the elapsed and CPU times scale rather well as the number of machines is increased. The elapsed time is reduced by a factor of 3.7 when 4-nodes are used: about 90% efficiency. The small communication overhead in part 1 does not increase with system size and almost perfect scaling would be achieved in larger calculations. The elapsed times during the SCF cycles are reduced by a factor of 3.3 on 4-nodes: about 80% efficiency. Two factors limit the efficiency of part 2. As noted above, the load balancing may not be ideal as the distribution of the Hamiltonian matrix construction tasks are determined by part 1. In this test, four identical computers were used and therefore, this problem is small; it would be more pronounced on a heterogeneous workstation cluster. Another factor limiting the efficiency is the communication of the partial Hamiltonian matrices to all processors. This step scales as N^2 where N is the number of basis functions. The CPU-time required in diagonalisation scales as N^3 and therefore, the communication overhead becomes less significant in studies of larger systems. For larger systems, we would also expect the efficiency of part 2 to improve. Unfortunately, for very large systems, the number of **k**-points required is reduced, and thus, the algorithm will not be able to make efficient use of a large number of processors. The other feature of part 2 is that the elapsed and CPU times are rather different. This is due to the large amount of disk access operations performed; as the stored integrals are read from disk, the CPU lies idle. On the workstation cluster, each processor has local disk storage, and as a result, the aggregate disk-access speeds improve with CPU performance during parallel operation. A scalable disk system is a vital component of any parallel computer and is not provided by the massively parallel machines currently available.

4. A Data-Parallel Algorithm

In this section, the implementation of the CETEP code [2] achieved by the groups at the University of Cambridge and the Edinburgh Parallel Computer Centre is described. The algorithm is designed for use on purpose-built parallel computers with high communication speeds between processors. As an alternative to a self-consistent field procedure, one can solve the SCF problem using direct minimisation (see Chapter 3, section 5). The basic idea of this method is that the total energy of the system may be written as a function of the set of single particle orbitals, that is:

$$E[\psi_i] = \sum_i \int d\mathbf{r} \psi_i(\mathbf{r}) H \psi_i(\mathbf{r}) \tag{2}$$

and that the total energy is minimised by the orbitals which solve the Kohn-Sham equation (see Chapter 3, Section 2.2). If the orbitals are expanded in a set of basis functions, and this time we choose plane waves, so that:

$$\psi_i(\mathbf{r}) \equiv \psi_{n\mathbf{k}}(\mathbf{r}) = \sum_{\mathbf{K}} C_{\mathbf{K}}^{n\mathbf{k}} \exp[\imath(\mathbf{k}+\mathbf{K})\cdot\mathbf{r}] \tag{3}$$

then the energy is a complicated function of the many expansion coefficients, C. Subject to the constraint that the single particle orbitals are orthogonal, we may now determine the coefficients by direct minimisation of the energy. A

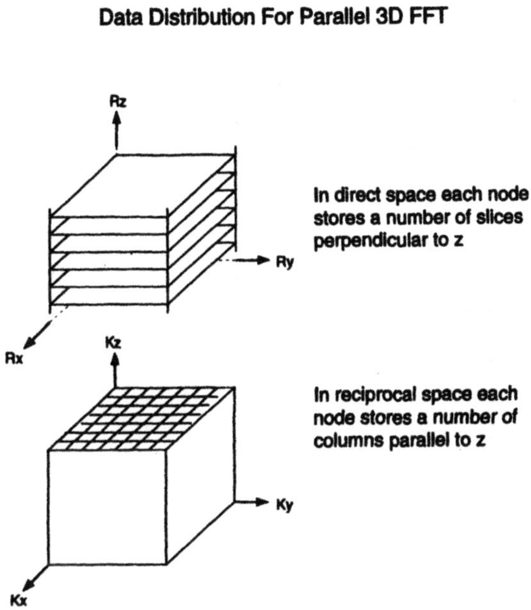

Figure 3. Data distribution for the parallel 3D FFT

plane wave basis set is only adequate for representing rather smooth functions. In calculations of this type, the core electrons are removed from the problem and replaced by a pseudo-potential; therefore, only the more gently varying valence wave functions are represented by the plane waves. Even so, many basis functions are required. Typical values for the number of plane waves ($N_{\mathbf{K}}$), number of bands (N_b) and number of **k**-points ($N_{\mathbf{k}}$) for a system containing N_{atoms} atoms are:

$N_{\mathbf{K}}$	$1000 * N_{atoms}$
N_b	$10 * N_{atoms}$
$N_{\mathbf{k}}$	$10 / N_{atoms}$

If the Hamiltonian matrix is generated in this representation and diagonalised, the computational cost would be of order $N_{\mathbf{K}}^3$ operations. In the 1980's, powerful algorithms were developed which allowed the minimisation process to be achieved in a number of operations of the order $N_b^2 N_{\mathbf{K}}$ [3]. The saving of roughly 4 orders of magnitude in the number of operations quickly led to these and similar algorithms being used in all calculations using large plane wave basis sets. The full details of these methods are given in the literature [3]. Here, we describe a data parallel algorithm which allows such a calculation to be performed efficiently on a parallel computer (for further details, see reference [4]).

In a plane wave basis set, the calculation $H\psi$ in equation 2 may be performed without ever storing the full H matrix. This is true because H may be split into two parts: one diagonal in (direct) **R**-space and the other diagonal in (reciprocal) **K**-space. This means that the product, $H\psi$, may be built efficiently if ψ is available in both reciprocal and direct space representations. The main computational task in the minimisation process is, therefore, to change

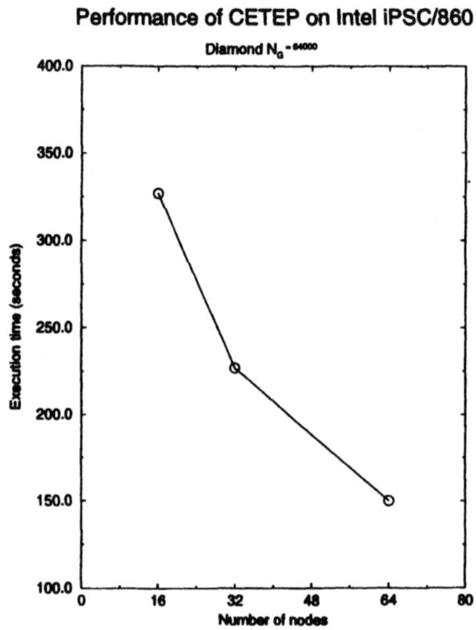

Figure 4. The performance of the CETEP package on the Intel iPSC/860

the representation of the wavefunction from reciprocal space to direct space using a 3-dimensional Fast Fourier transform (3D FFT). As the 3D transform is built from the repeated application of 1D FFTs, it is possible to perform this operation with the wave function distributed over many processors.

One possible algorithm, similar to that used in CETEP, is as follows. In direct space, $\mathbf{R} \equiv (R_x, R_y, R_z)$, the wavefunction is stored as (R_x, R_y) **planes** of data with each processor responsible for a certain range of R_z. In reciprocal space, $\mathbf{K} \equiv (K_x, K_y, K_z)$, the data is stored in **columns**; each node is responsible for a column of K_z values for some range of (K_x, K_y). This is illustrated in Figure 3. The parallel 3D FFT from direct space to reciprocal space proceeds by the following steps:

1. Each processor performs 2D FFTs on the (R_x, R_y) stored locally. It now has data (K_x, K_y, R_z) for some range of R_z.
2. Data is communicated between processors until each has a set of complete columns of data in the z-direction, that is, (K_x, K_y, R_z) for some range of (K_x, K_y).
3. Each node performs a 1D FFT, in the z-direction, for the columns it stores and the 3D FFT is complete.

The performance of the CETEP code for a calculation of the cubic 8-atom unit cell of diamond using 64000 plane waves is displayed in Figure 4. The calculation was performed on an Intel iPSC/860. The parallel efficiency is not perfect as the communication of the data is a significant overhead. The computational cost scales with system size somewhat faster than the communication time so greater efficiency would be expected for larger calculations. For this application, the parallel computer also represents a large step forward in the amount of memory which can be dedicated to a single task. On the Intel iPSC/860, for instance, each processor has 16 MB[2] of memory; on 64 nodes over 1000MB of memory is available. The Intel Paragon machine has 2048 nodes.

The data parallel algorithm has the advantage that the memory of the parallel computer is used efficiently. The communication time involved in passing data between processors is significant and for a given system, size will limit the scalability of the algorithm. The algorithm is best suited to closely coupled processors with fast communications, found in purpose-built parallel computers. It is worthy of note that the speed of inter-processor communication is increasing much more rapidly than processor speed. Applications which suffer from communication bandwidth problems will benefit from this in the near future.

5. Visualization of results

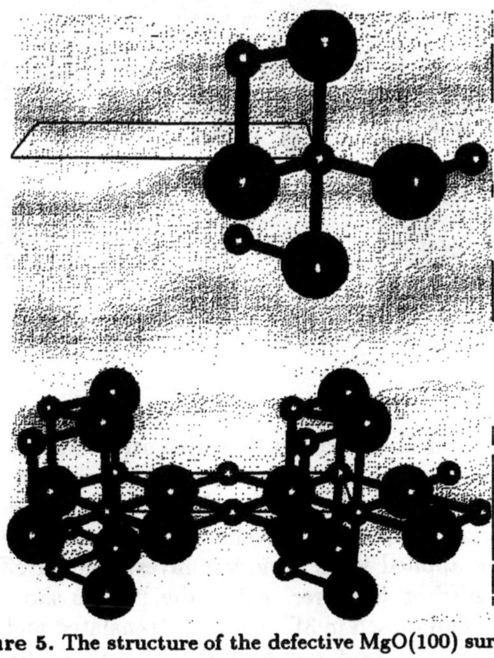

Figure 5. The structure of the defective MgO(100) surface

In this section, we briefly review some of the software and hardware available for the visualization of scientific data.

It is not only the processing power of computers which has grown rapidly over the past ten years; the rate at which complex images can be rendered has

[2] 1 MB = 10^6 bytes

also risen dramatically. In addition, the software used to generate images from scientific data has improved so that now, for a small investment of effort, it is possible to visualize complex data structures.

The hardware in a relatively inexpensive graphics workstation (Silicon Graphics Ltd Indy or equivalent) is quite remarkable. Most of the data that we wish to interpret can be visualized as coloured lines or polygons. The complexity involved in rendering a single coloured polygon to a screen is enormous and yet it is typical for purpose built hardware to be able to render several million such polygons per second. This allows one to generate spectacular images that move in real time.

Figure 6. An isovalue surface of the charge density above the defective MgO(100) surface: $\rho(\mathbf{r}) = 0.01|e|/(au)^3$

A limitation on our ability to do this with scientific data is the difficulty in mapping the data to suitable images. The data presented in this section was computed using the CRYSTAL code and the pictures were generated using DISPLAY [5] a visualization package developed at Daresbury Laboratory. Similar techniques may be applied with a modest investment of effort using the commercial packages AVS or Explorer. All of the images shown can be generated rapidly and then manipulated in 3D (rotated, translated and scaled) on a graphics workstation. By far the best way to understand this is to sit at a workstation and play.

The visualization of data becomes more difficult as we move to higher dimensions. Even 2D data, represented as contours and surface plots, may be hard to understand. For 3D data, there is a need for tools which allow one to apply many different rendering techniques to the same data in order to build up a mental image of its structure and relevance. It may also be vital to see the data *in context*; for our purposes, this often means plotted along side a structural model.

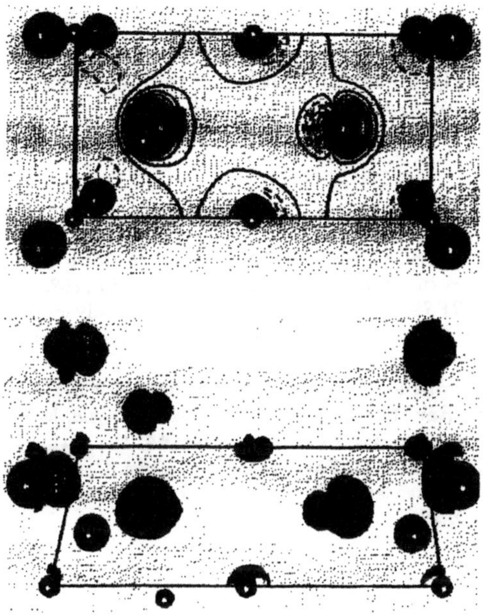

Figure 7. Isovalue and contour surface representations of the difference density: $\rho(\mathbf{r}) - \rho_{ions}(\mathbf{r})$ above the defective MgO(100) surface.

As examples from some recent research, I have included three slides. In the first, the structure of a surface is plotted (Figure 5). Complex crystal structures can be difficult to understand. A computer which allows us to render a simple ball and stick picture and spin it around on the screen can help a lot. In this case, the structure is rather simple. It is the surface of a cubic crystal; the (100) surface of MgO. The large spheres represent the O ions and the small spheres the Mg ions. The image is reproduced in grey scale from a colour original. The top panel shows the surface unit cell of this structure and the irreducible atoms. In the lower panel, the pattern has been repeated by a few lattice vectors to reveal more of the surface. This is not the ideal (100) surface; a "line" of MgO units has been removed to make a primitive model of a step defect. It takes a few minutes on a workstation to solve the SCF problem for this simple model. The resulting wavefunctions may then be used to compute various quantities of interest. The charge density is a 3D scalar field. We can represent this in many ways. In Figure 6, a surface of constant charge density has been evaluated and plotted. This is analogous to a 1D contour line cut from 2D data; it is a 2D contour surface cut from 3D data - an *isovalue surface*. The surface is at a rather low value of the charge density and essentially, follows the outline of each site. This helps to confirm the rather simple idea of large negative O^{2-} ions bound to smaller Mg^{2+} ions. The total density contains a large term due to the core electrons: a density more responsive to the local environment is obtained if we subtract the charge density of individual ions. A charge density difference map is plotted in Figure 7. In the upper panel, a 2D slice is taken from the 3D data and rendered using normal contour levels. The continuous lines indicate regions that have been depleted of charge and the dashed lines regions where charge has accumulated (relative to unperturbed ions). In the lower panel, the same data is rendered using 3D isovalue surfaces. It is clear that the O ions at the step edge are polarised strongly. From this plot, the regions of strong polarisation can be quickly identified and examined in more detail.

References

1. C. Pisani, R. Dovesi and C. Roetti, *Hartree-Fock ab initio Treatment of Crystalline Systems*, Lecture Notes in Chem. Vol. 48, Springer, Berlin (1988); R. Dovesi, C. Roetti, and V.R. Saunders, *CRYSTAL 92 User Documentation*, University of Torino, Torino (1992).
2. CETEP *user documentation* M.C. Payne, Cavendish Laboratory, University of Cambridge, UK.
3. R. Car and M. Parrinello, Phys Rev Lett 56, 2471 (1985); M.P. Teter, M.C. Payne and D.C. Allan, Phys. Rev. B 40, 12255 (1989); M.C. Payne, M.P. Teter, D.C. Allan, T.A. Arias and J.D. Joannopoulos, Rev. Mod. Phys. 64, 1045 (1992).
4. L.J. Clarke, I. Štich and M.C. Payne, Comput. Phys. Commun. 72, 14 (1992).
5. DISPLAY *User documentation* Dr P Sherwood, Daresbury Laboratory, Daresbury, Warrington, WA4 4AD, UK.

The CRYSTAL code

Carla Roetti

Department of Inorganic, Physical and Materials Chemistry, University of Torino, via Giuria 5, I-10125 Torino, Italy

May 28, 1996

Summary. The structure of the CRYSTAL code is analysed. This code permits the calculation of the wavefunction and properties of crystalline systems, using a Hartree-Fock Linear-Combination-of-Atomic-Orbitals (HF-LCAO) approximation. Both the theory and the practical implementation of the equations are discussed.

Key words: Hartree-Fock – Crystals – Coulomb Series – Exchange Series – Reciprocal Space Integration – Bipolar Approximation

1. Introduction

The CRYSTAL program permits the calculation of the wavefunction and properties of crystalline systems, within the Hartree-Fock Linear-Combination-of-Atomic-Orbitals (HF-LCAO) approximation.

The code was developed initially by the Theoretical Chemistry Group of the University of Torino at the end of the seventies [1, 2]. Since 1981 there has been a close collaboration on the project with V.R.Saunders (Daresbury Laboratory, UK) [3, 4]. The first public version was released in 1988 [5]. CRYSTAL92 [6] is now available as a FORTRAN source code, fully portable. The development of the current version of CRYSTAL has been greatly assisted by the constant support and interest of C. Pisani and by contributions from M. Causà, N. M. Harrison, E. Aprà and R. Orlando. The User's Manual includes 28 test cases, including both input and output, which provide examples of the wide range of program capabilities available.

A new release of the program is expected to appear in 1996. It is more efficient and has additional capabilities with respect to the current version, but the essential philosophy and techniques, which will be presented in this Chapter, have remained unchanged. A discussion of the present scheme in relation to other approaches, was presented in Chapter 3. In the following sections, the equations are presented for the closed-shell case; extension to the open-shell case is discussed in Chapter 6.

2. Theoretical framework

CRYSTAL is an *ab initio* Hartree-Fock CO-LCAO program for the treatment of periodic systems.

CO-LCAO, in the present case, means that each Crystalline Orbital (CO), $\psi_i(\mathbf{r};\mathbf{k})$, is a linear combination of Bloch functions (BF), $\phi_\mu(\mathbf{r};\mathbf{k})$, defined in terms of local functions, $\varphi_\mu(\mathbf{r})$ (here referred to as Atomic Orbitals, AOs).

$$\psi_i(\mathbf{r};\mathbf{k}) = \sum_\mu a_{\mu,i}(\mathbf{k})\phi_\mu(\mathbf{r};\mathbf{k}) \quad (1)$$

$$\phi_\mu(\mathbf{r};\mathbf{k}) = \sum_\mathbf{T} \varphi_\mu(\mathbf{r}-\mathbf{A}_\mu-\mathbf{T})\, e^{i\mathbf{k}\cdot\mathbf{T}} \quad (2)$$

$\mathbf{A}_\mu$ denotes the coordinate of the nucleus in the zero reference cell on which φ_μ is centred, and the $\sum_\mathbf{T}$ is extended to the set of all lattice vectors, $\mathbf{T}$.

The local functions are expressed as linear combinations of a certain number, n_G, of individually normalized Gaussian type functions (GTO) characterized by the same centre, with fixed coefficients, d_j and exponents, α_j, defined in the input:

$$\varphi_\mu(\mathbf{r}-\mathbf{A}_\mu-\mathbf{T}) = \sum_j^{n_G} d_j\; G(\alpha_j;\mathbf{r}-\mathbf{A}_\mu-\mathbf{T}) \quad (3)$$

The AOs belonging to a given atom are grouped into *shells*, λ. The shell can contain all AOs with the same quantum numbers, n and ℓ, (for instance $3s$, $2p$, $3d$ shells), or all the AOs with the same principal quantum number, n, if the number of GTOs and the corresponding exponents are the same for all of them (mainly *sp* shells; this is known as the *sp constraint*). These groupings permit a reduction in the number of auxiliary functions that need to be calculated in the evaluation of electron integrals and therefore, increase the speed of calculation.

A single, normalized, s-type GTO, G_λ, is associated with each shell (the *adjoined Gaussian* of shell λ). The α exponent is the smallest of the α_j exponents of the Gaussians in the contraction. The adjoined Gaussian is used to estimate the AO overlap and select the level of approximation to be adopted for the evaluation of the integrals.

The expansion coefficients of the Bloch functions, $a_{\mu,i}(\mathbf{k})$, are calculated by solving the matrix equation for each reciprocal lattice vector, $\mathbf{k}$:

$$\mathbf{F}(\mathbf{k})\mathbf{A}(\mathbf{k}) = \mathbf{S}(\mathbf{k})\mathbf{A}(\mathbf{k})\mathbf{E}(\mathbf{k}) \quad (4)$$

in which $\mathbf{S}(\mathbf{k})$ is the overlap matrix over the Bloch functions, $\mathbf{E}(\mathbf{k})$ is the diagonal energy matrix and $\mathbf{F}(\mathbf{k})$ is the Fock matrix in reciprocal space:

$$\mathbf{F}(\mathbf{k}) = \sum_\mathbf{T} \mathbf{F}^\mathbf{T}\, e^{i\mathbf{k}\cdot\mathbf{T}} \quad (5)$$

The matrix elements of $\mathbf{F}^{\mathbf{T}}$, the Fock matrix in direct space, can be written as a sum of one-electron and two-electron contributions in the basis set of the AO:

$$F_{12}^{\mathbf{T}} = H_{12}^{\mathbf{T}} + B_{12}^{\mathbf{T}} \tag{6}$$

The one-electron contribution is the sum of the kinetic and nuclear attraction terms:

$$H_{12}^{\mathbf{T}} = T_{12}^{\mathbf{T}} + Z_{12}^{\mathbf{T}} = \langle \varphi_1^{\mathbf{0}} \mid \hat{T} \mid \varphi_2^{\mathbf{T}} \rangle + \langle \varphi_1^{\mathbf{0}} \mid \hat{Z} \mid \varphi_2^{\mathbf{T}} \rangle \tag{7}$$

In core pseudopotential calculations, $\hat{Z}$ includes the sum of the atomic pseudopotentials.

The two-electron term is the sum of the Coulomb and exchange contributions:

$$B_{12}^{\mathbf{T}} = C_{12}^{\mathbf{T}} + X_{12}^{\mathbf{T}} = \tag{8}$$

$$\sum_{3,4} \sum_{\mathbf{Q}} P_{3,4}^{\mathbf{Q}} \sum_{\mathbf{S}} [(\varphi_1^{\mathbf{0}} \varphi_2^{\mathbf{T}} \mid \varphi_3^{\mathbf{S}} \varphi_4^{\mathbf{S}+\mathbf{Q}}) - \frac{1}{2} (\varphi_1^{\mathbf{0}} \varphi_3^{\mathbf{S}} \mid \varphi_2^{\mathbf{T}} \varphi_4^{\mathbf{S}+\mathbf{Q}})] \tag{9}$$

The Coulomb interactions, that is, those of electron-nucleus, electron-electron and nucleus-nucleus, are individually divergent, due to the infinite size of the system. The grouping of corresponding terms is necessary in order to eliminate this divergence.

The $\mathbf{P}^{\mathbf{Q}}$ density matrix elements in the AOs basis set are computed by integration over the volume of the Brillouin zone (BZ),

$$P_{3,4}^{\mathbf{Q}} = 2 \int_{BZ} d\mathbf{k} e^{i\mathbf{k}\cdot\mathbf{Q}} \sum_n a_{3n}^*(\mathbf{k}) a_{4n}(\mathbf{k}) \theta(\epsilon_F - \epsilon_n(\mathbf{k})) \tag{10}$$

where a_{in} denotes the i-th component of the n-th eigenvector, θ is the step function, ϵ_F, the Fermi energy and ϵ_n, the n-th eigenvalue.

The total electronic energy per unit cell is given by:

$$E^{elec} = \frac{1}{2} \sum_{1,2} \sum_{\mathbf{T}} P_{12}^{\mathbf{T}} (H_{12}^{\mathbf{T}} + B_{12}^{\mathbf{T}}) \tag{11}$$

Chapter 11 presents a discussion of the different contributions to the total energy.

2.1. Remarks on the evaluation of the integrals

The approach adopted for the treatment of the Coulomb and exchange series is based on a few simple ideas and on a few general tools, which can be summarized as follows:

1. Where possible, terms of the Coulomb series are aggregated so as to reduce the number of integrals to be evaluated;
2. Exchange integrals which will combine with small density matrix elements are disregarded;
3. Integrals between non-overlapping distributions are approximated;
4. Approximations for large integrals must be very accurate; for small integrals large percentage errors can be accepted;
5. Selection must be very efficient, because a large number of possible terms must be checked (adjoined Gaussians are very useful from this point of view);

2.2. Treatment of the Coulomb series

For the evaluation of the Coulomb contributions to the total energy and Fock matrix, correct coupling of electron-nucleus and electron-electron interactions is essential. The computational technique for doing so was presented by Dovesi *et al.* [3] and by Saunders *et al.* [7]. It may be summarized as follows.

Consider the Coulomb bielectronic contribution to the Fock matrix ($C^{\mathbf{T}}_{12}$) and to the total energy :

$$E^{coul}_{ee} = \frac{1}{2}\sum_{1,2}\sum_{\mathbf{T}} P^{\mathbf{T}}_{12}\sum_{3,4}\sum_{\mathbf{Q}} P^{\mathbf{Q}}_{3,4}\sum_{\mathbf{S}}[(\varphi^{0}_{1}\varphi^{\mathbf{T}}_{2} \mid \varphi^{\mathbf{S}}_{3}\varphi^{\mathbf{S+Q}}_{4}) \qquad (12)$$

Seven indices are involved in equation 12; four of them (1, 2, 3 and 4) refer to the AOs of the unit cell; in principle, the other three (**T**, **Q** and **S**) span the infinite set of translation vectors: for example, $\varphi^{\mathbf{T}}_{2}(\mathbf{r})$ is AO number 2 in cell **T**. P is the density matrix; the usual notation is used for the bielectronic integrals. Due to the localized nature of the basis set, the total charges, q_1 and q_2, associated with the two overlap distributions: $\{G_{10}\ G_{2\mathbf{T}}\}$ and $\{G_{3\mathbf{S}}\ G_{4\mathbf{S+Q}}\}$, decay exponentially to zero with increasing $|\mathbf{T}|$ and $|\mathbf{Q}|$ (for example, G_1 is the adjoined Gaussian of the shell to which φ_1 belongs).

A *Coulomb overlap* parameter, S_c, can be defined in such a way that when either q_1 or q_2 are smaller than S_c, the bielectronic integral is disregarded, and the sum over **T** or **Q** truncated. The ITOL1 input parameter is defined as **ITOL1**$=-log_{10}S_c$. The same parameter value is used for selecting overlap, kinetic and multipole integrals.

The problem of the **S** summation in equation 12 is more delicate, **S** being related to the distance between the two interacting distributions. The multipolar expansion scheme illustrated below is particularly effective when large unit cell or low dimensionality systems are considered.

The electron-electron and electron-nuclei series ($C^{\mathbf{T}}_{12}$ and $Z^{\mathbf{T}}_{12}$) can be rearranged as follows:

1. Mulliken shell *net* charge distributions are defined as :

$$\rho_\lambda(\mathbf{r}-\mathbf{S}) \equiv \{\lambda\}' \equiv \{\lambda\} - Z_\lambda = \sum_{3\in\lambda}\sum_{4\mathbf{Q}} P^{\mathbf{Q}}_{34}\,\varphi_3(\mathbf{r}-\mathbf{S})\,\varphi_4(\mathbf{r}-\mathbf{S}-\mathbf{Q}) - Z_\lambda \qquad (13)$$

where Z_λ is the fraction of nuclear charge formally attributed to shell λ, and $\{\lambda\}$ is the electronic charge distribution of shell λ.

2. Z and C contributions are reordered:

$$C^{\mathbf{T}}_{12} + Z^{\mathbf{T}}_{12} = \sum_{\lambda}\sum_{\mathbf{S}}\int d\mathbf{r}\,d\mathbf{r}'\varphi^{0}_{1}(\mathbf{r})\ \varphi^{\mathbf{T}}_{2}(\mathbf{r})\ |\mathbf{r}-\mathbf{r}'-\mathbf{S}|^{-1}\ \rho_\lambda(\mathbf{r}'-\mathbf{S}) \qquad (14)$$

3. For a given shell λ, there is a finite set of **S** vectors, B_λ, which the two interacting distributions overlap; in this B_λ zone (*bielectronic zone*), all the bielectronic integrals are evaluated explicitly. In the outer, infinite region which we define as M_λ, complementary to B_λ (the *monoelectronic zone*), ρ_λ can be expanded in multipoles and the series can be evaluated to infinity analytically, using Ewald's method combined with recursion formulae [7].

The resulting expression for the Coulomb contribution to the Fock matrix is:

$$\begin{aligned} C_{12}^{\mathbf{T}} + Z_{12}^{\mathbf{T}} &= \sum_{\lambda}\{\sum_{\mathbf{S}}^{B_\lambda}[\sum_{3\in\lambda}\sum_{4}\sum_{\mathbf{Q}} P_{34}^{\mathbf{Q}}(\varphi_1^{\mathbf{0}}\varphi_2^{\mathbf{T}} \mid \varphi_3^{\mathbf{S}}\varphi_4^{\mathbf{S}+\mathbf{Q}}) \\ &\quad - \sum_{\ell,m} \gamma_\ell^m(\mathbf{A}_\lambda;\{\lambda\})\Phi_\ell^m(12\mathbf{T};A_\lambda+\mathbf{S})] + \\ &\quad + \sum_{\mathbf{S}}\sum_{\ell,m} \gamma_\ell^m(A_\lambda;\{\lambda\}')\Phi_\ell^m(12\mathbf{T};\mathbf{A}_\lambda+\mathbf{S})\} \end{aligned} \tag{15}$$

where:

$$\gamma_\ell^m(\mathbf{A}_\lambda;\{\lambda\}) = \int d\mathbf{r}\, \rho_\lambda(\mathbf{r}-\mathbf{A}_\lambda) N_\ell^m X_\ell^m(\mathbf{r}-\mathbf{A}_\lambda) \quad \text{and} \tag{16}$$

$$\Phi_\ell^m(12\mathbf{T};\mathbf{A}_\lambda+\mathbf{S}) = \int d\mathbf{r}\varphi_1^{\mathbf{0}}(\mathbf{r})\varphi_2^{\mathbf{T}}(\mathbf{r}) X_\ell^m(\mathbf{r}-\mathbf{A}_\lambda-\mathbf{S})|\mathbf{r}-\mathbf{A}_\lambda-\mathbf{S}|^{-2\ell-1} \tag{17}$$

The X_ℓ^m functions entering in the definition of the multipoles and field terms are real, solid harmonics and N_ℓ^m, the corresponding normalization coefficients.

The advantage of using equation 15 is that many four-centre (long-range) integrals can be replaced by a smaller number of three-centre integrals.

The attribution of the interaction between $\rho_1 = \{10\,2\mathbf{T}\}$ and ρ_λ either to the *exact*, short-range or to the *approximate*, long-range zone is performed by comparing the penetration between ρ_1 and ρ_λ with the ITOL2 input parameter (if **ITOL2**$> -\log S_{1\lambda}$, then ρ_λ is attributed to the *exact* B_λ zone).

The multipolar expansion, in the approximate zone, is truncated at $L = \ell^{max}$. The default value of L is 4; the maximum possible value is 6, the minimum suggested value, 2.

2.3. The exchange series

The exchange series does not require particular manipulations of the kind discussed in the previous subsection, but needs a careful selection of the terms contributing appreciably to the Fock operator and to the total energy [4]. The exchange contribution to the total energy can be written as follows:

$$E^{ex} = \frac{1}{2}\sum_{12}\sum_{\mathbf{T}} P_{12}^{\mathbf{T}}\, [-\frac{1}{2}\sum_{34}\sum_{\mathbf{Q}} P_{34}^{\mathbf{Q}} \sum_{\mathbf{S}} (\varphi_1^{\mathbf{0}}\varphi_3^{\mathbf{S}} \mid \varphi_2^{\mathbf{T}}\varphi_4^{\mathbf{S}+\mathbf{Q}})] \tag{18}$$

where the term in square brackets is the exchange contribution to the 12T element of the direct-space Fock matrix. E^{ex} has no counterpart of opposite sign as the Coulomb term has; hence, it must converge by itself.

The **S** summation can be truncated after a few terms, since the $\{\varphi_1^{\mathbf{0}}\varphi_3^{\mathbf{S}}\}$ overlap distribution decays exponentially as **S** increases. Similar considerations apply to the second charge distribution. In CRYSTAL, the **S** summation is, therefore, truncated when the charge associated with either $\{G_1\mathbf{0}\; G_3\mathbf{S}\}$ or $\{G_2\mathbf{T}\; G_4\mathbf{S}+\mathbf{Q}\}$ is smaller than $10^{-\mathbf{ITOL3}}$.

The situation is more complicated when **T** and **Q** summations are analysed. Let us consider the leading terms at large distance, corresponding to $\varphi_1{=}\varphi_3$, $\varphi_2{=}\varphi_4$, $\mathbf{S} = \mathbf{0}$ and $\mathbf{Q} = \mathbf{T}$:

$$e^{\mathbf{T}}_{12} = -1/4(P^{\mathbf{T}}_{12})^2(10\,10|2\mathbf{T}\,2\mathbf{T}) = -(p^{\mathbf{T}})^2/(4|\mathbf{T}|) \tag{19}$$

(Here, $p^{\mathbf{T}}$ indicates the dominant P matrix element at long-range). Since the number of terms per unit distance of this kind increases as $|\mathbf{T}|^{d-1}$, where d is the dimensionality of the system, it is clear that the convergence of the series depends critically on the long-range behaviour of the bond order matrix.

Cancellation effects, associated in particular with the oscillatory behaviour of the density matrix in metallic systems, are not predominant at long-range. Even if the actual behaviour of the P matrix elements cannot be predicted because it depends, in a complicated way, on the physical nature of the compound [8], on orthogonality constraints and on basis set quality, the different range of valence and core elements can be exploited by adopting a *pseudoverlap* criterion. This consists in truncating **T** summations when the $\int d\mathbf{r}\varphi^0_1\varphi^{\mathbf{T}}_2$ overlap is smaller than a given threshold, defined as $P^{\mathbf{T}}_{ex}$ (where **ITOL4** = -log_{10} ($P^{\mathbf{T}}_{ex}$)) and also truncating the **Q** summation when $\int d\mathbf{r}\varphi^0_3\varphi^{\mathbf{Q}}_4$ overlap is smaller than the threshold, $P^{\mathbf{Q}}_{ex}$ (**ITOL5** = -log_{10} ($P^{\mathbf{Q}}_{ex}$)).

Despite its partially arbitrary nature, this criterion presents some advantages with respect to other, more elaborate schemes: it is similar to the other truncation schemes (ITOL1, ITOL2 and ITOL3) and so the same classification tables can be used; it is, in addition, reasonably efficient in terms of space occupation and computer time.

This truncation scheme is symmetrical with respect to the **T** and **Q** summations. However, if account is not taken of the different role of the two summations in the SC (Self-Consistent) stage, distortions may be generated in the exchange field, as felt by charge distributions $\varphi_1\varphi^T_2$, where T labels the largest (in modulus) **T** vector taken into account according to ITOL4. This distortion may be variationally *exploited* and unphysically large density matrix elements build up progressively along the SC stage, eventually leading to catastrophic behaviour (see Chapter II.5 of reference [2] for a discussion of this point). In order to overcome this problem, the threshold, $P^{\mathbf{Q}}_{ex}$ (**ITOL5**) for **Q** summation must be more severe than that for **T** summation (**ITOL4**). In this way, all the integrals whose second pseudocharge $\int d\mathbf{r}\varphi^0_3\varphi^{\mathbf{Q}}_4$ is larger than $P^{\mathbf{Q}}_{ex}$ are taken into account. A difference in the two thresholds ranging from three to eight orders of magnitude is sufficient to stabilize the SC behaviour in most cases.

2.4. Bipolar expansion approximation of Coulomb and exchange integrals

We may now return to the partition of the **S** summation in the Coulomb series, shown in equation 12. Consider one contribution to the charge distribution of electron 1, centred in the reference cell: $\rho^0 = \varphi^0_1\varphi^{\mathbf{T}}_2$; now consider the charge distribution $\rho_\lambda(\mathbf{S})$ of shell λ, centred in cell **S** (equation 13). For small $|\mathbf{S}|$ values, ρ_λ and ρ^0 overlap, so that all the related bielectronic integrals must be evaluated exactly, one by one; for larger values of $|\mathbf{S}|$, ρ_λ is external to ρ^0, so that all the related bielectronic integrals are grouped and evaluated, in an approximate way, through the multipolar expansion of ρ_λ .

However, in many instances, although ρ_λ is not external to ρ^0, the two-centre $\varphi^{\mathbf{S}}_3\varphi^{\mathbf{S}+\mathbf{Q}}_4$ contributions to ρ_λ are external to $\rho^0 = \varphi^0_1\varphi^{\mathbf{T}}_2$; in this case, instead of exactly evaluating the bielectronic integral, a two-centre, truncated, bipolar expansion can be used (see Chapter II.4.c in reference [2] and references therein).

In order to decide to which zone a shell may be ascribed, we proceed as follows: when, for a given pair of shells $\lambda_1^0\lambda_2^T$, shell λ_3^S is attributed to the B (*bielectronic*) zone, the penetration between the products of adjoined Gaussians $G_1^0 G_2^T$ and $G_3^S G_4^{S+Q}$ is estimated: the default value of the penetration parameter is 14, and the block of bielectronic integrals is attributed accordingly to the b_e (*exact*) or to the b_b (*bipolar*) zone. The set of $\mathbf{S}$ vectors, defining the B zone of $\rho^0 = \{12T\}$ and $\rho_\lambda \equiv \{\lambda_3\}$, is then split into two subsets, which are specific for each partner λ_4^l of λ_3.

A similar scheme is adopted for the selected exchange integrals (see previous section) whose pseudocharges do not overlap appreciably. The default value of the penetration parameter is 10.

The total energy change due to the bipolar expansion approximation should not be greater than 10^{-4} Hartree/atom; exact evaluation of all the two-electron integrals (obtained by setting the penetration parameter value > 20000) increases the computational cost by a factor of between 1.3 and 3. Multipolar expansion is very efficient, because the following two conditions are fulfilled:

1. A general algorithm is available for reaching high ℓ values easily and economically [3, 7]. The maximum allowed value is ℓ=6.
2. The multipolar series converges rapidly, either because the interacting distributions are nearly spherical (shell expansion) or because their functional expression is such that their multipoles are zero above a certain (low) ℓ value.

2.5. Exploitation of symmetry

Translational symmetry allows the factorization of the eigenvalue problem in periodic calculations, because the Bloch functions are a basis for irreducible representations of the translational group.

In periodic calculations, point symmetry is exploited to reduce the number of points for which the matrix equations are to be solved. Point symmetry is also explicitly used in the reconstruction of the Hamiltonian, which is totally symmetric with respect to the point group operators of the system.

In the HF-CO-LCAO scheme, the very extensive use of point symmetry allows us to evaluate bieletronic and monoelectronic integrals with saving factors as large as h, in the number of bielectronic integrals to be computed or h^2, in the number of those to be stored for the SCF part of the calculation, where h is the order of the point group. The main steps of the procedure [11] can be summarized as follows:

- The set of Coulomb and exchange integrals, whose indices 3 and 4 (3, 4) identify translationally equivalent pairs of AOs, so that the associated element of the density matrix P_{34} is the same, are summed together to give D_{1234} elements:

$$D_{1,2T;3,4Q} = \sum_{\mathbf{Q}}[(\varphi_1^0\varphi_2^T \mid \varphi_3^S\varphi_4^{S+Q}) - 1/2(\varphi_1^0\varphi_3^Q \mid \varphi_2^T\varphi_4^{S+Q})] \tag{20}$$

- The products of AOs $\varphi_1\varphi_2$ (and $\varphi_3\varphi_4$) are classified in symmetry-related sets; using the fact that the Fock matrix is totally symmetric, only those quantities are evaluated whose indices 1 and 2 (1, 2) refer to the first member of a symmetry set. The corresponding saving factor is as large as h.

- Using the symmetry properties of the density matrix, DM, quantities referring to 3, 4 couples belonging to the same symmetry set (and with the same 1, 2T index) can be combined, after multiplication by appropriate symmetry matrices, so that a single quantity for each 3, 4 symmetry set is to be stored, with a saving factor in storage of the order of h.
- The symmetry $\mathbf{P}_{34}^{\mathbf{Q}} = \mathbf{P}_{43}^{-\mathbf{Q}}$ is exploited.
- The symmetry $\mathbf{F}_{12}^{\mathbf{T}} = \mathbf{F}_{21}^{-\mathbf{T}}$ is exploited.

2.6. Reciprocal space integration

The integration in reciprocal space is an important aspect of *ab initio* calculations for periodic structures. The problem arises at each stage of the self-consistent procedure, when determining the Fermi energy, ϵ_F, when reconstructing the one-electron density matrix and, after self-consistency is reached, when calculating the density of states (DOS) and a number of observable quantities. The P matrix in direct space is computed following equation 10. The technique adopted to compute ϵ_F and the P matrix in the SCF step is described in reference [1]. The Fourier-Legendre technique presented in Chapter II.6 of reference [2] is adopted in the calculation of total and projected DOS. The Fermi energy and the integral in equation 10 are evaluated starting from the knowledge of the eigenvalues, $\epsilon_n(\kappa)$ and the eigenvectors, $a_{\mu n}(\kappa)$, at a certain set of sampling points, $\{\kappa\}$. In 3D crystals, the sampling points belong to a lattice (called the *Monkhorst net*, [9]) with basis vectors $\mathbf{b}_1/s_1$, $\mathbf{b}_2/s_2$, $\mathbf{b}_3/s_3$, where $\mathbf{b}_1, \mathbf{b}_2$ and $\mathbf{b}_3$ are the ordinary reciprocal lattice vectors; s_1, s_2 and s_3 (input as IS1, IS2 and IS3) are integer *shrinking factors*. Unless otherwise specified, IS1=IS2=IS3=IS. In 2D crystals, IS3 is set equal to 1; in 1D crystals both IS2 and IS3 are set equal to 1. Only points of the Monkhorst net belonging to the irreducible part of the Brillouin Zone (BZ) are considered, having associated geometrical weights, w_i.

In the selection of the κ points for non-centrosymmetric crystals, time-reversal symmetry is exploited ($\epsilon_n(\kappa) = \epsilon_n(-\kappa)$).

The number of inequivalent sampling points, κ_i, is given asymptotically by the product of the shrinking factors divided by the order of the point group. In high symmetry systems and with small s_i values, it may be considerably larger because many points lie on symmetry planes or axes.

Two completely different situations (which are automatically identified by the code) must now be considered, depending on whether the system is an insulator (or zero-gap semiconductor) or a conductor. In the former case, all bands are either fully occupied or vacant. The identification of ϵ_F is elementary and the Fourier transform expressed by equation 10 is reduced to a weighted sum of the integrand function over the set $\{\kappa_i\}$ with weights w_i, the sum over n being limited to occupied bands.

The case of conductors is more complicated: two additional parameters, ISHF and ISP, enter into play. ISHF is the number of symmetrized plane waves, $A_m(\mathbf{k})$, used for representing the $\mathbf{k}$ dependence of the eigenvalues. ISP (or ISP1, ISP2, ISP3) are *Gilat shrinking factors* which define a *Gilat net* [10], completely analogous to the Monkhorst net. The value of ISP is larger than IS (by up to a factor of 2), giving a denser net. Each point, κ_j', has a small sphere, S_j, attached, centred at κ_j' and with volume $\mathrm{v} = V_{BZ}/(s_1 s_2 s_3)$, that is, equal to the inverse density of Gilat points: the spheres are, therefore, partially overlapping. In a conductor, one or more bands are partially occupied. Suppose, for simplicity, that there is only one such band (the n-th, say). A truncated Fourier expansion of $\epsilon_n(\mathbf{k})$ is considered:

$$\epsilon_n(\mathbf{k}) \approx \tilde{\epsilon}_n(\mathbf{k}) = \sum_{m=1}^{ISHF} \tilde{\epsilon}_n A_m(\mathbf{k}) \tag{21}$$

with expansion coefficients, $\tilde{\epsilon}_n$, obtained from the knowledge of $\epsilon_n(\kappa)$ at the Monkhorst sampling points. We next consider a further approximation: $\hat{\epsilon}_n(\mathbf{k})'$ of $\epsilon_n(\mathbf{k})$, corresponding to linearly expanding $\tilde{\epsilon}_n(\mathbf{k})$ around each Gilat point κ_j' within its own sphere, S_j. Using this approximation for $\epsilon_n(\mathbf{k})$, it is possible to obtain ϵ_F easily and then to define appropriate weights, w_{in}', for each Monkhorst point and each band. The integral [10] is expressed, as before, as a weighted sum, but with weights w_{in}' instead of w_i. The accuracy of this procedure increases with increasing IS, ISP and ISHF.

In high symmetry systems, it is convenient to assign IS *magic* values such that all low multiplicity (high symmetry) points belong to the Monkhorst lattice. Although this choice does not correspond to maximum efficiency, it gives a safer estimate of the integral.

The values assigned to ISHF and ISP are irrelevant for non-conductors. However, a non-conductor may give rise to a conducting structure at the initial stages of the SCF cycle, owing, for instance, to a very unbalanced initial guess at the density matrix. The ISHF and ISP parameters must therefore be defined in all cases.

2.7. Basis set selection

The problem of selecting appropriate basis sets of contracted GTOs, for HF calculations of crystals, has many features in common with the corresponding one for molecules; yet, some aspects are markedly different in the two cases.

Diffuse Gaussian orbitals (exponent coefficient < 0.1) play a critical role in HF-CO-LCAO calculations of periodic structures, especially three-dimensional ones.

The number of integrals to be calculated increases dramatically with decreasing exponent. The risk of pseudo-linear dependence increases rapidly, demanding higher precision in order to avoid catastrophic behaviour. On the other hand, very diffuse AOs are much less important in three-dimensional, densely packed crystals than in atoms or molecules, where they serve to describe the tails of the electronic distribution towards the vacuum. Another effect of the much higher density of atoms in crystals, with respect to molecules, is that relatively poor basis sets provide larger variational freedom in the former case, since valence orbitals are, in fact, shared among a large number of neighbours and reaching the HF limit becomes easier.

Problems may arise when the binding energy is estimated by subtracting the total crystal energy per unit cell from the sum of HF energies of the isolated atoms. Using the same AOs for the isolated atoms as are used in the crystal necessarily overestimates the HF binding energy (Basis Set Superposition Error, BSSE). In order to allow for this, counterpoise corrections may be used.

As a general rule, atomic basis sets optimized for molecules can be the best starting point to build a basis set for an atom in a given crystal and in a given chemical bonding scheme. For instance, three different basis sets must be used (and optimized) for oxygen in MgO and H_2O (ice) [12, 13].

2.8. Calculation of observable quantities

The total electronic energy is the fundamental observable. However, comparing energy data requires some care. It is customary to extrapolate formation energies to zero temperature, to subtract nuclear-zero-point-motion effects and to add isolated atom energies to obtain values which may be compared with the experimental total electronic energy of the crystals. This procedure implicitly assumes the validity of the Born-Oppenheimer separation of nuclear and electronic motion. Such an approximation is a relatively minor one, when compared with the other errors that affect calculated HF energies, due to basis set incompleteness and numerical approximations. These errors strongly affect the absolute value of the total energy, but they cancel, to a large extent, when the energies of systems containing the same atoms in different configurations are compared.

The Hartree-Fock energy can be corrected for the neglect of correlation terms *a posteriori*, through a functional of the charge density [14, 15].

CRYSTAL can compute the following quantities, from a HF-CO-LCAO SCF wavefunction:

- Band structure; density of states
- Electronic charge density; charge density gradient; x-ray structure factors
- Atomic multipole and spheropole; Mulliken population analysis; topological analysis of charge density [16]
- Electron momentum distribution; Compton profiles and related quantities
- Electrostatic potential; electric field and electric field gradient

3. Structure of the code

The CRYSTAL code consists of a suite of three programs:
integrals, **SCF** and **properties** . The current version allows direct SCF execution.

The structure of the code closely follows the theory and is similar to those of molecular programs. In the flow diagram on the following page,the peculiar features of periodic Hartree-Fock are in bold-face.

Program communication is provided by means of a common filing system. Unformatted sequential data sets are used. Interfaces to molecular graphics programs and quantum molecular programs are supplied.

The s, p (in the order x, y and z) and d (in the order $2z^2-x^2-y^2$, xz, yz, x^2-y^2, xy) shells of GTF can be used. Shells of the type sp are also available (s and p shells, sharing the same set of exponents). The use of sp shells can give rise to considerable savings in CPU time.

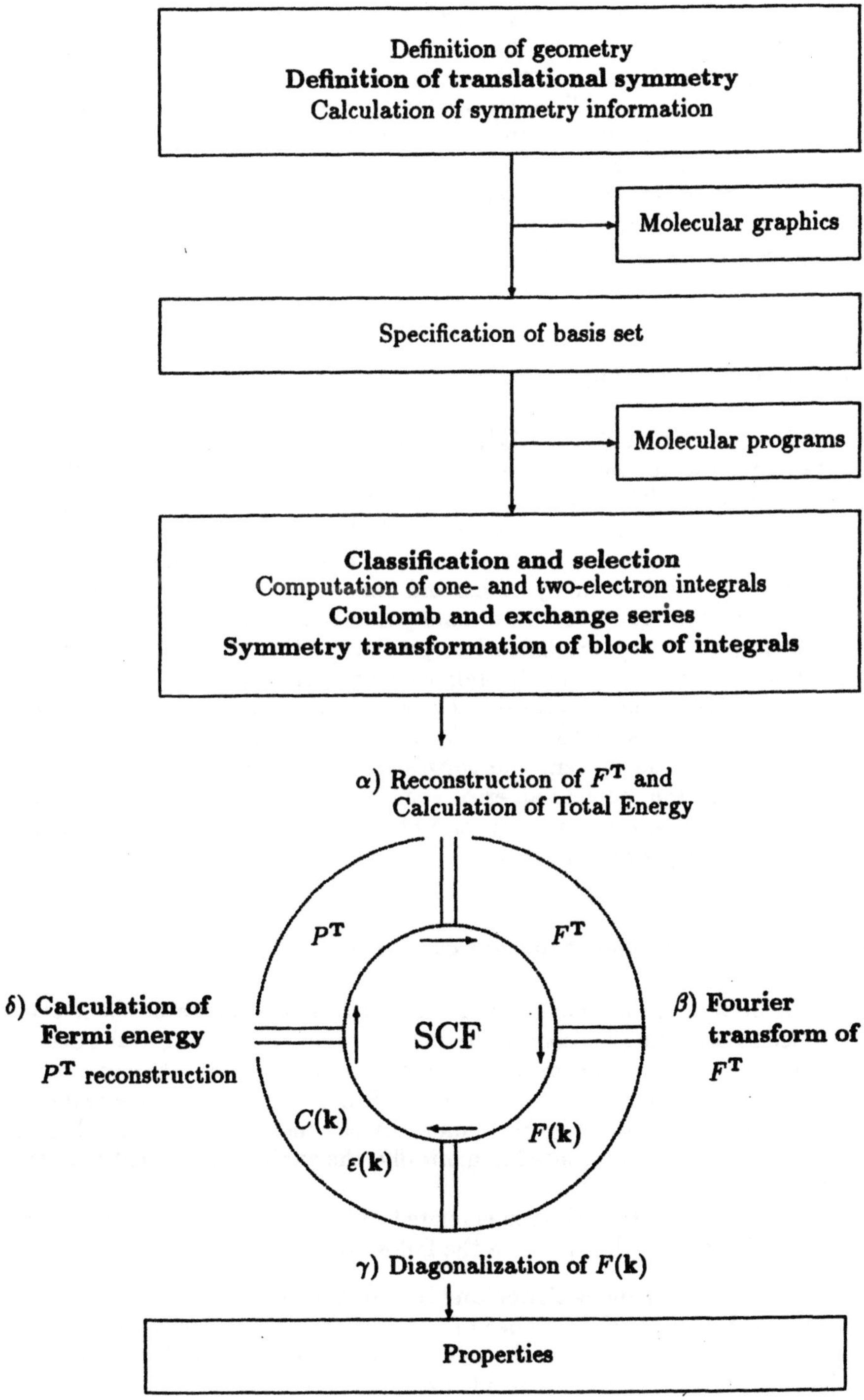

Figure 1. Flow chart of the CRYSTAL code.

The program can automatically handle space symmetry (230 space groups, 80 two-sided plane groups, 99 line groups, 45 point groups are available). In the case of polymers, it cannot treat helical structures (translation followed by a rotation around the periodic axis). However, when commensurate rotations are involved, a suitably large unit cell can be adopted. Point symmetries *compatible with translation symmetry* are provided for molecules.

The program can perform closed-shell, restricted Hartree-Fock, Restricted Open Shell (ROHF) and Unrestricted Open shell (UHF) calculations.

All-electron and valence-only basis sets with core pseudo-potentials are allowed.

3.1. Limits of applicability

The code is written in FORTRAN, standard ANSI77 and it has been tested on a large number of computer systems, from PCs to work-stations and CRAYs. It is readily portable. The distribution tape contains a makefile for Unix systems, to generate the executable files. No external libraries are required.

All the key dimensions are parameterized; in principle, it should be possible to run a system with as many atoms and basis functions as allowed by numerical accuracy, with good computational parameters. However, there are many dimensions which are very difficult to predict, as they depend in a correlated way on the number of symmetry-related atoms, the size of the basis set, the number of points in the reciprocal lattice used to compute Fermi energy and the required accuracy as specified by various tolerances. There are many checks on the dimensions of the system and suggestions are printed before the program stops, if such checks indicate an error condition. It is, therefore, difficult to define, in a few words, the maximum size of the systems that can be handled by CRYSTAL (see [6], Appendix F).

4. The cost of the wavefunction calculation

The cost of the calculation depends on many factors. Some of them can be controlled by adjusting the computational parameters, according to the precision required in the results. Information, such as formation energy, charges and band structure, do not require severe conditions. When looking for phase transitions or calculating the elastic properties, more severe conditions are required as total energy derivatives are computed numerically (the analytical gradient code is not yet implemented).

The cost of the **integrals** step is related to the number of integrals evaluated exactly and depends, therefore, on the following parameters:

- **The number of basis functions in the unit cell.**
  Given a basis set, the number of two-electron integrals evaluated exactly, depends almost linearly on the cell size, while the number of monoelectronic integrals has a quadratic dependence.
- **Symmetry of the system.**
  In high symmetry systems only a small fraction of integrals are actually evaluated, all the others being generated by symmetry transformations.

- **Exponents of the valence Gaussians.**
  The selection criteria of the integrals to be computed is based on the overlap of the lowest exponent Gaussians. Avoid very diffuse functions (exponents 0.1 $Bohr^{-2}$ or less).
- **Type of basis functions.**
  d orbitals are much more expensive to use than *s* and *p* orbitals; *sp* shells are much less expensive than split *s* and *p* shells.

The disk space to be allocated can be estimated by using a test-run option; the **SCF** direct program eliminates the storage of the bielectronic integrals that are computed at each SC cycle.

In the **SCF** step, most of the time is spent in matrix algebra operations, such as matrix multiplication and matrix diagonalization, which scale as the third power of the number of basis functions, when traditional numerical techniques are adopted. SCF time is proportional to i) the number of SCF cycles; ii) the number of reciprocal lattice points, κ, at which the Fock matrix is diagonalized (for the calculation of Fermi energy and the density matrix).

Description of an LAPW DF Program (WIEN95)

Karlheinz Schwarz and Peter Blaha

Institut für Technische Elektrochemie, Technische Universität Wien, A-1060 Vienna, Getreidemarkt 9/158, Austria

May 28, 1996

Summary. Band structure calculations based on density functional (DF) theory using the full-potential linearized augmented plane wave (FP-LAPW) method as implemented in the WIEN95 code are discussed, mainly in terms of available features but including a few examples. Recent improvements over the local density approximation (LDA), in the form of generalized gradient approximations (GGAs), are mentioned. The calculation of atomic forces allows relaxation of atomic positions and molecular dynamics simulations, in systems with complicated crystal structures, including metallic systems.

Key words: Density functional theory – Bandstructure – Local Density Approximation (LDA) – Generalized Gradient Approximation (GGA) – Linearized Augmented Plane Wave (LAPW) – Atomic forces

1. Introduction

The basic concepts of *ab initio* calculations for solids were discussed in Chapter 3, which gives an overview of different approaches. Here, we focus on applications within density functional theory (DFT) which is the basis of many first principles calculations. In DFT, the complicated many-body problem is replaced by effective one-electron equations which can be solved efficiently and lead to reliable electron densities and derived quantities, such as the total energy or forces. In this paper, the emphasis is on the full-potential linearized augmented plane wave (FP-LAPW) method, as implemented in the program package WIEN95, which was developed in our group. Two commonly used approximations (LDA and GGA) are discussed in section 2 and a few representative results (all corresponding to T=0) will be given in section 4.1.

Correspondence to: K.Schwarz - E-mail: kschwarz@email.tuwien.ac.at

2. Density functional theory (DFT)

The most important quantum mechanical (QM) schemes that can be used to study solids have already been discussed in Chapter 3, where DFT is compared with other QM treatments, such as HF theory, so that one can judge the advantages or disadvantages of the different approaches. Here, only a short summary of DFT is given. The foundation of DFT was laid by Hohenberg and Kohn [1], who showed that the total energy is a functional of the electron density. This means that one does not need to know the complicated many-electron wavefunction, but only the electron density, in order to determine the total energy of a system or other ground state properties. Therefore, the density, $\rho(\mathbf{r})$, is the key quantity in density functional theory. Although this is an extremely important theorem, one cannot solve the corresponding equations, since the functional is not known. For that purpose, Kohn and Sham [2] expressed the total energy (written for an atom, with an obvious generalization to molecules and solids):

$$E_{tot} = T_0[\rho] + \int V_{ext}(\mathbf{r})\, \rho(\mathbf{r})\, d\mathbf{r} + \frac{1}{2}\int \frac{\rho(\mathbf{r})\, \rho(\mathbf{r}')}{|\mathbf{r}-\mathbf{r}'|}\, d\mathbf{r}\, d\mathbf{r}' + E_{xc}[\rho] \qquad (1)$$

in terms which are easy to compute, namely: 1) T_0, the kinetic energy of non interacting particles; 2) the interaction between nuclei and electrons given by the external potential, $V_{ext}(\mathbf{r})$, and the electron density $\rho(\mathbf{r})$, respectively; 3) the Coulomb interaction between electrons at $\mathbf{r}$ and $\mathbf{r}'$ represented by their respective densities, $\rho(\mathbf{r})$ and $\rho(\mathbf{r}')$; and finally, 4) a term that contains everything else, namely, $E_{xc}[\rho]$, the exchange-correlation energy, which is defined by this equation.

2.1. Local density approximation (LDA)

Kohn and Sham (KS) used available information on the homogeneous electron gas [2] method, which treats the actual inhomogeneous system as being locally homogeneous with an exchange-correlation energy, ε_{xc}, that is accurately known (the leading term is proportional to $\rho^{1/3}$), giving:

$$E_{xc}^{LDA}[\rho] = \int \rho(\mathbf{r})\, \varepsilon_{xc}[\rho(\mathbf{r})]\, d\mathbf{r} \qquad (2)$$

Thus Kohn and Sham [2] introduced the so-called local (spin) density approximation (LDA or LSDA) where, in magnetic systems, spin polarization is allowed and the exchange-correlation energy (E_{xc}) becomes a functional of the local electron-spin densities, $\rho_\uparrow$ and $\rho_\downarrow$:

$$E_{xc}^{LSDA}[\rho_\uparrow, \rho_\downarrow] = \int \rho(\mathbf{r})\, \varepsilon_{xc}[\rho_\uparrow(\mathbf{r}), \rho_\downarrow(\mathbf{r})]\, d\mathbf{r} \qquad (3)$$

By writing E_{tot} in this form, one can now apply the variational principle and derive the following one-electron KS equations:

$$\{-\frac{1}{2}\nabla^2 + V_{ext}(\mathbf{r}) + V_c[\rho(\mathbf{r})] + \mu_{xc}[\rho(\mathbf{r})]\}\, \Phi_i(\mathbf{r}) = \varepsilon_i\, \Phi_i(\mathbf{r}) \qquad (4)$$

where μ_{xc} is given by the functional derivative of E_{xc} with respect to the density. In this method, one replaces the many-body problem by a set of effective one-electron KS equations which can be solved. In the KS scheme, the electron density is obtained by summing over all occupied states, i.e. by filling the KS orbitals according to the Aufbau principle (with increasing energy):

$$\rho(\mathbf{r}) = \sum_{i\ occupied} |\Phi_i(\mathbf{r})|^2 \tag{5}$$

This version of the DFT leads to a (spin) density that is close to the exact density provided the LDA is sufficiently accurate.

2.2. Generalized gradient approximation (GGA)

Despite its simplicity, LDA calculations often yield results in close agreement with experiment but there are also cases where they fail or show large discrepancies. In such cases, it is quite natural to take the next step beyond the homogeneous electron gas (the basis for LDA) and include gradient corrections, so that E_{xc} is a function of the spin densities and their gradients:

$$E_{xc}^{GGA}[\rho_\uparrow, \rho_\downarrow] = \int f[\rho_\uparrow(\mathbf{r}), \rho_\downarrow(\mathbf{r}), \nabla\rho_\uparrow(\mathbf{r}), \nabla\rho_\downarrow(\mathbf{r})]\, d\mathbf{r} \tag{6}$$

By imposing certain conditions, which, for example, the exchange (correlation) hole density must satisfy, generalized gradient approximations (GGAs) have been developed to improve the quality of LDA results; the GGA version suggested by Perdew and Wang [3] and Perdew [4] is one example. A series of tests on various systems [5] showed promising results, some of which will be illustrated in section 4.1.

3. Full-potential Linearized Augmented Plane Wave method

There are several methods that can solve the KS equations within LDA or GGA, where especially the latter requires high numerical precision. One of the most accurate schemes is the full-potential linearized augmented plane wave (FP-LAPW) method but other schemes of comparable quality are available too, e.g. modern pseudopotentials or other full-potential methods. In this context, the reader is referred to an excellent book by D. Singh [6]. There are also simplified versions of electronic structure calculations, such as the linear-muffin-tin-orbital (LMTO) or augmented-spherical-wave (ASW) method in which, in addition to LDA, the atomic sphere approximation (ASA) is made, i.e. within the self-consistency cycle, a spherically-averaged potential and charge density is assumed around each atomic site. Although these latter schemes are computationally faster, they often do not provide high enough accuracy to study such points as small energy variations as they occur, for instance, in phase transitions or diffusion.

The FP-LAPW method, as embodied in the WIEN95 code, has been developed in our group [7] and is frequently used worldwide. The original version (WIEN) was the first LAPW code that was published and thus, made available for other users. It is a scalar-relativistic version without spin-orbit coupling, where exchange-correlation is treated within DFT, using LDA or GGA.

3.1. Basis functions

In the LAPW method, the unit cell (as sketched in Figure 1) is partitioned into (non-overlapping) atomic spheres centered around the atomic sites (type I) and an interstitial region (II). For the construction of the basis functions, the

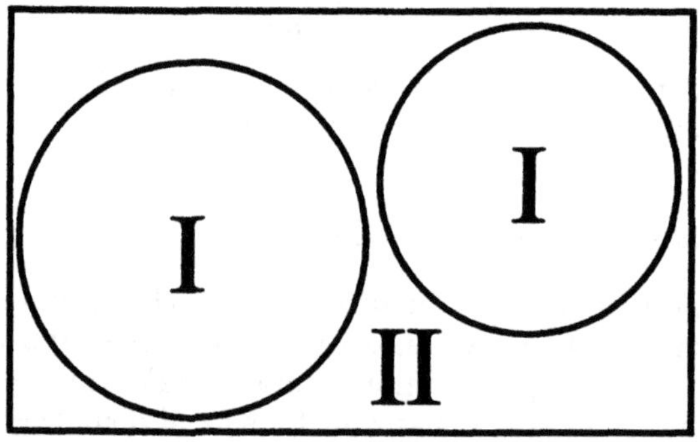

Figure 1. Schematic partitioning of the unit cell into atomic spheres (I) and the interstitial region (II).

muffin-tin approximation is used (Figure 2), according to which the potential is assumed to be spherically symmetric within the atomic spheres (I) but constant outside (in region II). Different basis sets are used in these respective regions.

In the interstitial region, plane waves form a complete basis set:

$$\Phi_{k_n} = \frac{1}{\sqrt{\Omega}}\, e^{i\mathbf{k}_n\mathbf{r}} \;; \quad \text{with } \mathbf{k}_n = \boldsymbol{\kappa} + \mathbf{K}_n \tag{7}$$

where $\boldsymbol{\kappa}$ is a wave vector in the Brillouin zone, $\mathbf{K}_n$ is a reciprocal lattice vector and Ω is the volume of the unit cell.

Inside each atomic sphere, the solutions for a spherically symmetric potential are atomic basis functions which consist of radial functions, $u_l(r, E)$, multiplied by spherical harmonics (equation 8).

For a typical valence state with azimuthal quantum number, l, the radial function is regular at the origin and varies with energy (Figure 3).

At the bottom of the band, E_{bottom}, $u_l(r.E)$ has a zero slope (bonding state) but it has a zero value (antibonding state) at the top of the band, E_{top}. This energy dependence is linearized by selecting an expansion energy, E_l, near the center of the band and writing the atomic function as:

$$\Phi_{k_n} = \sum_{lm}[A_{lm}u_l(r, E_l) + B_{lm}\dot{u}_l(r, E_l)]Y_{lm}(\hat{r}) \tag{8}$$

where u_l is the solution of the radial Schrödinger equation and $\dot{u}_l$ is its energy derivative, both of which are computed numerically. For each atom and lm, there are two free coefficients, A_{lm} and B_{lm}, which are chosen such that (at the sphere boundary) the plane wave with wave vector, $\mathbf{k}_n$, joins continuously onto the atomic function in value and slope.

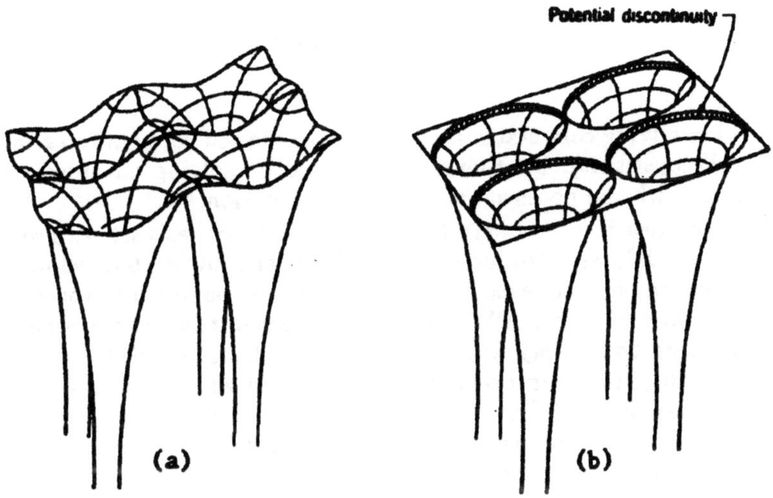

Figure 2. Crystal potential of a square two-dimensional lattice: a) actual (full) potential; b) *muffin tin* potential.

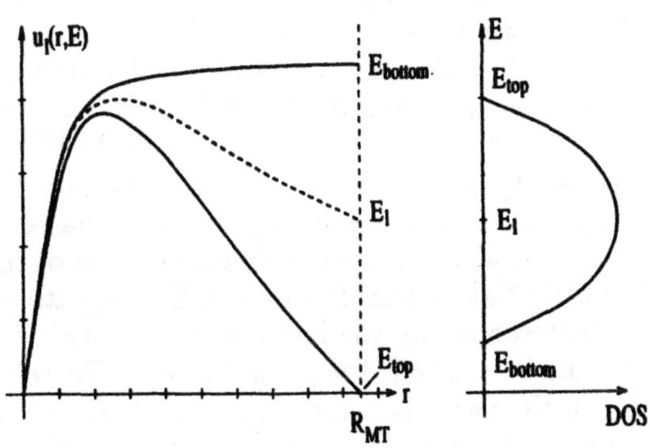

Figure 3. Schematic energy dependence of $u_l(r, E)$ and the corresponding density of states (DOS). The radius, r, goes up to the muffin tin radius, R_{MT}, which defines the atomic sphere.

These linearized augmented plane waves (LAPWs) form the basis for expanding the crystalline orbitals (Bloch states) and provide a rapidly converging series (50 to 100 plane waves per atom in the unit cell):

$$\Psi_{\mathbf{k}} = \sum_{k_n} C_{k_n} \Phi_{k_n} \tag{9}$$

All low-lying core states are included as thawed core states (**k**-independent, but self-consistent), i.e. they are computed by solving an atomic DFT Dirac equation with the potential of the current iteration.

Sometimes, it is necessary to extend the LAPW basis set with so-called local orbitals (LOs), as introduced by Singh [8] and illustrated below. With such an extension, higher level semi-core states can be treated, together with the valence states, in *one* LAPW calculation instead of *two* (for valence and semi-core states, using two energy windows). This LO scheme avoids the problem of non-orthogonality, that can occur in calculations in which the semi-core states are either frozen or treated in a separate energy window. An LO is chosen such that it vanishes in value and slope at the muffin tin radius and includes an additional radial function, corresponding to a different principal quantum number:

$$\Phi_{lm}^{LO} = [A_{lm} u_l(r, E_{1,l}) + B_{lm} \dot{u}_l(r, E_{1,l}) + C_{lm} u_l(r, E_{2,l})] Y_{lm}(\hat{r}) \tag{10}$$

The LO is then added to the usual LAPW basis set in equation 9. For example, one needs to include both the $3p$ semi-core, as well as the $4p$ valence orbitals, in order to describe Ti wavefunctions accurately. The use of LOs is illustrated in Figure 4, taken from Dufek [9], who showed that the $3p$ function at about -1.9 Ryd is not completely confined within the atomic sphere of R_{Ti}=2 a.u. However, by constructing an LO as a proper linear combination of the standard LAPW radial functions of $4p$ character (expanded around E_1= +1 Ryd) and a new function at E_2 = -1.9 Ryd: $a_{LO} * u(r, E_1) + b_{LO} * \dot{u}(r, E_1) + c_{LO} * u(r, E_2)$, a corresponding LO is formed, which has mainly $3p$ character and vanishes at R_{MT}. The inclusion of LOs makes the computation of matrix elements somewhat more difficult but one has a better model and saves computer time, since no semi-core states need to be calculated in a separate LAPW run.

In contrast to LCAO type schemes, where the basis functions are centered on atomic sites and may have a far reaching tail, methods based on atomic spheres (muffin tin or full potential) use a spatial decomposition of the wave functions. For example, if the tail of an orbital from a neighboring atom enters an atomic sphere, it must be represented by the basis set used within that sphere (in the LAPW method, it must be expanded in partial waves). Therefore, basis sets have a different meaning in the two cases and consequently quantities, which depend on the decomposition of the wave functions, such as partial charges, are affected by this conceptual difference. For example, a d-like charge in an LAPW calculation corresponds to an integrated charge density, found in the corresponding atomic sphere, which is derived from a d-like partial wave inside that sphere, while in LCAO, the analogous quantity represents the weight (Mulliken population) of a d-like function, centered on that atom, irrespective of its spatial extent.

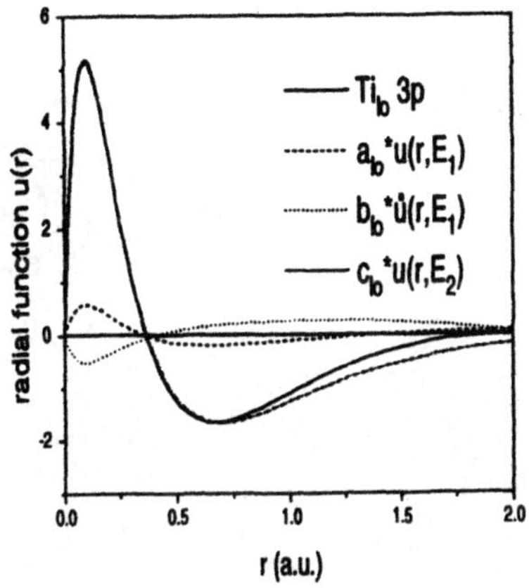

Figure 4. Construction of a local orbital (LO) resembling a Ti-3p state. The LAPWs of the 4p states, u and $\dot{u}$, are expanded at E_1=1 Ryd, while for the LO E_2 = -1.9 Ryd and a_{LO}= 0.128, b_{LO}= 1.566 and c_{LO}= 1.200.

3.2. Full potential and general charge density

The *muffin tin* approximation (MTA) was frequently used in the seventies and works reasonably well in highly coordinated (metallic) systems, such as face-centered cubic (fcc) metals. However, for covalently bonded solids or layered structures, MTA is a poor approximation and leads to serious discrepancies with experiment. In every case, however, a full-potential treatment is essential. In the FP-LAPW method, the potential and charge density are expanded (a) into lattice harmonics (inside each atomic sphere) and (b) as a Fourier series (in the interstitial region) and thus, they are completely general, so that such a scheme is termed a full-potential calculation:

$$V(\mathbf{r}) = \begin{cases} \sum\limits_{LM} V_{LM}(r)\, Y_{LM}(\hat{r}) & (a) \\ \sum\limits_{\mathbf{K}} V_{\mathbf{K}} \exp(\imath \mathbf{K}\cdot\mathbf{r}) & (b) \end{cases} \tag{11}$$

In order to have few LM values in the lattice harmonics expansion of equation (11), a local coordinate system for each atom sphere is defined, according to the point symmetry of the corresponding atom. This specifies a rotation matrix that relates the local to the global coordinate system of the unit cell.

Figure 5 shows the density of the spin-up electrons in the anti-ferromagnetic compound FeF_2 which crystallizes in the rutile structure. This diagram illustrates that a density, which is represented analogously to equation (11), is continuous, although different representations are used inside the atomic spheres and in the interstitial region. In this example, the spin density around Fe1 and F is spherically symmetric, while that at Fe2 has a large anisotropy.

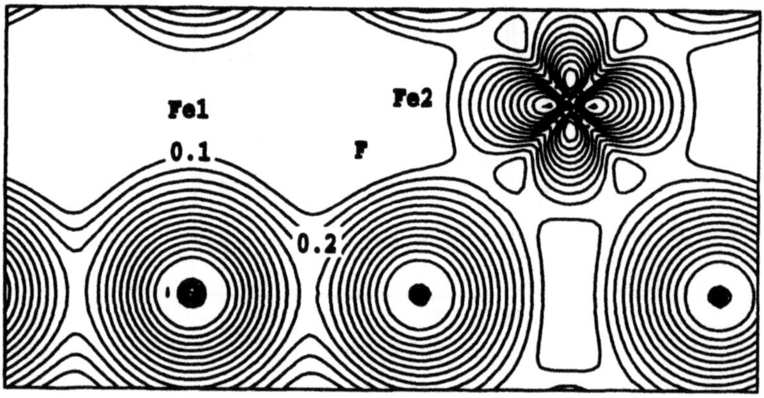

Figure 5. Spin-up density of FeF_2 in the (110) plane (from Dufek [9]).

3.3. SCF cycle

The effective potential in the KS equations (equation 4) can be found, provided the density is known, but conversely, the density can be obtained according to equation 5, by solving the KS equations which requires the knowledge of the potential. Such a case can only be solved iteratively by the well known SCF scheme, which is illustrated in Figure 6 for the LAPW scheme.

The WIEN95 package consists of several independent programs which perform different tasks and are linked via C-shell scripts, in an Unix environment, where communication is achieved by use of common files. One starts the LAPW calculation with an atomic calculation that generates the starting electron densities from atomic data. Then a self-consistency cycle is initiated and repeated until certain convergence criteria are met. The cycle consists of the following steps:

→	LAPW0 (POTENTIAL)	-generates potential from density
|	LAPW1 (BANDS)	-calculates valence bands (eigenvalues and eigenvectors)
|	LAPW2 (RHO)	-computes valence densities from eigenvectors
|	CORE	-computes core states and densities (fully relativistically)
←	MIXER	-mixes input and output densities

Either during the SCF cycle or after self-consistency is reached, additional quantities are obtained, such as total energy, forces, density of states (DOS), charge (or spin) densities, partial charges, structure factors, electric field gradients (EFG), among others.

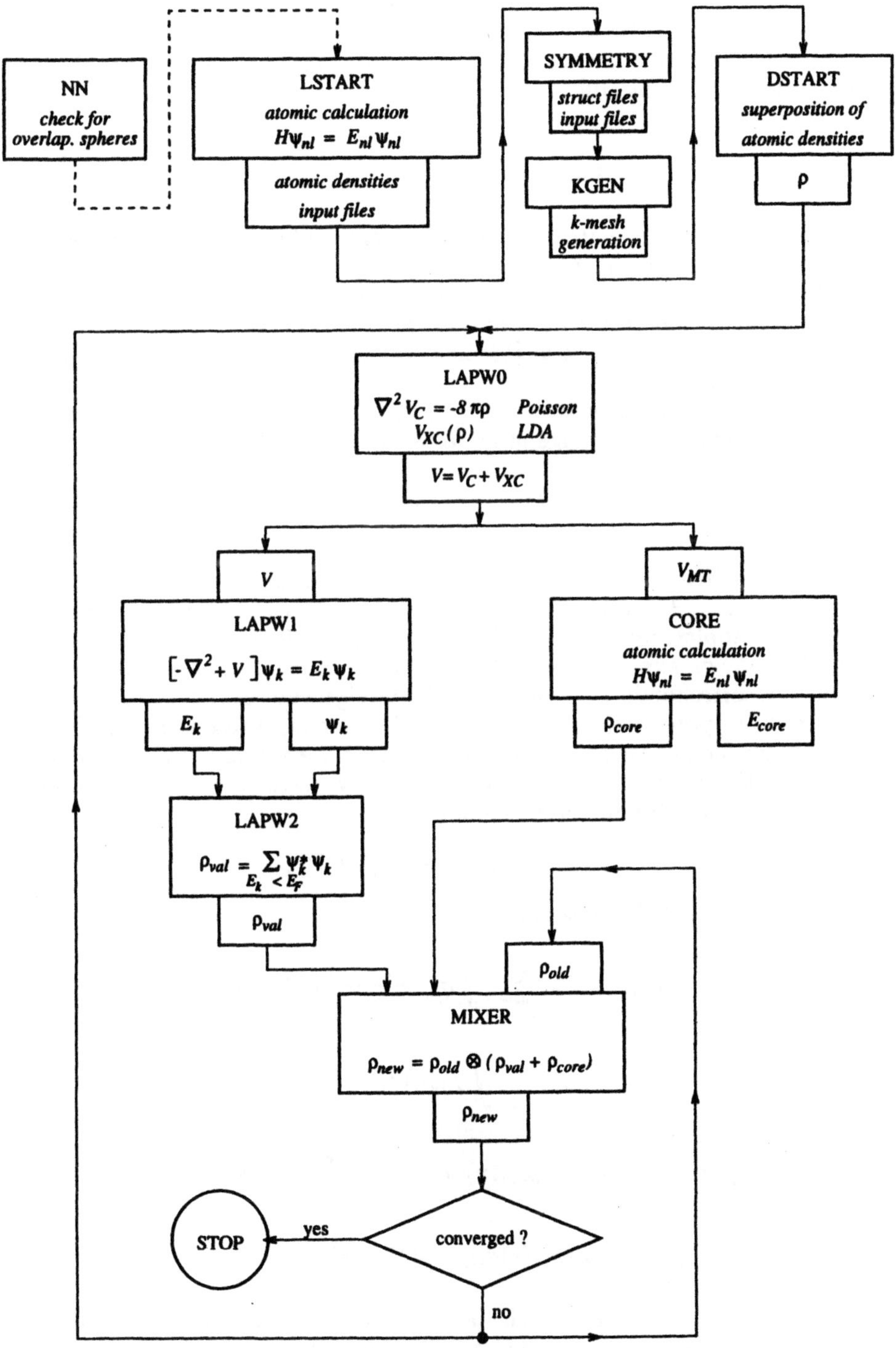

Figure 6. LAPW self-consistency cycle.

For magnetic systems, spin-polarized calculations can be performed, where some steps are performed for spin-up and for spin-down electrons separately, controlled by C-shell scripts for ferromagnetic, anti-ferromagnetic or fixed spin moment calculations [10].

3.4. Forces within the FP-LAPW Method

In the literature, there are two different ways to obtain forces within the LAPW method, namely, one by Yu, Singh and Krakauer [11] and the other by Soler and Williams [12]; both have been successfully implemented in WIEN. Here, we will briefly describe the basic equations of the first approach, which was originally programmed by Kohler et al [13] and was then modified and implemented in WIEN95 by us.

To find the atomic forces on atom α, we displace this atom by a small amount, $\delta\tau_\alpha$ and calculate the change in the total energy:

$$\mathbf{F}^\alpha = -\frac{\delta E}{\delta\tau_\alpha} = \mathbf{F}^\alpha_{HF} - \frac{1}{\delta\tau_\alpha}(\sum_i n_i\delta\epsilon_i - \int \rho(\mathbf{r})\delta V_{eff}\mathbf{r}) \tag{12}$$

F_{HF} is the Hellmann-Feynman force, which is equal to the electrostatic force on the nucleus. The other terms represent corrections, necessary because of basis set incompleteness, to the Hellmann-Feynmann force, the existence of which was first shown explicitly by Pulay [15]. In equation 12, n_i are the occupation numbers corresponding to the Kohn-Sham eigenvalues, ϵ_i and V_{eff} is the effective total potential. Within the LAPW method, where the basis functions depend on the atomic positions and the second derivative of the basis functions is not continuous across the sphere boundary, we have both a core and a valence correction to the Hellmann-Feynmann-force:

$$\mathbf{F}^\alpha = \mathbf{F}^\alpha_{HF} + \mathbf{F}^\alpha_{core} + \mathbf{F}^\alpha_{val} \tag{13}$$

The different contributions are:

$$\mathbf{F}^\alpha_{HF} = Z_\alpha \sum_{m=-1}^{1} \lim_{\alpha->0} \frac{V^{es}_{1m}(r_\alpha)}{\mathbf{r}_\alpha} \nabla_\alpha[\mathbf{r}_\alpha Y_{lm}(\hat{r})] \tag{14}$$

$$\mathbf{F}^\alpha_{core} = -\int \rho_{core}(\mathbf{r})\nabla_\alpha V_{eff}(r)dr \tag{15}$$

$$\mathbf{F}^\alpha_{val} = \int_\alpha V_{eff}(\mathbf{r})\nabla_\alpha\rho_{val}(\mathbf{r})dr + \sum_{k,i} n_i \sum_{\mathbf{K},\mathbf{K}'} C^*_i(\mathbf{K}')C_i(\mathbf{K}) \times$$
$$[(K^2-\epsilon_i)\oint \Phi^*_{\mathbf{K}'}(\mathbf{r})\Phi_{\mathbf{K}}(\mathbf{r})dS_\alpha - i(\mathbf{K}-\mathbf{K}')\langle\Phi_{\mathbf{K}'}|H-\epsilon_i|\Phi_{\mathbf{K}}\rangle_\alpha] \tag{16}$$

The first two terms are calculated at almost no extra cost but the third term requires careful programming, since a summation of matrix elements over all occupied states is involved. However, in our implementation, the extra time spent for computing the forces is almost negligible with respect to the SCF cycle. In several test cases, we compared the calculated forces with the respective derivative of the total energy and found high numerical stability. There was no need to increase internal convergence parameters.

Now we are in a position to minimize the total energy of a system with respect to the internal atomic coordinates within the unit cell and thus, can determine the atomic equilibrium positions. The simplest scheme for this purpose is the method of steepest descent in which the next geometry step is taken along the direction of the forces. There are more sophisticated schemes such as the variable metric method or a damped Newton dynamics which may be more efficient [13].

Another area where forces are indispensable is the field of molecular dynamics simulations, which can be performed within the LAPW method, even for metallic and magnetic systems, containing transition metals or $4f$ elements.

4. Applications and examples

4.1. LDA versus GGA calculations

During the last twenty years, numerous solid state calculations have demonstrated that the LDA yields reliable results in many cases, but, sometimes, they are in contradiction with experiment. One example is iron, where the spin-polarized generalization of LDA, i.e. the local spin-density approximation (LSDA), finds that nonmagnetic *fcc*-Fe has a lower total energy (about 6 mRy) than ferromagnetic *bcc*-Fe, which is the experimentally observed ground state. GGA calculations, however, led to the correct ground state [16]. The improved quality of GGA total energies changed the attitude of quantum chemists to using DFT-GGA calculations as a cost-effective, general procedure for studying physical properties of large molecules [5], instead of the traditional Hartree-Fock (HF) plus post-HF correlation procedures (e.g. second-order Moeller-Plesset theory). Recently, we have demonstrated ([17]) that GGA in the PW-II version (see [3] and [4]) can also partially cure another well-known difficulty of LSDA calculations, namely, that some anti-ferromagnetic (AFM) solids, for example CoO, are calculated to be metallic, instead of insulating, as experiment finds. This has led to the notion that such oxides are Mott or charge-transfer insulators, which cannot be described within band theory. Our recent work has shown that GGA enhances angular correlations according to orbital polarizations and thus, causes a band splitting that leads to insulating behaviour. One case where that works well is the AFM rutile compound FeF_2, which GGA calculations show to be insulating, agreeing with experiment, while LSDA studies imply that it is a metal. Another case is NiI_2 which, under pressure, undergoes a phase transition from an AFM insulator to a non-magnetic metal [18]. Presently, the improvement of DFT functionals or the combination of HF with DFT is an active field of research.

4.2. Total energy and phase transitions

Perovskites are very common among minerals and show a variety of phase transitions, which can also be studied by DFT calculations [19]. The basic building block is an octahedron which can have various distortions. One example of present interest is $BaBiO_3$, since it can be considered as a parent compound for the high Tc superconductors but one that contains no copper (see e.g. Blaha *et al* [20] and references therein). By partially replacing Ba with K (or Bi with Pb), $Ba_{1-x}K_xBiO_3$ becomes superconducting. The undoped material is a semiconductor in which the perovskite structure is unstable with respect to a breathing-like distortion of the oxygen octahedra (corresponding to a disproportionation of Bi^{4+} into the two inequivalent Bi valences Bi^{3+} and Bi^{5+}) and a rotation of these octahedra, leading to a monoclinic structure [20]. Substitution removes these distortions and leads to an ideal perovskite (simple cubic structure) and to superconductivity. Several *ab initio* calculations have been performed but the stability of $BaBiO_3$, with respect to the breathing distortion, was found to vary, depending on the method used and on the computational details, mainly in connection with the treatment of the Bi and Ba semi-core states. Such results illustrate two things: firstly, that such a delicate balance between two valences can be described within DFT, but, secondly, that an accurate representation of the electronic structure (treating the semi-core states with LOs) is required in order to reliably compute such small energy differences.

4.3. Electric field gradients and chemical bonding

The electric-field gradient (EFG) can be measured by various experimental methods which yield the nuclear quadrupole-interaction constant, which involves the product of the nuclear quadrupole moment, Q, multiplied by the EFG. For a better understanding of experimental results, a theoretical determination of the EFG is highly desirable, but, until recently, experimentalists had to interpret their results mainly on the basis of simple point charge models with additional Sternheimer antishielding corrections. We can determine the EFG, in solids, from first principles, since FP-LAPW calculations provide the electronic charge density, including all polarization effects. By numerically solving Poisson's equation, we obtain the Coulomb potential, in a form that allows computation of the EFG directly.

The general expression for the principal component of the EFG tensor, arising from a (nuclear plus electronic) charge density, $\rho(\mathbf{r})$, is defined as:

$$V_{zz} = \int \rho(\mathbf{r}) \, \frac{2P_2(\cos(\theta))}{r^3} \, d\mathbf{r} \tag{17}$$

where P_2 is the second-order Legendre polynomial. Once the charge density of a system is known to high precision, the EFG can be obtained numerically from equation (11) without further approximations (such as Sternheimer antishielding factors, assumptions about ionicities or specific charge distributions).

This new approach has been applied to various solids with very different chemical bonding and the relationship between the EFG and bonding is discussed in the following examples:

- Insulators and fairly ionic systems, such as Li_3N [21], Cu_2O [22], TiO_2 [23] or mercury(I) and (II) halides [24].
- Metals, such as *hcp* metals [25], and the pressure (c/a ratio) dependence of Be [26] or *bcc*-In, *hcp*-Zn and *hcp*-Ti [27].
- High-temperature superconductors, such as $YBa_2Cu_3O_7$ [28] and $YBa_2Cu_4O_8$ [29].

- Antiferromagnets, such as MnF_2 and NiF_2 [30]; or NiI_2 [18].
- Molecular crystals, such as Cl_2, Br_2, and I_2 with covalent bonds within the molecule, where we studied solid state effects [31].

Recently, we summarized the main EFG results for a representative example of each class [32]. For further details (computation, references and interpretations), the reader is referred to the original literature. Good overall agreement between our theory and experimental data has been obtained for the principal EFG component, the orientation and the asymmetry parameter, although these systems have extremely different chemical bonding. The LAPW calculations gave not only good results, but provided new insight into the origin of the EFG. It was found that the asymmetry of the valence electron distribution close to the nucleus, rather than core polarizations, causes the EFG. Sometimes, this asymmetry can be visualised by difference-electron densities or, after spatial integration, by symmetry-decomposed partial charges, quantities directly related to chemical bonding. The EFG is much more closely related to covalent bonding or polarizations than to charge transfer, so that an interpretation of EFGs, in terms of the point charge model (which relies on assumed ionicities), is highly questionable. Further analysis of the EFG allowed a distinction to be made between contributions from *p*- and *d*-like wavefunctions, where the former were found to contribute substantially to the EFG, even though they often represent only a small fraction of all the valence electrons, as, for example, in transition metals or their compounds. The reason for this is the r^3 denominator, which enhances the region close to the nucleus and the r^l behaviour of the corresponding radial wavefunctions, which favours contributions stemming from *p* over those from *d* functions.

The overall good agreement between theory and experiment gave us enough confidence to determine, recently, the nuclear quadupole moments by comparing our theoretical EFG results with experimental quadrupole splittings, e.g. for ^{57}Fe, the most important Moessbauer isotope [33].

The *ab initio* method presented here requires high precision quantum-mechanical calculations and large computational effort, but is capable of calculating reliable EFGs in crystalline solids and provides insight into the physical origin of the EFG.

4.4. *WIEN95 package*

Our computer program was first published as WIEN-code, where additional references to the LAPW method can be found. In 1993, we released a first update to this program package, WIEN93 and, two years later, we, in collaboration with other groups, have implemented new features and have developed a user-friendly, improved version WIEN95 [7], which runs on Unix workstations, as well as on host computers. In addition, its use by non-experts is made easier by a written User's Guide. Interested users should contact the author, preferably by e-mail. WIEN95 can handle crystal structures with unit cells containing up to about 30 atoms (on a fast computer with 128Mb or more memory).

5. Future outlook

In the present paper, several points were illustrated, which play a role in quantum mechanical calculations based on density functional theory. It is important to know both the merits and limitations of the LDA or GGA computations, in order to judge the accuracy that can be achieved with such calculations. Since small energy differences occur in many cases, an accurate computational method with good convergence is needed. Otherwise, basis set effects dominate the results, which do not correspond to the DFT limit and thus, cannot help us to understand the underlying physical principles. One of the most accurate schemes is the FP-LAPW method which presently can handle systems up to about thirty atoms per unit cell. The advantage of an *all electron* method and the use of *numerical* basis functions has been mentioned. These allow, in particular, the computation of quantities close to the atomic nucleus like EFGs, hyperfine fields or isomer shifts, where pseudopotential methods fail or schemes using Gaussian basis functions may not be sufficiently accurate.

Computational aspects will affect the future development of such schemes by improving computer codes that make use of efficient methods, such as fast Fourier transforms (FFT), iterative schemes versus diagonalization of large matrices, use of workstations, clusters, multiple processor machines or even massively parallel computing. With increasing computer power and the development of adapted algorithms, the *ab initio* DFT calculations, by the FP-LAPW method, should be able to bring new insight into many areas of solid state physics, chemistry or material sciences.

Acknowledgments

Work at the TU Wien was supported by the Austrian Science Foundation FWF Projects P8176, P9213 and M134. We want to thank all members of the group who contributed to the present paper, especially P. Dufek, a co-author of WIEN95, who was also involved in many LAPW (LDA or GGA) calculations. For all computational aspects, the help by R. Augustyn is greatly acknowledged.

References

1. P.Hohenberg and W. Kohn, Phys.Rev. **136**, B864 (1964).
2. W.Kohn and L.J. Sham, Phys.Rev. **140**, A1133 (1965).
3. J.P.Perdew and Y.Wang , Phys.Rev. B **33**, 8800 (1986).
4. J.P.Perdew, in*Electronic Structure in Solids* (P.Ziesche and H. Eschrig, eds), Akademie Verlag, Berlin, pp.11-20 (1991).
5. B.G. Johnson, P.M.W. Gill and J.A.Pople, J.Chem.Phys. **98**, 5612 (1993).
6. D.J.Singh, Planewaves, pseudopotentials and the LAPW method, Kluwer Academic Publishers, Boston, Dordrecht, London (1994).
7. P.Blaha, K.Schwarz, P. Dufek, and R.Augustyn, *Computer Code WIEN95* (1995): an improved and updated Unix version of the originally copyrighted WIEN code, which was published by P.Blaha, K.Schwarz, P.I.Sorantin and S.B.Trickey, Comput.Phys.Commun. **59**, 399 (1990).
8. D.J.Singh, Phys.Rev.B **43**, 6388 (1991).
9. P.Dufek, PhD Thesis, Technical University Vienna (1994).
10. K.Schwarz and P Mohn, J.Phys.F:Metal Phys. **14**, L129 (1984).
11. R.Yu, D.J. Singh, and H. Krakauer, Phys. Rev. B. **43**, 6411 (1991).
12. J.M Soler, and A.R. Williams, Phys. Rev. B. **40**, 1560 (1989).

13. B. Kohler, S. Wilke, M. Scheffler, R. Kouba, and C. Ambrosch-Draxl, Comp. Phys. Commun. **94**, 31 (1996).
14. H.G. Krimmel, J. Ehmann, C. Elsässer, M. Fähnle, and J.M. Soler, Phys. Rev. B. **50** 8846 (1994).
15. P. Pulay, Mol. Phys. **17**, 197 (1969).
16. P.Bagno, O.Jepsen and O.Gunnarsson, Phys.Rev. B **40**, 1997 (1989).
17. P.Dufek, P.Blaha, V.Sliwko and K.Schwarz, Phys. Rev. B **49**, 10170 (1994).
18. P.Dufek, P.Blaha, and K.Schwarz, Phys. Rev. B **51**, 4122 (1995).
19. K.Schwarz, Phase Transitions **52**, 109 (194).
20. P.Blaha, K.Schwarz, P.Dufek, G.Vielsack, and W.Weber, Z.Naturf. **48a**, 129 (1993).
21. P.Blaha, K.Schwarz and P.Herzig, Phys.Rev.Lett. **54**, 1192 (1985).
22. P.Blaha and K.Schwarz, Hyperf. Interact. **52**, 153 (1989).
23. P.Blaha, D.J.Singh, P.I.Sorantin and K.Schwarz, Phys.Rev. B **46**, 1321 (1992).
24. W.Tröger, T.Butz, P.Blaha and K.Schwarz, Hyperf. Interact. **80**, 1109 (1993).
25. P.Blaha, K.Schwarz and P.H.Dederichs, Phys.Rev B **37**, 2792 (1988).
26. P.Blaha and K.Schwarz, J.Phys.F: Met.Phys. **17**, 899 (1987).
27. P.Blaha, Hyperf. Interact. **60**, 773 (1990).
28. K.Schwarz, C.Ambrosch-Draxl, and P.Blaha, Phys.Rev B **42**, 2051 (1990).
29. C.Ambrosch-Draxl, P.Blaha and K.Schwarz, Phys.Rev B **44**, 5141 (1991).
30. P.Dufek, K.Schwarz and P.Blaha, Phys.Rev. B **48**, 12672 (1993).
31. P.Blaha and K.Schwarz, J.Mol.Struct. (Theochem) **261**, 355 (1992).
32. K.Schwarz and P.Blaha, Z.Naturforsch. **47a**, 197 (1992).
33. P.Dufek, P.Blaha, and K.Schwarz, Phys. Rev. Lett. **75**, 3545 (1995).

A Pseudopotential Plane Waves Program (PWSCF) and some Case Studies

Andrea Dal Corso

Institut Romand de Recherche Numérique en Physique des Matériaux (IRRMA), IN-Ecublens, 1015 Lausanne, Switzerland

May 28, 1996

Summary. We analyze the PWSCF code. This code solves the self-consistent Kohn and Sham equations, obtained for a periodic solid, in the framework of density functional theory (DFT) using the local density approximation (LDA). The orbitals are expanded in a plane wave basis set and the cores are described by norm-conserving pseudopotentials. The theory and the implementation of the equations are discussed. We present three examples of applications to solids: a semiconductor, silicon; an insulator, NaCl; and a metal, aluminium. For each system, we compute the total energy, the band structure and the electronic charge density. Examples of calculations of the lattice constants and of the bulk modulus are also given. Several practical issues which were encountered in these calculations are discussed.

Key words: Density Functional Theory (DFT)– Local Density Approximation (LDA)– Plane Waves (PW) Code – Pseudopotentials (PP)

1. Introduction

The program PWSCF was originally developed by S. Baroni and P. Giannozzi; it has received, over the years, important additions and variants by several authors, including S. de Gironcoli, P. Pavone, A. Dal Corso, A. Debernardi, C. Bungaro, K. Stokbro and R. Valente. One of the latest, simplified versions of the code is available on the public server `itncpl.science.unitn.it` in the directory `/pub/PHONON` and can be obtained by anonymous `ftp` [1]. Together with the code, there is a complete documentation. Part of this chapter is based on the unpublished notes which are distributed with the program.

Correspondence to: A. Dal Corso - e-mail dalcorso@eldpa.epfl.ch

Several advanced features have been implemented over the main body of the program. These will not be discussed here: a non-exhaustive list includes calculation of phonon spectra [2], of macroscopic stress [3], of linear-response properties [4], of the macroscopic polarization by a Berry phase [5]; implementations of the nonlinear-core correction [6, 7] and of the gradient correction are available [8, 9]. Only one part of the publicly available version will be discussed here, namely the part which uses PPs in the Bachelet, Hamann and Schlüter (BHS) form. The use of fully separable PPs is also possible but it is not described here.

The PWSCF code computes the electronic band structure, the electronic charge density and the total energy of a periodic crystal, with a given Bravais lattice and a given space group symmetry. The algorithm is based on DFT in the local density approximation (LDA) [10]. It computes the solution of the self-consistent KS equations [11], which describe a valence electron in the potential created by the periodic lattice of the pseudonuclei (described by norm-conserving PPs [13]) and by all the other valence electrons. The KS orbitals are expanded over a finite plane wave basis set [14] which allows the transformation of the partial differential KS equations into an algebraic eigenvalue problem which is solved by iterative techniques [15]. The program exploits the point group symmetry of the solid to reduce the number of operations necessary to compute the charge density and the total energy. The computation of these quantities allows the study of structural (lattice constants, bulk modulus and elastic constants) and dynamical (zone-centre phonon frequencies) properties, the study of structural phase transitions and of the effect of pressure on the solid. Furthermore, by using the supercell approach, defects, interfaces and surfaces can be studied.

Here, we give three examples of the application of the code: a semiconductor, silicon; an insulator, NaCl; and a metal, aluminium. These systems are sufficiently simple to allow a complete control of the numerical noise, allowing us to test the physical approximations used in the code on some real examples. The computations performed on these systems should allow the reader to compare between the density functional theory [11] (DFT) in the local density approximation (LDA), implemented with PPs and a PW basis set [12], with other methods which are currently used in the study of solids.

In our computations, there are two physical approximations. One is the LDA, which is used to give an explicit form to the exchange-correlation energy, the other is the frozen core approximation [16] which is used to substitute the core electrons with a fixed PP. Only the ground state energy of the valence electrons is computed by expanding the Kohn and Sham orbitals [11] in a PW basis set. In many interesting cases, these two approximations are well justified and the use of the code allows us to predict many quantities to within a few percent of the experimental values.

The LDA approximation is a simple physical approximation of the exchange-correlation energy which does not require any empirical parameter, being based on the exact quantum Monte Carlo calculation of the exchange-correlation energy of the uniform electron gas [17]. It gives very good results in covalent and metallic systems, and, in general, it is one of the most powerful tools of the *ab initio* investigations of condensed systems whether in the liquid, amorphous or crystalline phases. The frozen core approximation is a necessary hypothesis to describe the charge density with PWs. This basis has several numerical advantages but it is not possible to represent the strong oscillations of the all-electron wavefunctions close to the nucleus with a reasonable number of basis functions. It is necessary to substitute the nuclear potential with a different potential which represents both the nucleus and the core electrons. The remaining valence charge is a smooth function which, in several solids, can be described by a treatable number of PWs. The number of electrons which are considered as valence electrons, depends only on the physical problem, so that, in principle, it is always possible to introduce, into the valence states, all the electrons which are important in a given computation. In practice, it is necessary to avoid very localized charges which are difficult to describe with PWs.

There are, however, some cases where the systematic errors due to physical approximations are more important and physically relevant. For example, it is known that if the Kohn and Sham eigenvalues are interpreted as energy bands and compared with photoemission data, the LDA gap of semiconductors and insulators is underestimated. The errors of the theoretical lattice constants are systematic: all the bond strengths are overestimated in LDA. This error is usually very low in covalent and metallic systems, but it can be substantial in solids where the Van der Waals bonds or the hydrogen bonds [18] are important. The weak bonds which are responsible for cohesion in molecular crystals are too strong when computed using the LDA [8, 19]. Being a mean-field approximation, the LDA cannot provide correct information in those cases where strong correlation effects are significant.

It is important not to add spurious effects, due to the numerical noise, to the physical errors. There are two major sources of numerical noise which must be accurately monitored in each calculation:

a) **Incomplete basis set.**

The number of PWs used in the calculation is not sufficient to represent correctly the wavefunctions and the charge density. The number of PWs is a critical parameter in solids which contain localized charge (see the example of NaCl) or which have a very large unit cell. In these cases, a compromise must be made between the CPU-time, memory requirements and the precision of the calculation.

b) **Poor sampling of the Brillouin zone (BZ)**

The number of special points is not sufficient to compute accurately the BZ integrals. Usually this problem manifests itself in the computation of particular physical quantities (such as the dielectric constant) where the integrand function is not smooth. In the case of metals, the special-point technique has to be used with care: it is important to describe correctly the Fermi energy surface.

We performed our tests on small systems, where it is possible to control the convergence of all the results. In the case of aluminium, we also checked the convergence with respect to the smearing parameter [20, 21] which is used to deal with the presence of a Fermi surface.

This chapter is organized as follows. In section 2, we present the theoretical framework of DFT and the plane wave formulation of the KS equations. In section 3, we summarize the organization of the code, giving a short description of the main routines. In section 4, the dimension parameters are explained and the CPU-time and memory requirements are discussed. A description of the input parameters and an example input file are given in section 5. In section 6, we describe the computations performed on silicon. We start by analyzing the convergence of several properties with respect to the number of PWs and of special points used. We compute the total energy, the theoretical lattice constant and the bulk modulus. We show examples of the band structure computed at the experimental and at the theoretical lattice constant; finally, we show some plots of the electronic charge density computed on the (100) and (110) planes. In section 7, we present similar results for NaCl. In section 8, we discuss the case of aluminium. Some problems related to the description of the Fermi surface are addressed in this section. Section 9 contains the conclusions drawn from this work.

2. Theoretical framework

In the formulation of DFT by KS, the total energy of an interacting electronic system, in an external potential $V_{\mathrm{ext}}(\mathbf{r})$, is given by the following relationship:

$$\begin{aligned} E_{\mathrm{tot}} &= -\frac{1}{2}\sum_i \int \psi_i^*(\mathbf{r})\nabla^2\psi_i(\mathbf{r})d\mathbf{r} \\ &+ \int \rho(\mathbf{r})V_{\mathrm{ext}}(\mathbf{r})d\mathbf{r} + \frac{1}{2}\int \frac{\rho(\mathbf{r})\rho(\mathbf{r}')}{|\mathbf{r}-\mathbf{r}'|}d\mathbf{r}d\mathbf{r}' + \int \rho(\mathbf{r})\epsilon_{\mathrm{xc}}(\rho(\mathbf{r}))d\mathbf{r} \end{aligned} \tag{1}$$

where $\rho(\mathbf{r})$ is the electronic charge density and ϵ_{xc} is the exchange-correlation energy which, within the LDA, is a function of the density [17]. The kinetic energy of the system is written as the kinetic energy of a gas of independent electrons, with the same density as the interacting system. This is achieved by the introduction of KS orbitals, $\psi_i(\mathbf{r})$, which fulfill the orthonormality constrains:

$$\int \psi_i^*(\mathbf{r})\psi_j(\mathbf{r})d\mathbf{r} = \delta_{ij} \tag{2}$$

These orbitals are obtained by minimizing the total energy of the system. The minimization problem is equivalent to the solution of the KS equations:

$$\left[-\frac{1}{2}\nabla^2 + V_{\mathrm{ext}}(\mathbf{r}) + \int \frac{\rho(\mathbf{r}')}{|\mathbf{r}-\mathbf{r}'|}d\mathbf{r}' + \mu_{\mathrm{xc}}(\rho(\mathbf{r}))\right]\psi_i(\mathbf{r}) = \epsilon_i\psi_i(\mathbf{r}) \tag{3}$$

where the exchange-correlation potential, μ_{xc}, is given by:

$$\mu_{\mathrm{xc}}(\rho(\mathbf{r})) = \frac{d}{d\rho}(\rho\epsilon_{\mathrm{xc}}(\rho)) \tag{4}$$

The charge density is related to the KS orbitals by the relationship:

$$\rho(\mathbf{r}) = \sum_i |\psi_i(\mathbf{r})|^2 \tag{5}$$

with the constraint:

$$\int \rho(\mathbf{r})d\mathbf{r} = N \tag{6}$$

where N is the number of electrons of the system. The number of occupied states can be computed by using equation 6 and the Fermi statistics. In an insulator or a semiconductor, the number of occupied states is $n/2$; in a metal, equation 6 defines the Fermi energy of the system and hence, the occupied orbitals. The PWSCF code solves equation 3 to compute the KS orbitals, uses equation 5 to compute the charge density and finally, computes the total energy of the system. For this purpose, it uses a formula equivalent to equation 1, which is more convenient from a numerical point of view. This formula can be obtained by multiplying the KS equations by $\psi_i^*(\mathbf{r})$:

$$E_{\text{tot}} = \sum_i \epsilon_i - \frac{1}{2}\int \frac{\rho(\mathbf{r})\rho(\mathbf{r}')}{|\mathbf{r}-\mathbf{r}'|} d\mathbf{r}d\mathbf{r}' + \int \rho(\mathbf{r})\Big(\epsilon_{\text{xc}}(\rho(\mathbf{r})) - \mu_{\text{xc}}(\rho(\mathbf{r}))\Big) d\mathbf{r} \quad (7)$$

where the sum over i is performed over the filled states.

2.1. The self-consistent loop

The KS equation for $\psi_i(\mathbf{r})$ is formally equivalent to a Schrödinger equation for one electron, where the external potential is an effective potential which depends, in a self-consistent way, on the solutions of all the other equations. The solutions of these equations are obtained by iteration. Starting from an arbitrary choice of initial potential, $V_{\text{in}}^{(0)}(\mathbf{r})$, the code solves the following iteration:

$$\left[-\frac{1}{2}\nabla^2 + V_{\text{in}}^{(n)}(\mathbf{r})\right]\psi_i^{(n)}(\mathbf{r}) = \epsilon_i^{(n)}\psi_i^{(n)}(\mathbf{r}) \quad (8)$$

$$\rho^{(n)}(\mathbf{r}) = \sum_i |\psi_i^{(n)}(\mathbf{r})|^2 \quad (9)$$

$$V_{\text{out}}^{(n)}(\mathbf{r}) = V_{\text{ext}}(\mathbf{r}) + \int \frac{\rho^{(n)}(\mathbf{r}')}{|\mathbf{r}-\mathbf{r}'|} d\mathbf{r}' + \mu_{\text{xc}}(\rho^{(n)}(\mathbf{r})) \quad (10)$$

From this solution, a new initial potential is created, $V_{\text{in}}^{(n+1)}$, so that the sequence of $V_{\text{in}}^{(n)}$ and $V_{\text{out}}^{(n)}$ converges to the self-consistent potential, V_{scf}, which is defined as the potential for which $V_{\text{in}} = V_{\text{out}}$. The simplest iteration $V_{\text{in}}^{(n+1)} = V_{\text{out}}^{(n)}$, usually makes the problem unstable and no self-consistent solution is reached. A better solution is to *mix* input and output potentials:

$$V_{\text{in}}^{(n+1)} = (1-\beta)V_{\text{in}}^{(n)} + \beta V_{\text{out}}^{(n)} \quad (11)$$

where β is a number between 0 and 1. Even if the linear mixing is usually sufficient to cause convergence, a more efficient strategy is the Anderson method [22], where more than one preceding iteration is used. In the simplest form, the iteration is the following:

$$V_{\text{in}}^{(n+1)} = (1-\beta)\Big((1-\theta_n)V_{\text{in}}^{(n)} + \theta_n V_{\text{in}}^{(n-1)}\Big) + \beta\Big((1-\theta_n)V_{\text{out}}^{(n)} + \theta_n V_{\text{out}}^{(n-1)}\Big) \quad (12)$$

where θ_n is defined by the condition that the norm:

$$\left\|(1-\theta_n)\Big(V_{\text{in}}^{(n)} - V_{\text{out}}^{(n)}\Big) + \theta_n\Big(V_{\text{in}}^{(n-1)} - V_{\text{out}}^{(n-1)}\Big)\right\|^2 \quad (13)$$

is a minimum. The parameter β can be quite large for small systems ($\simeq 0.7$) but it must be reduced in systems where the convergence is more difficult.

2.2. The KS equations

We consider, from now on, a periodic system with N cells and n electrons per cell. At each iteration of the self-consistent loop, the KS equations at fixed potential must be solved. This problem can be transformed into an algebraic eigenvalue problem on a finite basis set. The code PWSCF implements these equations in the case of a periodic crystal with a lattice $\mathbf{R}$, a unit cell of volume Ω which contains N_{at} atoms of N_{typ} different types in the positions $\mathbf{d}_s$. Each type of atom is characterized by a valence charge, $Z_{\mu(s)}$ and a PP, $\widehat{V}_{\mu(s)}$. We use $\mathbf{K}$ to indicate the reciprocal lattice vectors. The Bloch theorem allows us to classify the electronic states by the vectors $\boldsymbol{\kappa}$ in the BZ. These states are expanded in plane waves $|\boldsymbol{\kappa}+\mathbf{K}\rangle$ which have the appropriate translational symmetry:

$$\psi_i(\mathbf{r}) \rightarrow \psi_v(\mathbf{r};\boldsymbol{\kappa}) = \sum_{\mathbf{K}} \widetilde{\psi}_{\mathbf{K},v}(\boldsymbol{\kappa})|\boldsymbol{\kappa}+\mathbf{K}\rangle \quad \text{where } |\boldsymbol{\kappa}+\mathbf{K}\rangle = \mathcal{V}^{-\frac{1}{2}}e^{\iota(\boldsymbol{\kappa}+\mathbf{K})\mathbf{r}} \tag{14}$$

and where $\mathcal{V}$ is the total volume of the crystal. We assume that there are $N \rightarrow \infty$ unitary cells with periodic boundary conditions. The total volume of the solid is $\mathcal{V} = N\Omega$, the total number of electrons is $N\sum_{i=1}^{N_{at}} Z_{\mu(i)}$ and there are N discrete values of $\boldsymbol{\kappa}$ in the BZ. The index v is a band index which, in the case of an insulator or a semiconductor, identifies $N_b = n/2$ bands, each of them occupied by two electrons.

The wavefunctions are expanded in a finite PW basis set. The number of PWs, N_{pw}, is computed introducing a kinetic energy cutoff, E_{cut} and using the PWs that fulfill the relationship $(\boldsymbol{\kappa}+\mathbf{K})^2 \leq E_{cut}$, where E_{cut} is sufficiently large to assure a good representation of the wave functions. The solution of the KS equation for a given $\boldsymbol{\kappa}$ is equivalent to an *eigenvalue equation* of dimension N_{pw}:

$$\sum_{\mathbf{K}'} H(\boldsymbol{\kappa}+\mathbf{K},\boldsymbol{\kappa}+\mathbf{K}')\widetilde{\psi}_{\mathbf{K}',v}(\boldsymbol{\kappa}) = \epsilon_{\boldsymbol{\kappa},v}\widetilde{\psi}_{\mathbf{K},v}(\boldsymbol{\kappa}) \tag{15}$$

where H is the following matrix:

$$\begin{aligned} H(\boldsymbol{\kappa}+\mathbf{K},\boldsymbol{\kappa}+\mathbf{K}') &\equiv <\boldsymbol{\kappa}+\mathbf{K}|H|\boldsymbol{\kappa}+\mathbf{K}'> \\ &= \frac{1}{2}(\boldsymbol{\kappa}+\mathbf{K})^2\delta_{\mathbf{K}\mathbf{K}'} + \frac{4\pi}{\Omega}\frac{|\widetilde{\rho}(\mathbf{K}-\mathbf{K}')|}{|\mathbf{K}-\mathbf{K}'|^2} \\ &\quad +\widetilde{\mu}_{\mathrm{xc}}(\mathbf{K}-\mathbf{K}') + \sum_i e^{-\iota(\mathbf{K}-\mathbf{K}')\mathbf{d}_i} \times \\ &\quad \left(\widetilde{V}^{loc}_{\mu(i)}(\mathbf{K}-\mathbf{K}') + \sum_l \widetilde{V}^{nl}_{\mu(i),l}(\boldsymbol{\kappa}+\mathbf{K},\boldsymbol{\kappa}+\mathbf{K}')\right) \end{aligned} \tag{16}$$

and where we have defined the Fourier coefficients as:

$$\widetilde{a}(\mathbf{K}) = \frac{1}{\Omega}\int_{\Omega} a(\mathbf{r})e^{-\iota\mathbf{K}\mathbf{r}}d\mathbf{r} \quad \text{and} \quad a(\mathbf{r}) = \sum_{\mathbf{K}} \widetilde{a}(\mathbf{K})e^{\iota\mathbf{K}\mathbf{r}} \tag{17}$$

It is worthwhile to note that E_{cut} is fixed independently of $\boldsymbol{\kappa}$, so that N_{pw} is a function of $\boldsymbol{\kappa}$. In the last term, i.e. in the matrix elements of the PP, we have separated a local term from a non-local contribution. The exact expression of the matrix elements depends on the PP type. The program uses semilocal PPs of the Bachelet, Hamann and Schlüter (BHS) type:

$$V_{\text{ext},\mu}(\mathbf{r}) \equiv V_{\mu}^{loc}(r) + \sum_{l=0}^{l_{MAX}} \hat{P}_l V_{\mu,l}^{nl}(r) \tag{18}$$

where $\hat{P}_l$ is the projector upon the states of l angular momentum:

$$\hat{P}_l = \sum_{m=-l}^{+l} Y_{ml}(\mathbf{r}) Y_{ml}^{*}(\mathbf{r}')\delta(r - r') \tag{19}$$

2.3. The self-consistent potential

The sum over the BZ needed for the computation of the charge density, $\rho(\mathbf{r})$, at each iteration, is approximated very efficiently by a sum over a discrete set of points. This technique is known as the ***special-points*** technique and has been discussed in more detail in Chapter 4. Given a set, $N_{\boldsymbol{\kappa}}$, of special points $\{\boldsymbol{\kappa}\}$ in the irreducible part of the BZ (IBZ), with the correct weights, $w_{\boldsymbol{\kappa}}$ (equal to the number of equivalent points in the total BZ), the total charge density is obtained in two steps. First an unsymmetrized charge density is obtained by adding all the points in the IBZ:

$$\tilde{\rho}(\mathbf{r}) = \sum_{\boldsymbol{\kappa} \in IBZ} \sum_{v} w_{\boldsymbol{\kappa}} |\psi_v(\mathbf{r}; \boldsymbol{\kappa})|^2 \tag{20}$$

and then this object is symmetrized:

$$\rho(\mathbf{r}) = \frac{1}{N_S} \sum_{m=1}^{N_S} \tilde{\rho}((S^m)^{-1}\mathbf{r} - \mathbf{f}^m) \tag{21}$$

where $(S^m|\mathbf{f}^m)$ are the N_S symmetry operations of the space group of the crystal.

A very efficient way to compute the charge density and the LDA potential uses the discrete Fourier Transform, in particular the algorithm of the Fast Fourier-Transform (FFT). For this purpose, a mesh in $\mathbf{K}$ space is introduced:

$$\mathbf{K}_{m_1,m_2,m_3} \equiv m_1\mathbf{b}_1 + m_2\mathbf{b}_2 + m_3\mathbf{b}_3 \tag{22}$$

where $\mathbf{b}_1, \mathbf{b}_2$ and $\mathbf{b}_3$ are the principal reciprocal lattice vectors and where:

$$m_1 = -\frac{N_1}{2}, \ldots, \frac{N_1}{2} - 1\ ;\ m_2 = -\frac{N_2}{2}, \ldots, \frac{N_2}{2} - 1\ ;\ m_3 = -\frac{N_3}{2}, \ldots, \frac{N_3}{2} - 1 \tag{23}$$

with even values of N_j. Then we introduce a mesh in real space within a unit cell:

$$\mathbf{r}_{m_1,m_2,m_3} \equiv \frac{m_1 - 1}{N_1}\mathbf{a}_1 + \frac{m_2 - 1}{N_2}\mathbf{a}_2 + \frac{m_3 - 1}{N_3}\mathbf{a}_3 \tag{24}$$

where $\mathbf{a}_1, \mathbf{a}_2$ and $\mathbf{a}_3$ are three primitive vectors which generate the direct lattice and:

$$m_1 = 1, \ldots, N_1\ ;\ \ m_2 = 1, \ldots, N_2\ ;\ \ m_3 = 1, \ldots, N_3 \tag{25}$$

The integer numbers N_1, N_2 and N_3 define the mesh in real space and equivalently, the corresponding one in reciprocal space. They must be sufficiently large so that the vectors $\mathbf{K}_{m_1,m_2,m_3}$ contain all the vectors $\mathbf{K} - \mathbf{K}'$, defined by the basis set.

Given a function in $\mathbf{K}$ space, defined on the mesh points $\widetilde{f}(m_1, m_2, m_3) \equiv f(\mathbf{K}_{m_1,m_2,m_3})$, its three dimensional Fourier transform can be obtained by a 3D-FFT:

$$f(m_1, m_2, m_3) = \sum_{l_1} \sum_{l_2} \sum_{l_3} \widetilde{f}(l_1, l_2, l_3) \times e^{\imath 2\pi l_1 m_1/N_1} \times e^{\imath 2\pi l_2 m_2/N_2} \times e^{\imath 2\pi l_3 m_3/N_3} \tag{26}$$

The use of FFT to switch from real to reciprocal space is very efficient and is used whenever possible. To compute the charge density, one computes the KS orbitals, $\psi_{\mathbf{K},v}(\boldsymbol{\kappa})$, on the mesh $\widetilde{\psi}_v(m_1, m_2, m_3, \boldsymbol{\kappa})$ in reciprocal space, then with an FFT one converts to real space, where the square of the wavefunctions and the total charge are computed. Then μ_{xc} is computed in real space, while the Hartree potential is obtained by converting back to reciprocal space.

3. Structure of the code

The structure of the code closely follows the theory just explained. The main routines of the code are the following:

routine	*calls*	*descriptions*
readin		reads input data and PP
setup		variable initialization
	latgen	generates the vectors $\mathbf{a}_1, \mathbf{a}_2, \mathbf{a}_3$, computes Ω
	recips	generates $\mathbf{b}_1, \mathbf{b}_2, \mathbf{b}_3$
	ggen	generates the vectors $\mathbf{K}$ and the FFT indices
	hexsym	generates symmetry operations
	cubicsym	generates symmetry operations
	sgama	check the compatibility among special points, atomic positions and point group symmetry
openfil		opens necessary files
summary		writes information on input data and initial variables
hinit		initializes the Hamiltonian
	setlocal	computes the local part of the PP
	setab	computes a bidimensional table for BHS PP
	setnlpot	computes $\boldsymbol{\kappa} + \mathbf{K}$ and puts the BHS PP into the Hamiltonian

... continued overleaf

routine	*calls*	*descriptions*
`potinit`		computes the initial potential
	`sumatom`	computes the charge as a sum of atomic charges
`wfcinit`		initializes the wavefunctions
	`h_fill`	fills the Hamiltonian with a small matrix
	`cdiagd`	diagonalizes with EISPACK [23] routines
`pwscf`		controls the self-consistent loop
	`c_bands`	computes the wavefunctions
	`sum_band`	computes the charge density
	`v_of_rho`	computes the new potential
	`delta_e`	computes some terms of the total energy
	`dmixp`	computes the new input potential
`punch`		save results on disk
`closfil`		closes all the files

There is an initial part where the input parameters are read from a file, and several useful quantities are computed. These are the direct and reciprocal lattice, the list of the **K** vectors and the local and non-local PP. Then a guess at the initial wavefunctions and at the Hartree and exchange-correlation potential are obtained from the Hamiltonian at a small cutoff energy. After these preparatory computations, the self-consistent cycle is repeated up to the convergence of the potential. At each iteration, the total energy is computed. The routine `c_bands` uses the routine `cegter` for the iterative diagonalization. This routine calls `h_psi` to compute the products $H\psi$. At the end, the self-consistent potential, the KS wavefunctions and eigenvalues are saved on the disk for further analysis.

4. Use of the program

The code is composed of several files. The file `pwscf.f` contains the main program and all the routines which depend on the *commons* and on the dimensions of arrays and matrices. The routines which do not depend on the commons and on the array dimensions are contained in the file `pwlib.f`. The commons and the definitions of the global variables are contained in the file `pwcom.inc`. The dimensions of the matrices and array are contained in the file `pwpar.inc`.

At run time, the program also requires a file which contains the parameters of the PP. The file `pseudop.dat`, distributed with the code, contains several examples of PPs for the most common atomic species. Ref. [13] contains a table with the parameters of the PPs of all atoms.

4.1. Dimensions

The file `pwpar.inc` contains the array and matrix dimensions which are used to define the memory size of the program. These dimensions depend on the system under study. Therefore, this file must be written before the compiling phase. A typical format for the file `pwpar.inc` is the following:

```
 parameter ( mxhc=300, mxvc= 32, mxh0=100, nax= 2, ntypx= 2,
&            lcorx=1, npsx=2, nrx1=16, nrx2=16, nrx3=16,
&            maxter= 50, npk=60)
 parameter ( qmax =  6.0,  dq=  .034)
```

where the parameters have the following meaning:

MXHC	maximum number of plane waves (MXHC $\geq N_{pw}$)
MXVC	a number proportional to the number of bands N_b. It must obey MXVC $\geq 4N_b$ (for good performance, MXVC $\simeq 7-8N_b$)
MXHO	maximum dimension of the initial hamiltonian
NAX	maximum number of atoms in the unit cell (NAX $\geq N_{at}$)
NTYPX	maximum number of different types of atoms (NTYPX$\geq N_{typ}$)
NPSX	maximum number of different PPs
LCORX	maximum l in the non-local PPs
NRX1,NRX2,NRX3	maximum values in N_1, N_2, N_3 in the FFT mesh
MAXTER	maximum number of self-consistent iterations
NPK	maximum number of $\boldsymbol{\kappa}$ points for integration on the IBZ (NPK$\geq N_{\boldsymbol{\kappa}}$)
QMAX	maximum length, in a.u., of the vectors $\boldsymbol{\kappa}+\mathbf{K}$: QMAX $> \lvert\boldsymbol{\kappa}+\mathbf{K}\rvert$
DQ	step for the bidimensional table of the non-local PPs.

4.2. Libraries, memory and time requirements

The linking of the program requires some auxiliary routines which can be found in the libraries BLAS, BLAS2, BLAS3, EISPACK and LINPACK. These libraries are available on many machines, however, an auxiliary file `library.f` contains the Fortran sources of the required routines. If this file is used, the performance of the code may be slightly reduced because these routines are not optimized for any machine.

The CPU time, T_{cpu}, is proportional to the number of atoms in one unit cell and can be approximately estimated as

$$T_{cpu} \sim N_{iter}N_{\boldsymbol{\kappa}} \times \left(\mathcal{O}(N_bN_{pw}^2) + \mathcal{O}(N_bN_{pw}\log N_{pw}) + \mathcal{O}(N_b^2N_{pw})\right) \tag{27}$$

where N_{iter} is the number of iterations needed to achieve self-consistency (typically 5–20). The RAM memory can be estimated as (in bytes):

$$M_{ram} = 16\mathtt{MXHC}^2 + 16\mathtt{MXHC} * \mathtt{MXVC} + 9\mathtt{NRXX} \tag{28}$$

where NRXX=NRX1*NRX2*NRX3. The program uses some disk space in scratch files which are deleted at the end of the run. This space can be estimated, in bytes as:

$$M_{I/O} \simeq N_{\kappa} \times \left(16\texttt{MXHC}^2 + 16\texttt{MXHC}*N_b\right) + 64\texttt{NRXX} \tag{29}$$

5. Input Data

The file which contains the input data has the following format:

Title (max. 75 characters)
Name of the Crystal (max. 20 characters)
&input *variables in the namelist* ... **&end**
$\mathbf{d}_1, \mu(1)$
...
$\mathbf{d}_{N_{at}}, \mu(N_{at})$
$atom(1),\ n_{ps}(1), (j_{ps}(1,i), x_{ps}(1,i), i = 1, n_{ps}(1))$
...
$atom(N_{typ}),\ n_{ps}(N_{typ}), (j_{ps}(N_{typ},i), x_{ps}(N_{typ},i), i = 1, n_{ps}(N_{typ}))$
N_{κ}
κ_1, w_{κ_1}
...
$\kappa_{N_\kappa}, w_{\kappa_{N_\kappa}}$

where $atom(\mu)$ is the name of the atom of type μ (max 3 characters), j_{ps} and x_{ps} are used to specify the PP by which a given atom is described. Normally, $n_{ps} = 1$, $x_{ps} = 1$ and j_{ps} numbers the PP.

The set of κ-points should be appropriate to the point group of the Bravais lattice. If the input atomic positions are incompatible with the assumed symmetry, the program finds the true point group and adds other κ points updating the weight of each point. Therefore, the value of N_κ can be larger that the input value.

The variables which appear in the input namelist are divided into two types. In the first, there are the variables which do not have any default value and must be specified by the user:

ibrav	index of the Bravais lattice
indpg	index of the point group
celldm(i)	parameters of the lattice
nat	number of atoms, N_{at}, in the unit cell
ntyp	number of different types of atoms, N_{typ}
zpsx(i), $i = 1, N_{ps}$	atomic numbers of the PPs
nr1, nr2, nr3	FFT dimensions N_1, N_2, N_3
ecut(0)	cutoff E_{cut}, in Rydberg, for the initial diagonalization
ecut(i)	cutoff E_{cut}, in Rydberg, for the following iterations

The first **celldm**(1) is the lattice parameter a_0; **ibrav** and **celldm**(i), $i > 2$ are explained at the beginning of the program.

The second type of variable is optional and has a default value.

`npseu`	number of PPs N_{ps}. Default: `npseu=ntyp`.
`ngauss`	for metals, this gives the interpolation order in the *Gaussian broadening* [24]. `ngauss=0` (default) is the normal Gaussian broadening.
`degauss`	value, in Rydberg, of the Gaussian broadening.
`niter`	maximum number of iterations. Default: 20.
`beta`	mixing factor in the Anderson algorithm. Default: 0.7.
`id`	type of Anderson algorithm: `id=1` simple mixing; `id=2` uses the result of the previous iteration; `id=3` (default) uses the results of the two previous iterations.
`tr2`	precision of self-consistency. Default: 10^{-12}.
`nbnd`	number of bands to be computed. Default for a semiconductor is the number of filled bands. For a metal, two more bands are computed.
`sumat`	`.true.` (default) means that the starting potential is obtained from the sum of atomic charge densities. `.false.` sets the initial potential to zero.
`lscf`	`.true.` (default) causes self-consistent computation to be performed. `.false.` means band computation with a fixed potential.
`input_pot`	name of the file which contains the self-consistent potential. Default: no potential is read.
`output_pot`	name of the file where the SCF potential is written. Default: no potential is written.
`filpun`	file where all the other quantities are saved. Default: No quantity is written.
`filpsd`	file with the parameters of the PP. Default: `pseudop.dat`.

The following is an example of input data, for silicon in an *fcc* Bravais lattice, with two atoms per unit cell and two special points in the BZ.

```
Silicon at 12 Ry cutoff and 2 special points in the IBZ

Si

 &input ibrav=2, indpg=31,     celldm(1)=10.26, nat=2,

        ntyp=1,  npseu=1,      zpsx(1)=14.0,    nr1=16,

        nr2=16,  nr3=16,       ecut(0)=4.0,     ecut(1)=12.0,

        filpun='carica',       input_pot='sipotin',

        output_pot='sipotout', lforce=.false.,

        lstres=.false.,        kwrite=1,        tr2=1.0e-14,

        filpsd='PSEUDOP',      sumat=.f.
```

```
&end
    0.00    0.00    0.00 1
    0.25    0.25    0.25 1
 'Si', 1, 1, 1.0
2
    0.25  0.25  0.25   1.0
    0.25  0.25  0.75   3.0
```

6. Silicon

The electronic configuration of the silicon atom is $[Ne]3s^2 3p^2$. Crystalline silicon is a group IV covalent semiconductor, with an experimental energy gap of 1.17 eV. Each atom is tetrahedrally coordinated and connected to its neighbours by a covalent bond derived from sp^3 hybrid orbitals. The electronic valence charge density is concentrated on the bonds. The charge is described by a smooth function of the position which can be represented with a small number of PWs. The neon-like cores are tightly bound to the nuclei and do not relax appreciably when the atoms form the solid. The LDA approximation is particularly good in the case of sp^3 covalent bonds and the PP approximation does not introduce substantial errors. Silicon is the most studied material in the framework of DFT-LDA [25]. Nowadays, a large amount of well established theoretical results are available for most of its properties.

The crystal can be described by an *fcc* Bravais lattice with two atoms per unit cell. The atomic positions are fixed by the choice of the origin. If one atom is placed on the origin, the other is at $\mathbf{d} = a_0(0.25, 0.25, 0.25)$ where a_0 is the edge of the conventional cubic cell. The experimental value of a_0 is $a_0 = 10.26$ a.u. Silicon has 8 electrons per unit cell, which fill the four valence bands.

We start by computing the total energy of the system as a function of the kinetic energy cutoff and of the number of special points in the irreducible BZ (IBZ). We have explained in Chapter 4 how to derive sets of special points with the method of Monkhorst and Pack. We use here the points reported in Table 1 of that chapter.

In Table 1, we report the total energy of the system. All the computations were performed at the experimental lattice constant. The absolute value of the total energy is given (in parenthesis, in Rydberg) in the last row and column of the table. This value has converged to within 1 mRy; the other entries of the table are the differences with respect to this value. These differences are expressed in mRy. It is interesting to note that all the entries in the table are positive. The decrease of the energy, with an increase kinetic energy cutoff and therefore, of the dimension of the basis set, is due to the fact that we are looking for the minimum of the total energy functional: an increase of the subspace of the sampling functions improves the estimate of the minimum. On the other hand, the decrease of the energy with the increase of the number of sampling points is not a general feature: it is simply a property of silicon. A second observation concerns the efficiency of the special-point technique. Passing from the Γ point to 2 special points, the error is reduced by two orders of magnitude. To have a convergence error lower than 2 mRy per atom it is necessary to use 6 κ-points and a cutoff of 30 Ry. As the LDA and the PP approximations introduce errors of this order of magnitude, or higher, this precision is sufficient in all practical cases.

Table 1. Errors in the total energy of silicon due to different cutoffs and special-point sampling, at the experimental lattice constant. The differences are in mRy. The converged value of the total energy is also reported in Ry (in parenthesis).

E	Γ	2 pt.	6 pt.	10 pt.
4 Ry	1809.2	517.70	485.11	483.47
12 Ry	1306.7	52.87	40.78	40.11
20 Ry	1263.8	18.78	7.78	6.57
30 Ry	1256.4	14.93	3.87	2.65
40 Ry	1254.8	12.35	1.22	(−15.88870)

In real applications, the absolute convergence of the total energy is not very important. The most important quantities are the energy differences between different states of the solid. Generally, it is possible to obtain a precision higher than 2 mRy, with a cutoff lower than 30 Ry, because the energy differences converge more rapidly than the energy itself.

To illustrate this point, we studied the convergence of two structural properties: the lattice constant and the bulk modulus. These quantities are related to the form of the function $E(\Omega)$ around Ω_0. Here, Ω is the volume of the unit cell and Ω_0 is the volume which minimizes the total energy. The theoretical lattice constant has the value, a_0, which corresponds to Ω_0, while the bulk modulus is related to the curvature of the energy curve at the minimum:

$$B_0 = -\Omega \frac{\partial^2 E}{\partial \Omega^2}\bigg|_{\Omega=\Omega_0} \tag{30}$$

Both the lattice constant and the bulk modulus depend on the difference between total energies at different volumes but not on the absolute value of the energy. The total energy curves computed at various cutoffs differ from one another by a rigid shift but they have the same form. This is the reason why the energy differences converge much faster than the energy itself.

The values of a_0 and B_0 can be computed by evaluating the total energy for several values of a_0 and interpolating these values with the equation of state of the solid. When the volume is changed the number of PWs, which are used to describe the density and the wavefunctions, changes as well if the cutoff is kept constant. There are, therefore, two conceptually different ways of plotting the curve $E(\Omega)$. In one case, the number of PWs is kept constant; in the other case, the kinetic energy cutoff is kept constant. Experience shows that the latter method results in a faster convergence of the lattice constant and the bulk modulus and therefore, we used this method. This faster convergence can be explained by noticing that a constant kinetic-energy cutoff means, in fact, a constant resolution in real space. After the computation of $E(\Omega_i)$, it is necessary to choose an interpolating formula. There are many choices for the interpolating curve. Close to the minimum, the curve is a parabola, but the points deviate rapidly from the quadratic law and, at a larger distance from a_0, it is necessary to fit the points to something more sophisticated. One of the most popular fitting curves is the Murnaghan equation [26] which depends upon three parameters: Ω_0 (the volume of the cell), B_0 (the bulk modulus), B_0' (the derivative of the bulk modulus with respect to pressure):

$$E(\Omega) = \frac{\Omega_0 B_0}{B_0'}\left[\frac{1}{B_0'-1}\left(\frac{\Omega_0}{\Omega}\right)^{B_0'-1} + \frac{\Omega}{\Omega_0}\right] + const \qquad (31)$$

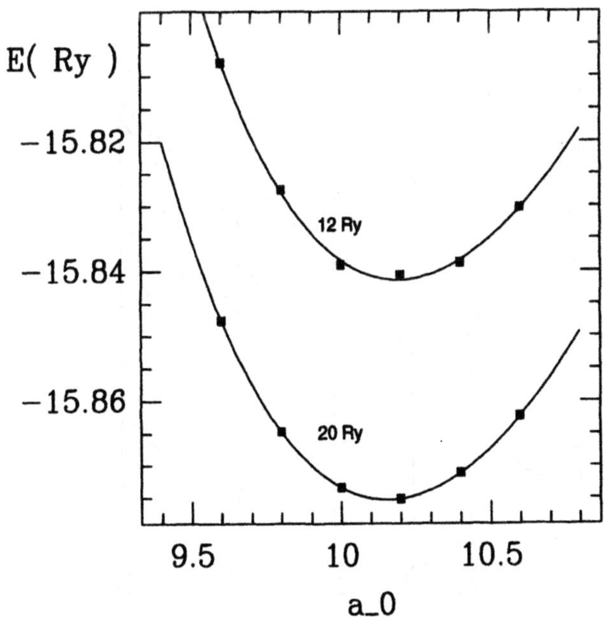

Figure 1. Total energy of silicon for different values of the lattice constant. The numerical values (squares) are interpolated with a Murnaghan equation (continuous lines). The two curves correspond to a cutoff of 12 and 20 Ry.

In Figure 1, we show the energy as a function of the lattice constant, for 6 κ-points, with 12 and 20 Ry cutoffs and the interpolation obtained by using equation 31. In Table 2, we report the values of a_0 and B_0 which have been obtained for silicon. The value at 6 κ-points and 20 Ry is already converged and therefore, we do not report the other calculations. The lattice constants are given in a.u. and the bulk moduli are given in kbar.

Table 2. Lattice constant and bulk modulus of silicon, $a_0 - B_0$ as a function of cutoff and number of special points (n pt.): the lattice constant is expressed in atomic units and the bulk modulus in kbar. The experimental values are 10.26 – 990.

cutoff	2 pt.	6 pt.
4 Ry	9.65 – > 2000	10.08 – 1586
12 Ry	10.16 – 952	10.19 – 976
20 Ry	10.17 – 953	10.16 – 960

This table shows that it is possible to obtain meaningful results for the structural properties, studying silicon with even a 12 Ry cutoff, a cutoff is sometimes used in simulations with large unit cell.

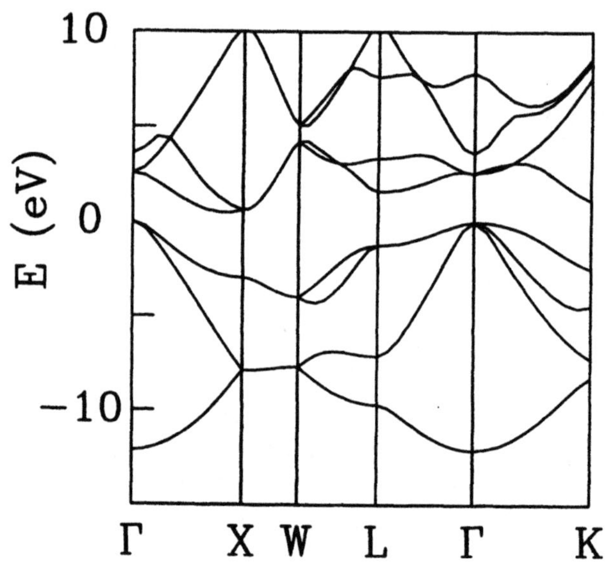

Figure 2. Band structure of silicon along several symmetry lines. The lowest 8 bands are plotted.

We now study the energy band structure of silicon. Although the eigenvalues of the Kohn and Sham equations are not to be considered as the energy of the electrons, it is a common practice to compare these eigenvalues with the results of photoemission experiments. In many cases, the results are good. While the energy gap of semiconductors and insulators is underestimated, the form of the bands and their width is correct. Experimentally, the energy bands of a solid are known to a precision of the order of 0.1 eV. In silicon, the eigenvalues converge within this error, as fast as the total energy. Our results have been obtained with 6 κ-points and 20 Ry.

In Figure 2, we show with a continuous line, the energy eigenvalues of the lowest 8 bands along several symmetry lines. The computations are performed at the theoretical lattice constant. The direct gap is 2.5 eV, the indirect gap is 0.4 eV, less than one half of the experimental value.

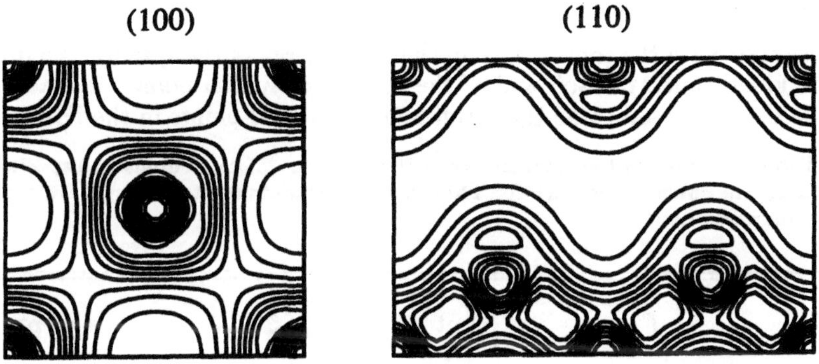

Figure 3. Electronic pseudo-charge density of silicon, in the (100) and (110) planes.

The pseudo-charge density of this system, in the (100) and (110) planes, is shown in Figure 3. This charge density has been plotted on a contour map by dividing its range of variation into 10 equal values. The charge in the (110) plane shows several maxima, corresponding to the covalent bonds between the atoms.

7. NaCl

NaCl is an ionic crystal, composed of an alkaline metal, sodium, whose atom has an electronic configuration $[Ne]3s^1$ and a halogen, chlorine, with electronic configuration $[Ne]3s^23p^5$. Each sodium atom gives an electron to a chlorine atom, thereby becoming a positive ion and transforming the chlorine into a negative ion. The ions attract each other by Coulomb forces. The equilibrium distance is given by the balance between the long-range Coulomb interactions and the short-range repulsion between closed-shell ions. These short-range forces are due to the interaction of the neon-like core of sodium with that of the chlorine ion.

The PP of sodium given in Ref. [13] is generated assuming that only the $3s^1$ is a valence electron. It is clear that, with this approximation, the PP calculation cannot provide a correct value for the lattice constant and bulk modulus, as it does for silicon. In fact, the short-range repulsion, which balances the long-range Coulomb forces, involves the $2p$ electrons of sodium, which are missing in this PP scheme. To study an ionic crystal composed of an alkaline metal and a halogen, it is necessary to use a PP for the metal atom which includes part of the core in the valence charge. With this technique, phonon spectra of some alkali halides have also been computed [27].

In this example, we limit ourselves to the use of the PP of Ref. [13]. We studied the convergence of the total energy, the band structure of the system and the electronic charge density. We performed our calculations at the experimental lattice constant.

The NaCl crystal can be described by an fcc Bravais lattice, with two atoms per unit cell. One atom is at the origin while the other is at $\mathbf{d} = a_0(0.5, 0.5, 0.5)$. We put the sodium atom at the origin. There are 8 electrons per unit cell, and they fill the four valence bands which are mainly derived from chlorine $3s$ and $3p$ levels. The experimental value of a_0 is 10.66 a.u..

We start with the convergence of the total energy of the system, as a function of the kinetic energy cutoff and of the number of special points. In Table 3, we report the results up to 80 Ry cutoff and up to 10 κ-points in the IBZ.

Table 3. Errors in the total energy of NaCl due to different cutoffs and special-point sampling. The differences are in mRy. The converged value of the total energy is also reported, in Ry (in parenthesis).

E	Γ	2 pt.	6 pt.	10 pt.
12 Ry	787.22	368.07	367.59	369.14
20 Ry	440.37	78.93	78.89	—
40 Ry	387.43	7.62	—	—
60 Ry	371.25	2.86	—	—
80 Ry	368.82	(-30.81859)	—	—

The total energy converges very slowly as a function of the kinetic energy cutoff. Note the different scale of the kinetic energy cutoffs, with respect to Table 1. At the experimental lattice constant used here, a cutoff of 80 Ry corresponds to about 3700 PWs for the wavefunctions and 27000 PWs for the charge density. On the other hand, NaCl is an ionic solid and the energy bands are very flat and close to atomic-like levels. Therefore, all the quantities converge rapidly with respect to κ-point sampling. In this table, we have not reported all the entries relative to 10 and 6 special points because they coincide with the results obtained with 2 κ-points.

The lowest 8 energy bands of the system, along several symmetry lines in the BZ, are shown in Figure 4. The bands have been produced with a cutoff energy of 60 Ry and 2 points in the IBZ.

As expected, the energy dispersion of the valence bands is very low, when compared with the dispersion of a covalent solid like silicon. The gap is direct at the Γ point. Its theoretical value is 4.7 eV.

In Figure 5, we show the electronic charge density in the (100) and (110) planes. As expected, the chlorine is perfectly spherical, while the sodium is not visible on the plot because no valence charge is left on it.

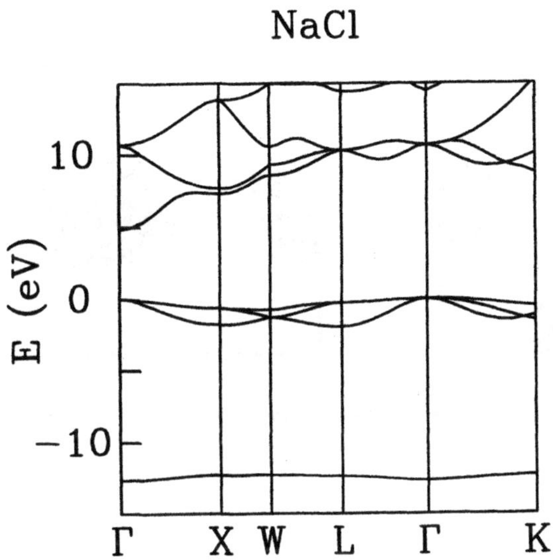

Figure 4. Band structure of NaCl along several symmetry lines. The lowest 8 bands are plotted.

8. Aluminium

The electronic configuration of the aluminium atom is $[Ne]3s^2 3p^1$. Crystalline aluminium is a group IIIA metal. Its Fermi surface is close to the free electron surface for an *fcc* Bravais lattice with three conduction electrons per atom. It has one filled *s* band and partially filled *p* bands. The electronic charge density can be described by a smooth function, which is represented quite efficiently by PWs. The neon-like core is tightly bound to the nucleus so that the frozen core approximation is justified. The LDA approximation works well in the case of metallic bonds and the PP-PW technique can be used efficiently to study this system. The main difficulty, here, is the description of the Fermi surface: the special-point technique must be used with care, in this case. However, this problem is common to all techniques used to study a metal.

The aluminium crystal can be described by an *fcc* Bravais lattice with one atom per cell. One parameter, the edge of the conventional cubic cell, is sufficient to specify the dimensions of the Bravais lattice. Experimentally, $a_0 = 7.66$ a.u. A good PP for aluminium can be obtained from the parameters of Ref. [13].

We illustrate this system with the same approach used for silicon. We start with the study of the convergence of the total energy as a function of the cutoff energy and of the special-point sampling. For a metal, it is necessary to use a smearing parameter to avoid the effects of the Fermi surface. We used the method explained in Ref. [21] and briefly summarized in Chapter 6. With this method, it is possible to obtain a converged value of the energy with a large smearing parameter, without adding any correction term to the total energy functional. Tables 4 and 5 give the convergence of the total energy. In both tables, we report the difference of the energy, in mRy, with respect to the converged value which

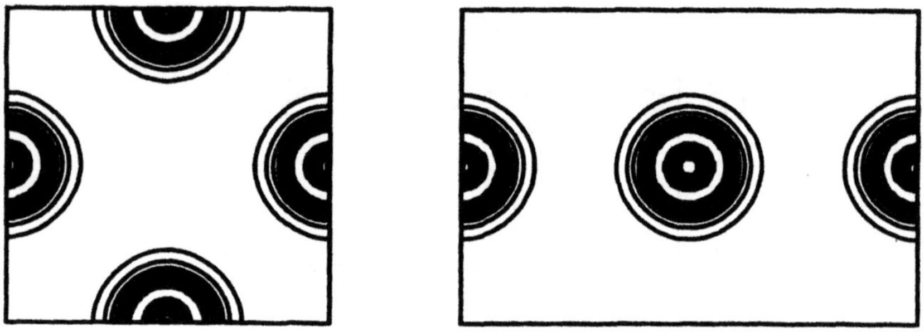

Figure 5. Electronic pseudo-charge density of NaCl in the (100) and (110) planes.

is given in Ry in the last row and column of each table. In the first case, we use a value of the smearing parameter Δ (see section 7 of Chapter 6) equal to 0.2 Ry (this value is also used in actual computations). In this case, the total energy converges rapidly with respect to the number of κ-points.

Table 4. Errors in the total energy of aluminium due to different cutoffs and special-point sampling. The differences are in mRy. The converged value of the total energy is also reported in Ry (in parenthesis). The calculations have been performed with Δ=0.2 Ry.

E	6 pt.	10 pt.	28 pt.	60 pt.
12 Ry	2.09	4.38	4.09	4.05
20 Ry	0.83	1.49	1.14	1.13
30 Ry	−1.98	0.34	0.00	(−4.20680)

Table 5. Errors in the total energy of aluminium due to different cutoffs and special-point sampling. The differences are in mRy. The converged value of the total energy is also reported in Ry (in parenthesis). The calculations have been performed with Δ=0.01 Ry.

E	6 pt.	10 pt.	28 pt.	60 pt.
12 Ry	−4.02	8.03	4.91	4.03
20 Ry	−6.89	5.19	1.99	1.14
30 Ry	−8.04	4.06	0.86	(−4.20725)

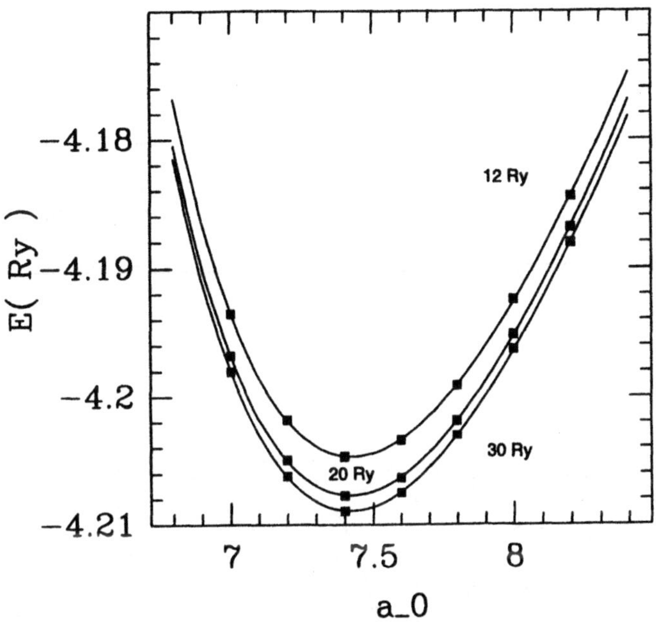

Figure 6. Total energy of aluminium for different values of the lattice constant. The numerical values (squares) are interpolated with a Murnaghan equation (continuous lines). The cutoffs are indicated on the figure.

In practice, 10 special points gave the same value for the energy as 28 or 60 points. For the first time, we found some negative values for the total energy differences. We have already noted that the variational properties of the total energy functional guarantee only that the energy decreases with increasing number of basis functions. Aluminium is a system where an increase in the number of special points may increase the energy.

In Table 5, we report the values of the total energy obtained with $\Delta = 0.01$ Ry. The energy is converged also with respect to the smearing parameter at this value. In this case, the convergence with respect to the number of κ-points is worse.

The converged values of the total energy, in the two cases, differ by 0.45 mRy but, using a large Δ, we can perform the calculations with only 10 special points.

We then studied the convergence of the lattice constant and of the bulk modulus of this system, performing all the calculations with $\Delta = 0.2$ Ry. We used the same method already explained for silicon.

The results are reported in Table 6. In Figure 6, we show the interpolations of the total energy as function of the lattice parameter with the Murnaghan equation.

Table 6. Lattice constant and bulk modulus of aluminium, $a_0 - B_0$ as a function of cutoff and number of special points (n pt.): the lattice constant is expressed in atomic units and the bulk modulus in kbar.

n pt	12 Ry	20 Ry	30 Ry
6	7.39 – 901	7.39 – 900	7.39 – 900
10	7.43 – 845	7.42 – 853	7.41 – 857

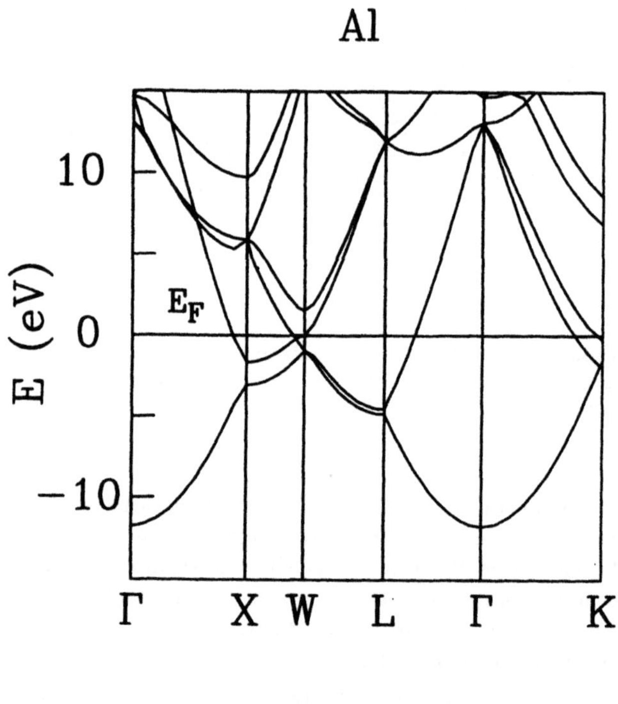

Figure 7. Band structure of aluminium along several symmetry lines. The lowest 10 bands are plotted.

The convergence of these quantities with respect to the cutoff energy is very fast, but at least 10 special points are necessary to obtain convergence of the bulk modulus. The agreement with respect to experiment is quite good. The error on the lattice constant is only 2%. The experimental values of the bulk modulus are in the range $720 < B_0 < 880$ kbar. Our result is within the experimental range.

Figure 8. Electronic pseudo-charge density of aluminium, in the (100) and (110) planes.

We then studied the band structure of the solid at the theoretical lattice constant. In Figure 7, we show the energy eigenvalues of the lowest 10 bands along several symmetry lines in the BZ. The energy zero is set at the Fermi energy. The bands have been obtained by a self-consistent calculation with 60 special points, $\Delta = 0.2$ Ry and 30 Ry cutoff.

The pseudo-charge density of this system, in the (100) and (110) planes, is shown in Figure 8. This charge density has been plotted on a contour map by dividing its range of variation into 10 equal values.

9. Conclusions

We have described the PWSCF code and we have given examples of its applications to three physically different solids: a semiconductor (Si), an insulator (NaCl) and a metal (Al). After describing the theoretical framework on which the code is based, we have presented a description of the main routines and of the input data. We have shown that, for all systems that we have considered, the code can compute the ground-state electronic properties. This is the starting point for the computation of other physical properties whose evaluation can be performed very efficiently with a PW basis set. We have shown that, by a careful study of the convergence, it is possible to reduce the numerical noise due to the finite basis set and the discrete sampling of the BZ, below the experimental error. All the remaining discrepancies between theory and experiment are due to the physical approximations, which are used in the theoretical formulation of the electronic problem. These errors must always be accounted for in the study of a given system, but the results of a DFT-LDA computation can be considered as a physically meaningful starting point which, without any empirical parameters, can give important insights into the physics and chemistry of materials.

References

1. A copy of the code can be obtained also by sending an e-mail to `dalcorso@eldpa.epfl.ch`.
2. P. Giannozzi, S. de Gironcoli, P. Pavone, and S. Baroni, Phys. Rev. B **43**, 7231 (1991).
3. S. de Gironcoli, S. Baroni, and R. Resta, Phys. Rev. Lett. **62**, 2843 (1989).
4. S. Baroni, P. Giannozzi, and A. Testa, Phys. Rev. Lett. **58**, 1861 (1987); *ibid.* **59**, 2662 (1987).
5. A. Dal Corso, R. Resta, and S. Baroni, Phys. Rev. B **47**, 16252 (1993).
6. S. de Gironcoli, S. Baroni, and R. Resta, Ferroelectrics **111**, 19 (1990).

7. A. Dal Corso, S. Baroni, R. Resta, and S. de Gironcoli, Phys. Rev. B **47**, 3588 (1993).
8. A. Dal Corso, S. Baroni, and R. Resta, Phys. Rev. B **49**, 5323 (1994).
9. A. Dal Corso and R. Resta, Phys. Rev. B **50**, 4327 (1994).
10. P. Hohenberg and W. Kohn, Phys. Rev. **136**, 864 (1964).
11. W. Kohn and L.J. Sham, Phys. Rev. **140**, 1133 (1965).
12. W.E. Pickett, Computer Phys. Reports **9**, 115 (1989).
13. G.B. Bachelet, D.R. Hamann, and M. Schlüter, Phys. Rev. B **26**, 4199 (1982).
14. J. Ihm, A. Zunger, and M. L. Cohen, J. Phys. C **12**, 4409 (1979).
15. E.R. Davidson, Computer Phys. Commun. **53**, 49 (1989).
16. U. von Barth and C.D. Gelatt, Phys. Rev. B **21**, 2222 (1980).
17. D.M. Ceperley and B.J. Alder, Phys. Rev. Lett. **45**, 566 (1980); J. Perdew and A. Zunger, Phys. Rev. B **23**, 5048 (1981).
18. C. Lee, D. Vanderbilt, K. Laasonen, R. Car, and M. Parrinello, Phys. Rev. B **47**, 4863 (1993).
19. D. Vanderbilt and J.D. Johannopoulos, Phys. Rev. B **27**, 6296 (1983).
20. See Chapter 4, this issue.
21. M. Methfessel and A.T. Paxton, Phys. Rev. B **40**, 3616 (1989).
22. D.G. Anderson, J. Assoc. Comput. Mach. **12**, 547 (1965).
23. B.T. Smith, *Matrix Eigensystem routines–EISPACK guide*, Berlin, (1979).
24. M. Methfessel and A.T. Paxton, Phys. Rev. B **40**, 3616 (1989).
25. M.T. Yin and M.L. Cohen, Phys. Rev. B **26**, 5668 (1982).
26. F.D. Murnaghan, Proc. Nat. Acad. Sci. USA **50**, 697 (1944).
27. M.B. Nardelli, S. Baroni and P. Giannozzi, Phys. Rev. Lett. **69**, 1069 (1992).
28. S. G. Louie, S. Froyen and M. L. Cohen, Phys. Rev. B **26**, 1738 (1982).
29. S. Froyen and M.L. Cohen, Phys. Rev. B **29**, 3770 (1984).

Total Energy and Related Properties

Roberto Dovesi

Department of Inorganic, Physical and Materials Chemistry, University of Torino, via Giuria 5, I-10125 Torino, Italy

September 6, 1996

Summary. Computational aspects related to the evaluation of total energy and energy-related properties with quantum mechanical programs are discussed; examples are reported, with reference to the Hartree-Fock periodic LCAO CRYSTAL code.

Key words: Total Energy – Equilibrium Geometry – Phase Transitions – Geochemistry – Elastic properties – Basis set – Numerical accuracy – Comparison with Experiment.

1. Introduction

The ground-state total energy and its derivatives are among the most important properties that can be obtained from *ab initio* quantum mechanical methods when applied to solids; many aspects of the crystalline behaviour can be understood when the total energy of the system is known, or, more often, when the energy difference between two or more electronic or nuclear, competitive configurations is known as a function of some parameters. The energy differences of interest are, as a rule, very small (of the order of 1 kcal/mol or less), whereas the total electronic energy of the system can be five to nine orders of magnitude larger; the problem of the accuracy in the determination of the total energy of the system (and its derivatives) is then of crucial importance. Some aspects of this problem will be discussed in the next section. In section 3, the use of the total energy for the calculation of many quantities of interest, such as the formation energy of a crystalline compound or the relative stability of the ferro- and antiferromagnetic phases of a magnetic insulator, is discussed. Derivatives of the total energy of the system with respect to some parameters are also very important. In section 4, results concerning the equilibrium geometry are reported;

section 5 contains a discussion on how to obtain the equation of state of a crystalline compound and how to compare the enthalpy $H = E + PV$ of two phases as a function of pressure; this section is of particular relevance for the simulation of problems of geological interest. Among the properties related to energy second derivatives, the elastic constants and vibrational spectrum are the most important. In section 6, examples of calculated bulk moduli and elastic constants are reported, while vibrational properties (phonons) will be discussed in Chapter 12. In section 7, some aspects of the correlation problem are discussed; section 8 is devoted to the basis set and section 9 to numerical problems encountered when performing calculations with the Hartree–Fock LCAO program CRYSTAL. Finally, in section 10, some problems involved in the comparison of calculated with experimental results are tackled.

The numerical results to be presented have been obtained with the codes CRYSTAL92 and CRYSTAL95 (see Chapter 8). Energy derivatives have been calculated numerically, point by point. It is possible to evaluate, analytically, energy derivatives within the Hartree–Fock (HF) or the density functional (DF) formalism and this is a standard rule in molecular quantum mechanics [1]. In periodic programs and in plane-wave DF codes, such as the PWSCF program discussed in Chapter 10, such an implementation is relatively trivial, while it is more difficult for codes which use a different basis set: this facility has just been introduced in WIEN (Chapter 9), and its implementation in CRYSTAL (Chapter 8) is still in progress.

2. The problem of numerical accuracy

The total energy of the system, as obtained with a quantum mechanical program, such as WIEN, PWSCF or CRYSTAL, is affected by errors that, schematically, can be classified into four types.

The first source of error is related to the choice of the Hamiltonian; in WIEN and PWSCF, where a DF scheme is adopted, different values of the total energy are obtained, according to the selected exchange-correlation functional: many different options are possible (see, for example, [2] for a list of these). The relative merits and limitations of the various functionals is still under discussion (see for example [3, 4] for applications to solids and [2, 5] for molecules) and the residual error with respect to the "exact" total energy in solids is not easily defined, due to the presence of other sources of error. In the case of the HF approach, the deficiency is due to the so-called *correlation effects*, that is, to lack of correlation in the motion of different electrons (in the HF scheme, each electron moves in the mean field created by the other electrons). This error affects all the computational quantities; it is particularly large when bonds are broken, that is, when a *closed-shell* configuration is compared with an *open-shell* one, as is the case for the evaluation of the binding energy of a molecule or a crystalline solid from atoms. Many-body techniques are used in molecular quantum chemistry [1, 6] for improving the HF results, which are not yet available for periodic systems. An *a posteriori* (that is after the HF-SCF cycle) estimate of the correlation energy with a *correlation-only* functional [7], using the HF charge density is usually adopted in CRYSTAL, which recovers most of the correlation effects in the crystalline formation energies [8, 9, 10]. As will be discussed below, this correction is less useful in the calculation of other energy-related properties, such as the equilibrium geometry or the bulk modulus [9].

The second source of error is related to the use of a finite variational basis set. Convergence of the total energy and related properties, as a function of the number of plane waves is discussed in Chapter 10, where the PWSCF code is presented. As expected, the convergence is very fast for metals (core electrons must be eliminated by using pseudopotentials), but very slow for ionic systems. However, when Bloch functions constructed with *atomic orbitals* (AOs) are used, a reasonable description of delocalized electrons is very expensive and, in some cases, difficult, due to problems related to the non-orthogonal nature of the basis functions (see the discussion in section II.8 of reference [11]), whereas localized electrons (including core electrons) are easily described by relatively small numbers of Gaussian functions. Some examples will be given concerning ionic and semi-ionic compounds.

The third source of error is related to the numerical approximations introduced in the solution of the HF or DF equations. In each computational scheme, numerical integrations are performed and series expansions are truncated that influence to some extent the results. In the CRYSTAL implementation of the HF-LCAO scheme, for example, terms of the infinite Coulomb and exchange series are approximated, in different ways according to their importance and small terms are neglected. The reciprocal space integration necessary for the calculation of the density matrix, at each cycle of the SCF process, is performed numerically, with reference to a finite set of reciprocal space points. A more explicit description of these aspects is given in Chapter 8, where the CRYSTAL code is discussed. The influence of the computational features on the calculated quantities is documented in section 9; other examples can be found in references [12, 11].

The last source of inaccuracy is related to real errors in the computer codes ("bugs"); this aspect is often underestimated, probably because it is the most trivial one. In this area, however, we are still not too far from the pioneering stage of "one researcher, one program"; most of the programs are not public, and are under continuous modification; among the (very few) public codes, cross checks are impossible, because they differ in the hamiltonian and/or type of basis set adopted, as is the case of the three codes discussed in this book.

The absolute error in the total energy, resulting from the above four possible sources is much larger than the relevant energy differences for many chemical and physical transformations. Fortunately, in many cases most of the errors cancel in performing these energy differences; the most trivial example is the error related to an approximate (but *reasonable*) description of inner (*core*) electrons, whose wavefunction remains essentially unaltered through any chemical modification involving the atoms to which they belong.

We report in Table 1 the lattice parameter and the bulk modulus (see below) of NaCl resulting from *ab initio* calculations performed in the last decade, as an example of the possible range of calculated values which may be produced, for simple properties of a model system, even when the same Hamiltonian is used.

Method	Functional	Ref.	year	$a(\AA)$	B(GPa)
LDA-PS-PW	Wigner	[13]	1986	5.52	31.2
LDA-FLAPW	Hedin-Lundqvist	[14]	1986	5.64	30.4
LDA-LAPW	Wigner	[15]	1987	5.54	28.7
LDA-LAPW	Hedin-Lundqvist	[15]	1987	5.47	31.4
LDA-Numerical	Ceperley-Alder	[16]	1991	5.48	31.5
LDA IONIC	Ceperley Alder	[17]	1991	5.75	23.2
LDA ASA+E	Hedin-Lundqvist	[18]	1992	5.49	33.0
HF - AIPI		[19]	1992	6.11	14.8
HF - LCAO		[20]	1993	5.80	22.8
HF - LCAO+P91		[20]	1993	5.54	30.1
Experimental		[21]	1972	5.57	28.5

Table 1. Calculated lattice parameter and bulk modulus of NaCl from recent papers. For the meaning of the acronyms, see the indicated references

3. The formation energy and other energy differences

In Table 2, some of the energy differences of interest are listed, with appropriate references, where examples and comparison with experimental data may be found. It should be noted that, in many cases, systems of different dimensionality (D) are involved: 3D and 2D for the evaluation of the surface energy, 2D and 0D for the chemisorption on a surface, 3D and 0D for the binding energy of a molecular or ionic crystal and so on. In performing calculations, the question must be raised concerning the above-mentioned cancellation of errors of different origin for the two systems to be compared.

With reference to the examples reported in Table 2, and to the CRYSTAL code, the following comments can be made:

i) The accuracy of the code is approximatively the same for 3D, 2D, 1D and 0D systems, so that errors of the third type (see previous section) are negligible.

Computed Energy	System 1	System 2	Example	Reference
Binding	bulk	atoms	ionics,covalent	[22], [8]
Binding	bulk	molecule	Urea, Ice	[23], [24]
surface	bulk	slab	MgO	[25]
surface relax	slab	slab	MgO	[25]
adsorption	slab+molecule	slab,molecule	CO on LiF	[26]
substitution	defective solid	solid, atoms	C in Si	[27]
superexchange	AFM bulk	FM bulk	NiO	[28]
solid state reaction	bulk	bulk	$MgO + Al_2O_3 \rightarrow MgAl_2O_4$	[29]

Table 2. Examples of total energy differences and related references. In all cases cited here, the CRYSTAL code has been used for the calculation. AFM and FM stand for antiferromagnetic and ferromagnetic, respectively.

		N		P		As		Sb	
		E	%	*E*	%	*E*	%	*E*	%
B	HF	9.1	-33.1	6.6	-36.7	5.6			
	HF+c	13.1	-3.8	10.2	-1.9	9.2			
	Expt.	13.6	--	10.4	--				
Al	HF			5.7	-32.4	4.9	-38.3	4.0	-39.5
	HF+c			8.7	+3.3	7.9	-1.3	6.8	+3.0
	Expt.			8.4	--	8.0	--	6.6	--
Ga	HF			4.2	-41.4	3.5	-47.3	3.0	-50.2
	HF+c			7.2	+0.0	6.5	-2.8	5.8	-3.2
	Expt.			7.3	--	+6.6	--	+6.0	--
In	HF			3.9	-41.5	3.5	-44.5	2.8	-49.6
	HF+c			6.7	+1.0	6.3	-0.2	5.5	-0.9
	Expt.			6.6	--	6.3	--	5.6	--

E	%	
10.4	-30.8	C
14.7	-2.4	
15.1	--	
6.3	-34.0	Si
9.5	-0.7	
9.5	--	
4.4	-43.5	Ge
7.5	-4.2	
7.8	--	
4.2	-34.2	Sn
7.1	+10.7	
6.4	--	

Table 3. Calculated and experimental binding energies (in eV) of the III-V (left) and IV-IV (right) semiconductors with the zincblende (or diamond) structure. Percentage errors with respect to experiment are given in the columns headed "%". HF+c is the Hartree-Fock energy corrected *a posteriori* with the P86 functional [7]. For the basis sets and other computational details see ref [22].

ii) Errors of the fourth type can reasonably be excluded.

iii) The total energy per formula unit is very stable when evaluated with unit cells of increasing size (for example, the total energy of MgO bulk is the same to the fifth decimal figure whether evaluated with unit cells containing 2, 8, 16, 32, 64 or 128 atoms [30]); this aspect is very important (but often underestimated) for the evaluation of defect formation energies with the supercell scheme [30, 27] and when the relative stability of different magnetic phases is computed [31, 28].

iv) The accuracy for a given set of computational parameters is NOT the same for ionic, covalent and metallic systems (see sections II.4 and II.5 of [11]).

v) The efficiency of a given basis set changes with the local symmetry and the dimensionality of the system; so, for example, in the evaluation of the binding energy of NaCl, two different basis sets are used for atoms and for bulk; as a rule [9], the same basis set is used for inner electrons, whereas extra functions must be added to the Na^+ bulk basis set for an accurate description of the valence electron of the isolated Na atom. This point is discussed in section 8.

vi) In each process involving the formation and breaking of bonds, correlation effects, disregarded at the HF level, entail systematic errors (about 30% for the formation energy of covalent systems) which may be corrected, as discussed above. More generally, the correlation error affects all the observables evaluated at the HF level; the error, as will be discussed below, is in some way systematic, although of varying importance for different properties and for different families of compounds.

In Tables 3, 4 and 5, the calculated formation energies of covalent and ionic crystalline compounds are compared with the corresponding experimental results. As anticipated, the HF formation energy is always smaller than the experimental one, the percentage error ranging from −20% to −50%. The error is larger for heavy atoms for two reasons. Firstly, because dispersion effects (not taken into account at the HF level) are larger, and secondly, electrostatic effects (correctly described at the HF level) are less important, due to larger lattice parameters. The *a posteriori* correlation correction with the P86 or P91 functionals (they provide very similar results) recovers most of the missing energy, and the final error is never larger than 10%.

		F		Cl		Br		I	
		E	%	*E*	%	*E*	%	*E*	%
Li	HF	6.74	−23.2	5.66	−20.5	4.95	−23.9	4.25	−23.4
	HF+c	8.47	−3.6	7.40	+4.0	6.59	+1.4		
	Expt.	8.78		7.12		6.50		5.55	
Na	HF	5.93	−23.3	5.41	−18.2	4.83	−20.6	4.23	−18.5
	HF+c	7.54	−3.8	6.93	+4.8	6.30	+3.6		
	Expt.	7.83		6.61		6.08		5.19	
K	HF	5.54	−27.0	5.54	−17.3	5.20	−16.7	4.56	−16.0
	HF+c	7.12	−6.2	6.97	+4.0	6.48	+3.8		
	Expt.	7.59		6.70		6.24		5.43	
Rb	HF	5.29	−30.4	5.41	−18.2	4.94	−20.8	4.48	−17.3
	HF+c	6.74	−11.2	6.80	+2.9	6.27	+0.5		
	Expt.	7.60		6.61		6.24		5.42	

Table 4. Calculated and experimental binding energies (in eV) of alkali halides with the NaCl structure. Percentage errors with respect to experiment are given in the columns headed "%". HF+c is the Hartree-Fock energy corrected *a posteriori* with the P86 functional [7]. For the basis sets and other computational details see references [9, 10].

System	Ref	Binding Energy				
		HF	%	HF + c	%	Expt.
BeO	[8]	9.22	-24	11.59	-4	12.14
MgO	[8]	7.32	-29	9.69	-6	10.28
CaO	[8]	7.59	-30	10.10	-7	10.86
$CaCO_3$	[32]	20.76	-30	27.35	-7	29.58
$MgCO_3$	[33]	19.48	-31	26.12	-8	28.24
Al_2O_3	[29]	23.65	-26	30.54	-5	32.14
$MgAl_2O_4$	[29]	31.67	-26	40.50	-5	42.71

Table 5. Calculated and experimental formation energies(in eV) of simple oxides. Percentage errors with respect to experiment are given in the columns headed "%". HF+c is the Hartree-Fock energy corrected *a posteriori* with the P91 functional [7].

4. The equilibrium geometry

In Tables 6 and 7, the calculated equilibrium geometry for the families of semiconductors and alkali halides, discussed in the previous section, are reported. In both cases, the energy minimization is simple, because only one parameter (the cell edge) must be optimized.

In the case of semiconductors, the lattice parameter is always overestimated; the error ranges between 0 and +2%; it is negligible for BN and diamond; then it remains roughly constant going down the periodic table from Al to In and from

		N		P		As		Sb	
		a	%	*a*	%	*a*	%	*a*	%
B	HF	3.62	0.0	4.60	+1.3	4.83	+1.0		
	Expt.	3.62		4.54		4.78			
Al	HF			5.52	+1.3	5.76	+1.8	6.22	+1.5
	Expt.			5.45		5.66		6.13	
Ga	HF			5.56	+2.0	5.76	+1.9	6.23	+1.8
	Expt.			5.45		5.65		6.12	
In	HF			5.93	+1.0	6.15	+1.5	6.56	+1.4
	Expt			5.87		6.06		6.47	

a	%	
3.58	+0.3	C
3.57		
5.46	+0.6	Si
5.43		
5.79	+2.3	Ge
5.66		
6.49	0.0	Sn
6.49		

Table 6. Calculated and experimental lattice parameter *a* (in Å) of the III-V (left) and IV-IV (right) semiconductors with the zincblende (or diamond) structure. Percentage errors with respect to experiment are given in the columns headed "%". For experimental references, basis sets and other computational details, see ref [22].

		F		Cl		Br		I	
		a	%	*a*	%	*a*	%	*a*	%
	HF	4.02	+0.8	5.28	+4.1	5.73	+5.7	6.32	+6.9
Li	HF+c	3.85	-3.5	5.03	-0.8	5.44	+0.4		
	Expt.	3.99		5.07		5.42		5.91	
	HF	4.63	+1.3	5.80	+4.1	6.23	+5.6	6.79	+6.3
Na	HF+c	4.44	-2.8	5.52	-0.9	5.92	+0.3		
	Expt.	4.57		5.57		5.90		6.39	
	HF	5.49	+3.0	6.59	+5.6	7.03	+7.7	7.53	+7.9
K	HF+c	5.20	-2.4	6.23	0.0	6.59	+0.9		
	Expt.	5.33		6.23		6.53		6.98	
	HF	5.92	+5.9	7.00	+6.7	7.43	+9.3	7.95	+9.7
Rb	HF+c	5.63	+0.7	6.65	+1.3	7.02	+3.2		
	Expt.	5.59		6.56		6.80		7.25	

Table 7. Calculated and experimental lattice parameters (in Å) of alkali halides with the NaCl structure. Percentage errors with respect to experiment are given in the columns headed "%". HF+c is the Hartree-Fock energy corrected *a posteriori* with the P86 functional [7]. For basis sets and other computational details, see ref [9, 10].

P to Sb. The lattice parameter is also overestimated, in the case of the alkali halides; the trend along the series from light to heavy atoms is however very different from that shown by semiconductors. The error increases systematically with increasing size of the cation and the anion (from +0.8% for LiF to +9.7% for RbI). The reason for this difference is the much higher *relative* importance of dispersion and polarization contributions in alkali halides with respect to semiconductors; the former systems are characterized by very flat energy versus cell volume curves, with bulk moduli (see section 6) ranging from 77 (LiF) to 13 (RbI) GPa, whereas for semiconductors, the extremes are at 440 (diamond) and 46 (InSb) GPa.

In semiconductors, the equilibrium geometry and the stiffness of the E(V) curve is determined by strong covalent bonds, which are reasonably well described at the Hartree-Fock level; the overestimation of 1-2% is in line with the Hartree-Fock results for molecular systems, characterized by essentially covalent bonds. In alkali halides, the main contribution to the attractive part of the $E(V)$

System	Ref.	*a*			*c*			fractionary coord.		
		calc.	exp.	%	calc.	exp.	%	type	calc.	exp.
MgO	[29]	4.191	4.195	−0.1						
CaO	[28]	4.864	4.811	+1.1						
MnO	[28]	4.526	4.445	+1.8						
NiO	[28]	4.265	4.168	+2.3						
$MgAl_2O_4$	[29]	8.056	8.060	-0.1				O_x	0.264	0.262
$MgCO_3$	[32]	4.618	4.632	-0.3	15.114	14.991	+0.8	O_x	0.275	0.278
$CaCO_3$	[32]	5.045	4.995	1.0	17.504	16.920	+3.5	O_x	0.252	0.257
α–Al_2O_3	[29]	4.742	4.750	-0.2	12.936	12.96	-0.2	O_x	0.306	0.306
								Al_z	0.353	0.352
α–Fe_2O_3	[34]	5.112	5.035	+1.5	13.820	13.747	+0.5	O_x	0.302	0.306
								Fe_z	0.354	0.355
SiO_2	[35]	4.126	4.177	-1.2	2.6896	2.6651	+0.9	O_x	0.305	0.306
$MgSiO_3$	[35]	4.713	4.720	-0.01	13.454	13.520	-0.3	Si_z	0.156	0.158
ilmenite								Mg_z	0.360	0.360
								O_x	0.320	0.321
								O_y	0.036	0.036
								O_z	0.240	0.241

Table 8. Calculated and experimental equilibrium geometry (in Å) of simple oxides. % is the percentage error; SiO_2 data refer to the stishovite phase.

curve comes from electrostatic interactions, which, however, are not extremely large, because only ±1 charges and relatively large distances are involved. As the atomic number of the ions increases, interelectronic and intraelectronic correlation effects, disregarded at the Hartree-Fock level, become more important due to:

- the larger polarizability of the ions, resulting from their larger dimension;
- the larger distance of the valence electrons from the nucleus, resulting from the screening of and orthogonality to inner shells, while electrostatic forces become less important, as a consequence of the large cation-anion distance.

In Table 8, the calculated equilibrium geometry of a number of ionic or semi-ionic oxides is reported; the error for linear parameters is, in all but two cases, smaller than 2%.The results for oxides agree much better with experiment than do those for alkali halides, for the following reasons:

a) Charges are larger (for example, ±2 in MgO, CaO and NiO), so that electrostatic forces, correctly described at the HF level, play a more fundamental role with respect to dispersion effects.
b) The ions present low polarizability (all the cations appearing in Table 8 belong to the first three rows of the periodic table).
c) Some of the systems of Table 8 show partially covalent character.

As a matter of fact, the geometrical structure of most important families of compounds, such as simple and transition metal oxides, silicates and semiconductors, can be described using the HF method, with a relatively high accuracy.

5. Equation of state and phase transitions.

In many cases, it is interesting to explore the behaviour of crystalline systems as a function of pressure or volume. In geophysics, for example, simulation (quantum mechanical or semi-classical) is used to investigate the relative stability and possible solid-state chemical reactions of many silicates, which are supposed to be the most abundant components of the earth's mantle [36], extending to a depth of about 2900 Km. Pressure increases with depth and is believed [37] to vary from 10 to 30 GPa in the very important ***transition zone*** between the ***upper*** and ***lower mantles***; at the lower mantle edge, it is supposed to be as large as 150 GPa. There is no direct experimental information on the chemical composition and phase-stability order in the transition zone (extending roughly from 400 to 700 Km) and lower mantle, because erupted materials come from depths of the order of 100 Km only. Most of our knowledge is, therefore, indirect and comes from high pressure experiments with diamond anvil cells and computer simulation. The two techniques, whose reliability is limited by many factors of different origin, must be considered complementary tools for the investigation of the high pressure behaviour of minerals. In high pressure experiments, P and T

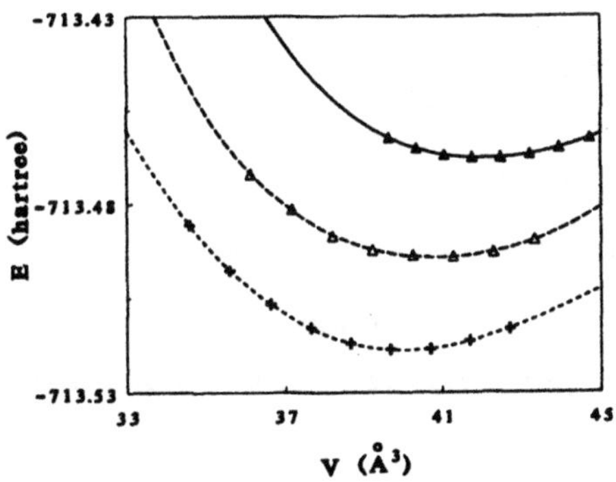

Figure 1. Hartree-Fock total energy of $MgSiO_3$ perovskite, as a function of the cell volume. *Crosses, open* and *full triangles* refer to the orthorhombic, tetragonal and cubic cells, respectively. Symbols indicate calculated energy points.

are the independent variables. In calculations, however, the independent variable is the volume V (and $T = 0$); for a given volume, V_i, the total energy of the system is minimized, by optimizing all the remaining geometrical parameters (cell parameters and fractionary coordinates). This process is then repeated for a limited number of volumes (6 to 15 in general); the E versus V curve is fitted to an analytic function (polynomial, or one of the many ***equations of state*** proposed in the literature), which is then differentiated:

$$P = -\partial E/\partial V \tag{1}$$

It is now possible to represent P as an analytic function of V, $P \equiv P(V)$ or, conversely, $V \equiv V(P)$ and $E \equiv E(P)$. Under experimental conditions, the controlled variables are N (number of atoms), T and P, so that the thermodynamic function of interest is the Gibbs free energy:

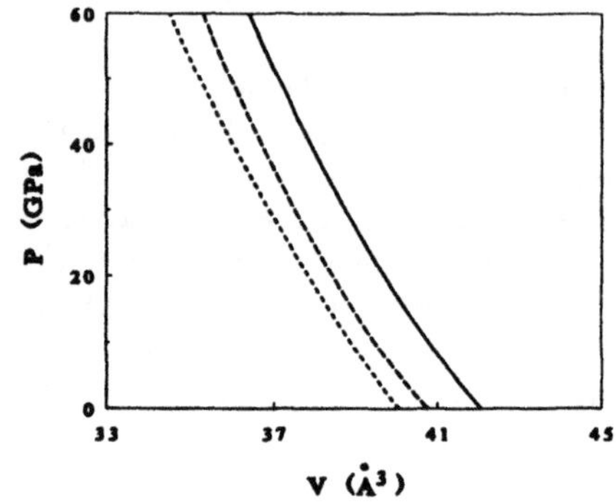

Figure 2. Equation of state for the three phases of $MgSiO_3$ perovskite. *Solid, long-dashed* and *short-dashed* lines are used for the cubic, tetragonal and orthorhombic cells respectively.

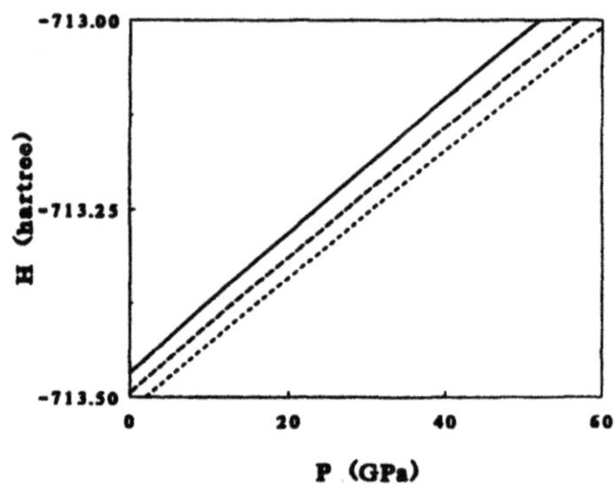

Figure 3. Enthalpy of the cubic, tetragonal and orthorhombic phases of $MgSiO_3$ perovskite. Lines as in Fig. 2.

$$G = TS + E + PV \tag{2}$$

The most stable phase is the one with the lowest Gibbs free energy at given values of N, T and P.

Calculated data refer to the *static limit* ($T = 0$ and frozen nuclear motion); G reduces then to H, the enthalpy ($H = E + PV$) and the function of interest is then:

$$H \equiv H(P) \tag{3}$$

which is easily obtained because both functions $E \equiv E(P)$ and $V \equiv V(P)$ have been obtained in an analytic form. It must be noted that in this process only the first derivative of the energy has been evaluated numerically. The numerical accuracy of the CRYSTAL code (see section 9) and of many periodic quantum mechanical codes, is high enough to give accurate numerical first derivatives of the energy, with respect to a geometrical parameter.

An example of this process is given in Figures 1, 2 and 3, which refer to three phases of $MgSiO_3$ perovskite. Perovskite is known to be the most stable phase of magnesio-silicates above 27.5 GPa at $T = 0$ [38, 36]. In the range of pressure and temperature explored experimentally, it is orthorhombic [39] and can be described as being formed by (nearly) regular octahedra rotated by about 8 degrees with respect to the **b** and **c** cell edges (see [40] for a discussion of the octahedral structure in perovskites). It has, however, been proposed [41] that tetragonal (null rotation with respect to **b**) or cubic (perfectly ordered, untilted arrays of regular octahedra) $MgSiO_3$ perovskite structures might exist in particular $T - P$ regions. A related question concerns the possible increase or decrease of orthorhombic distortion [40] in $MgSiO_3$ with pressure. Yagi *et al* [42] argued that the cubic form should be favoured at high pressure, whereas according to other authors [43], [44] [45] the distortion should increase with pressure.

The calculated energy points [35] for the three phases and for several volumes are shown in Figure 1. It must be noted that the unit cells of the cubic (C), tetragonal (T) and orthorhombic (O) phases contain 5, 10 and 20 atoms, respectively; in order to obtain the $E \equiv E(V)$ curve, for a given volume, the energy of the system has to be optimized with respect to all the geometrical parameters (there are 0, 2 and 9 of them for the C, T and O phases, respectively). Obtaining the $E(V)$ curve for the C phase is then very cheap (five atoms, just one parameter to explore); for the O phase (20 atoms, 10 geometrical parameters), however, the computational effort is extremely demanding and a medium size basis set had to be adopted [35]. Figures 2 and 3 show the $P(V)$ and $H(P)$ curves obtained from the $E(V)$ curve, as described above. The result is that the $H(P)$ curves for the three phases run nearly parallel in the explored pressure interval, the stability order being O > T > C. The question which can be asked is then: what is the influence of the many approximations implicit in the calculation on this result, namely:

a) basis set limitations;
b) lack of interelectronic correlation;
c) temperature effects and
d) zero point energy?

As will be shown later, each of these approximations can alter the relative stability of the various phases, so that a careful analysis of their importance must be performed. Single-energy-point test calculations and a comparative analysis of the behaviour of other systems indicate, however, that the errors associated with points a) to d) above cancel nearly exactly in the case of C, T and O phase perovskite, due to the fact that they are very similar structures. As regards the influence of pressure on the rotation of octahedra, in [35] it is shown that, for both the T and O phases, the rotation angles increase with increasing P.

As a second example of pressure-driven phase transition, let us consider the relative stability of $MgSiO_3$ perovskite (*Pv*; orthorhombic), ilmenite (*Il*; trigonal; 10 atoms per unit cell; 6 geometrical parameters to be optimized) and the assemblage of the two simple oxides, MgO periclase (*Pe*; cubic; 2 atoms per cell; 1 geometrical parameter) and SiO_2 stishovite (*St*; 6 atoms per cell; 3 geometrical parameters) [46]. As anticipated, *Il* is believed to be the most stable polymorph of $MgSiO_3$ between 23 and 27 GPa in the temperature range 0<T<1500 (see Figure 2 in [36]). The *Il* - *Pv* phase transition has been explored by Ito and Takahashi [38] with a diamond anvil apparatus at various temperatures. The phase boundary can be described by a straight line given by:

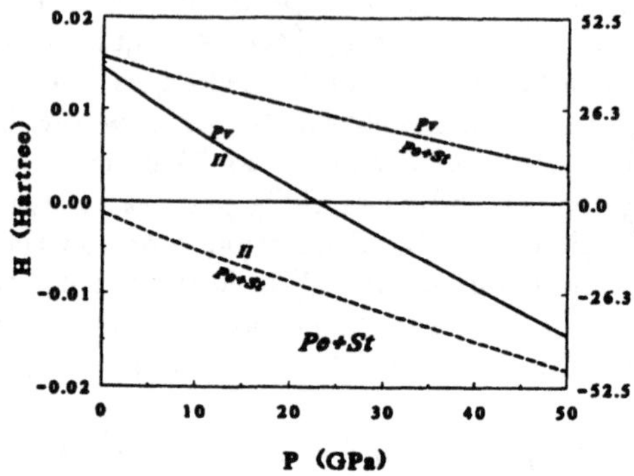

Figure 4. Enthalpy of the reactions *Il* → *Pv*, *Il* → *St*+*Pe*, and *Pv* → *St*+*Pe* with the *medium quality* basis set. *Solid*, *dashed* and *doubly-dashed* lines correspond to *Il* → *Pv*, *Il* → *St*+*Pe*, and *Pv* → *St*+*Pe* reactions, respectively. The calculated pressure of transition and the pressure domain of stability of the phases are reported.

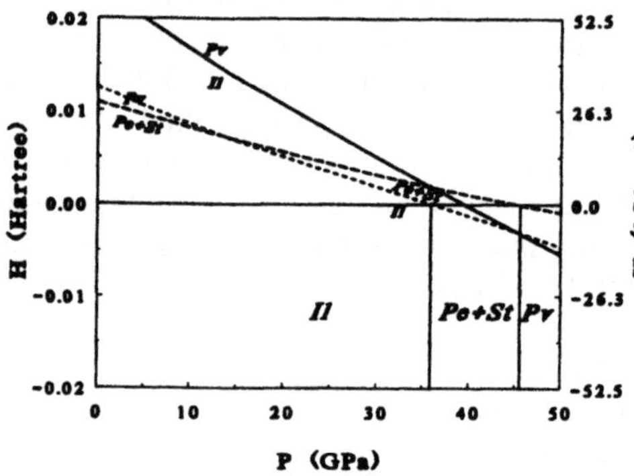

Figure 5. As in Fig. 4, with the *high quality* basis set.

$$P(\text{GPa}) = 26.8 - 0.0025T(^{\circ}C) \tag{4}$$

which gives the 0 K value of 27.5 GPa mentioned above. There is no experimental evidence of a stability zone for the *Pe*+*St* assemblage in the explored interval, to our knowledge. As in the previous example, the bottleneck of the calculation was represented by $MgSiO_3$ perovskite (20 atoms/cell; only 8 symmetry operators in the point group; 10 geometrical parameters to be optimized). The equation of state was then obtained with the same medium-size basis set (see Table 1 in [35]) that was shown to be sufficiently flexible for the description of the relative stability of the C, T and O perovskite phases. In the present case, however, additional variational freedom was exploited, in quite a different way, by the four systems (see Table 17), characterized by different point symmetry, atomic coordination, bond lengths and ionicity. As a matter of fact, the addition of a set of *d* polarization functions on Mg and O and of an *sp* valence shell on both atoms, lowered the total energy at calculated minima by 24.4, 33.6, 19.7

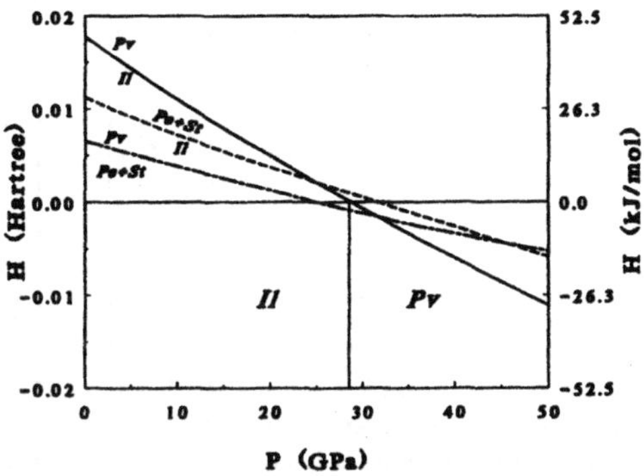

Figure 6. As in Fig. 4, with the *high quality* basis set and including correlation corrections.

millihartree for *Pv, Il* and *(Pe+St)*, respectively, altering the relative stability of the three phases obtained at $P = 0$ with the poorer basis set.

The interelectronic correlation energy, as estimated through the *a posteriori* correction mentioned above, was also shown to give different contributions to the three phases (at $P = 0$, *Pv* is stabilized by 5.7 millihartree with respect to *Il* and *Il* by 1.4 millihartree with respect to *Pe+St*). If it is supposed that the main effect of basis set improvements and inclusion of correlation is a nearly rigid downwards shift of each of the $E(V)$ curves for the four systems, the $H(P)$ curves shown in Figure 4 (and obtained with the medium size basis set, and full optimization of the geometry at each volume) can be corrected as shown in Figures 5 and 6. Figure 4 shows that, at the *medium quality* basis set level, the *Pe+St* assemblage is the most stable phase in all the explored interval. When the high quality basis set is used (Figure 5), *Il* is the stable phase below 36 GPa; between 36 and 46 GPa the assemblage is more stable than *Pv*. Finally, when correlation contributions are taken into account (Figure 6), the assemblage is always less stable than *Il* or *Pv*. The *Il* to *Pv* transition, is now at 29 GPa, not too far from the $T = 0$ experimental result (27.5 GPa).

6. Bulk modulus and elastic constants.

The properties considered so far refer to the total energy and its first derivatives with respect to volume, V (or pressure, P). Other important properties are related to second derivatives of the total energy of the system with respect to 1) volume; 2) the parameters of the unit cell (the lattice parameters **a**, **b**, **c** and the angles between them α, β, γ); and 3) the fractional coordinates of the atoms. The related observables are the bulk modulus, the elastic constants and vibrational properties, respectively.

Tables 9–11 report the bulk moduli (B) evaluated for the semiconductors, alkali halides and simple oxides already discussed in sections 3 and 4. Hartree-Fock results for semiconductors, in line with molecular experience on covalent systems, overestimate the experimental B value by 0–10%. For alkali halides, the error ranges from −1% (LiF) to −42% (RbI). This large underestimation is due to the increasingly large overestimation of the lattice parameter, when going

		N		P		As		Sb	
		B	%	*B*	%	*B*	%	*B*	%
B	HF	416	-	171	-1.2	149	-		
	Expt.			173					
Al	HF			96	+11.6	79	+2.6	61	+5.2
	Expt.			86		77		58	
Ga	HF			95	+4.4	77	0.0	59	+5.4
	Expt.			91		77		56	
In	HF			76	+5.6	64	+10.3	50	+8.7
	Expt			72		58		46	

B	%	
471	+6.6	C
442		
109	+10.1	Si
99		
85	+11.8	Ge
76		
56	+5.7	Sn
53		

Table 9. Calculated and experimental bulk modulus, B (in GPa) of the III-V (left) and IV-IV (right) semiconductors with the zincblende (or diamond) structure. Percentage errors with respect to experiment are given in the columns headed "%". For experimental references, basis sets and other computational details, see ref [22].

		F		Cl		Br		I	
		B	%	*B*	%	*B*	%	*B*	%
Li	HF	75.9	−1.3	30.0	−18.7	21.4	−28.9	15.4	
	HF+c	99.6	+29.5	40.8	+10.6	31.0	+3.0		
	Expt.	76.9		36.9		30.1			
Na	HF	51.1	−5.0	22.3	−22.0	16.6	−30.3	12.3	−33.8
	HF+c	69.9	+29.9	32.3	+12.9	23.5	−1.3		
	Expt.	53.8		28.6		23.8		18.6	
K	HF	30.0	−15.5	15.1	−27.4	11.8	−32.2	9.0	−35.3
	HF+c	45.0	+26.8	24.0	+15.4	17.0	−2.3		
	Expt.	35.5		20.8		17.4		13.9	
Rb	HF	23.4	−26.0	13.0	−30.5	10.5	−34.0	7.4	−42.6
	HF+c	38.0	+16.8	22.0	+17.6	15.1	−5.6		
	Expt.	31.6		18.7		15.9		12.9	

Table 10. Calculated and experimental bulk modulus (in GPa) of alkali halides with the NaCl structure. Percentage errors with respect to experiment are given in the columns headed "%". HF+c is the Hartree-Fock energy corrected *a posteriori* with the P86 functional [7]. For experimental data, basis sets and other computational details, see ref [9, 10].

from light to heavy atom systems, because the bulk modulus is evaluated from the second derivative of the energy at the calculated equilibrium geometry.

When smaller ions and larger electrostatic effects are involved, as in the case of the oxides (formal charges ±2 or 3; see Table 11), the calculated bulk moduli are more reasonable: as for semiconductors, in all but one case, there is an overestimation of between 7 and 12% with respect to experiment.

The bulk moduli reported in Tables 9–11 have been obtained as follows: the total energy of the system has been evaluated at various volumes, V_i, (usually 9–11 points are used); the energy points are interpolated with a polynomial curve, which is then differentiated in order to obtain B. All the geometrical parameters of the unit cell which can be varied, while keeping V_i constant, must be optimized, in each case; in the corundum case, for example, where the unit cell is completely defined by **a**, **c** (lattice parameters), x_O and z_{Al} (fractionary coordinates of oxygen and aluminium), the **c**/**a** ratio and the two fractional coordinates must be optimized. For $MgSiO_3$ perovskite in the orthorhombic

System	Ref	B		
		HF	%	Expt
BeO	[8]	253	+12	224
MgO	[8]	186	+11.4	167
CaO	[8]	128	+6.7	120
$CaCO_3$	[32]	75.3	−5.8	79.9
$MgCO_3$	[33]	125	+6.8	117
Al_2O_3	[29]	287	+10	261
$MgAl_2O_4$	[29]	227	+10	206

Table 11. Calculated and experimental bulk modulus (in GPa) of simple oxides; percentage errors with respect to experiment are given under %.

phase (see previous section), nine geometrical parameters must be optimized for each V_i volume.

Regarding elastic constants, we refer to a standard textbook [47] for definitions and to previous papers [12, 48, 49, 50] for the strategy adopted in CRYSTAL for their numerical evaluation. The quality of Hartree-Fock results for elastic constants is, roughly speaking, similar to that documented above for bulk moduli. In Table 12, we report examples, referring to three simple oxides. As expected, overestimation of about 10% with respect to experiment is usually observed.

System	Ref.	C_{11}			C_{12}			C_{44}		
		calc.	Expt.	%	calc.	Expt.	%	calc.	Expt.	%
Li_2O	[51]	234	216	+8.3	22	25	−12.0	68	67	+1.5
MgO	[52]	326	314	+3.8	111	94	+18.1	183	160	+14.4
CaO	[53]	245	240	+2.1	69	60	+15.0	95	83	+14.5

Table 12. Calculated and experimental elastic constants (in GPa) of simple oxides.

7. Some conclusions on the Hartree-Fock energy and energy derivatives and on their *a posteriori* correlation correction.

Table 13 summarizes the results of a statistical analysis performed on the three families of compounds considered here, namely semiconductors, alkali halides and simple oxides. The effect of an *a posteriori* correction on the binding energy, lattice parameter and bulk modulus (the last two properties for alkali halides only) through a density-functional, gradient-corrected formula is also analyzed. The statistics have been performed on the data reported in Tables 3, 6 and 9 for semiconductors (with results for SiC taken from reference [22]), in Tables 4, 7 and 10 for alkali halides, and in Table 5, 8 and 11 for oxides (note that for Table 8, only the a and c columns have been considered, because fractional coordinates are not dimensionally homogeneous).

The underestimation of the binding energy ranges from 30% to 50% for semiconductors, from 16% to 30% for alkali halides and from 24% to 30% for simple oxides; trends along the series are regular and systematic. Most of the error may be attributed to the HF Hamiltonian, rather than to basis set inadequacy or to

		Semiconductors			Alkali halides				Oxides		
		BE	a_0	B	BE	BEI	a_0	B	BE	a_0,c_0	B
HF	maxΔ%	−50.2	+2.3	+11.8	−30.4	−13.6	+9.7	−42.6	-31	+3.5	+12
	minΔ%	−30.3	0.0	0.0	−16.6	−1.2	+0.7	−1.3	-24	−1.2	-5.8
	$\langle\|\Delta\|\rangle$	39.2	1.2	6.5	21.2	7.2	5.7	26.6	28	0.9	9.0
	σ	6.7	0.7	3.6	3.8	3.1	2.4	11.4	2.7	1.2	27.8
	N	16	17	15	16	16	16	15	7	17	7
HF+c	maxΔ%	+10.7			+4.8	+5.7	+3.2	+29.9	−8		
	minΔ%	−4.2			−11.2	−5.7	−3.5	−5.6	−4		
	$\langle\|\Delta\|\rangle$	2.6			3.3	2.9	1.4	12.8	4		
	σ	2.5			1.5	1.8	1.2	10.0	1.4		
	N	16			12	12	12	12	7		

Table 13. Statistical analysis of the calculated Hartree-Fock results with (HF+c) and without (HF) the *a posteriori* correlation correction, for the III-V and IV-IV semiconductors and for alkali halides, when compared to experiment. BE and BEI are the binding energies evaluated with respect to neutral atoms and to ions, respectively; a_0 is the lattice parameter and B is the bulk modulus. Maximum (maxΔ%), minimum (minΔ%), mean ($\langle|\Delta|\rangle$) and standard deviation (σ) of the percentage errors are reported. N denotes the number of systems included in the statistics.

numerical inaccuracy. The error is larger in semiconductors, due to the greater importance of correlation effects in coupling electrons, during the formation of covalent bonds. It is always larger when heavy atoms are involved, because of the greater importance of intra-atomic and inter-atomic correlation effects.

It is interesting to compare the BE and BEI columns for alkali halides. Since BE is evaluated from atoms and BEI from ions, the difference is, therefore, a measure of the correlation energy involved in transferring one electron from the cation (where a single electron in an s shell has essentially no correlation with inner electrons) to the anion (where the new electron strongly correlates with the 5 electrons already present in the p shell). The table shows that the largest part of the BE error is due to this electron transfer; the BEI error is a measure of the *interionic* correlation, that is, the error due to dispersion forces, reciprocal polarization of the ions and increasing intra-atomic correlation, resulting from contraction of the electronic cloud in the formation of the crystal.

In the case of oxides, the error range is obviously narrower (24–31%); its mean value is larger than for alkali halides, because roughly two electrons are transferred from the cation to the anion; it is, however, substantially smaller than for semiconductors, because bonds remain essentially ionic. In all cases, the *a posteriori* correction improves the binding energy dramatically; the maximum error is now as small as 2.6, 3.3 and 4% for the three families of compounds.

The error in the lattice parameter ranges from 0 to 2%, and from -1.2 to +3.5% for semiconductors and oxides, respectively; for alkali halides, where dispersion forces play a more important role for the heavy terms of the series, the maximum error can be as large as +9.7%. The mean absolute error is 1.2, 0.9 and 5.7% for semiconductors, oxides and alkali halides respectively, a result which is, on the whole, satisfactory, considering the large set of compounds taken into account. The gradient-corrected, *a posteriori* correction has been applied systematically only in the case of alkali halides. In most cases, the calculated lattice parameter improves dramatically; there is a systematic reduction of distances, the correction being larger when the error is larger; when the HF result is essentially correct, the *a posteriori* corrected lattice parameter is shorter than the experimental one. The error for correlated alkali halide lattice parameters ranges from −3.5% to +3.2% and the absolute mean error is very small, 1.4%; the dispersion is also small (σ=1.2), about 50% of that obtained at the HF level.

The results for bulk moduli are, in part, influenced by the error in the geometry, because B is evaluated as the second derivative, at the calculated minimum. For semiconductors, the results are in line with those from molecular calculations: the error ranges from 0 to 12%, the mean value being 6.5%. In the case of oxides, the situation is similar, with errors ranging between −6 and 12%; for alkali halides, a systematically large underestimation is observed, as a consequence of the large overestimation of the lattice parameter. The improvement of lattice parameters due to the *a posteriori* correction entails an improvement of bulk moduli, where the mean absolute error decreases from 27% to 13%; the dispersion remains, however, essentially the same: σ=11.4 and 10.0, respectively.

In summary, the *a posteriori*, gradient-corrected formula in the Perdew 91 [7], Perdew 86 [7] or Colle-Salvetti forms [54] appears to be a very useful tool for improving systematically the HF binding energies. It can be used to correct the HF geometries, when dispersion forces play an important role in determinating the equilibrium geometry, as is the case of heavy cation/anion alkali halides, silver halides, molecular crystals and layered compounds (graphite, brucite [55] $MgCl_2$ [56]). When covalent bonds, small ions or large electrostatic effects are involved, its usefulness is less clear (sometimes the results improve, sometimes not) and the present author prefers to work at the uncorrected HF level. For second energy derivatives (bulk moduli, elastic constants, vibrational frequencies), a real improvement of the HF results can be obtained, in the author's opinion, only through a scheme improving the HF wavefunction, in a self-consistent way and not just the energy, as in the *a posteriori* scheme adopted here.

8. Basis set effects.

In this section, some examples of the effect of basis set on total energy, and related properties, are reported and discussed. The examples refer to ionic, semi-ionic and covalent situations. As anticipated in section 2, when Bloch functions built on localized functions (AOs) are used, convergence with the basis set size is faster for ionic, than for covalent, compounds; for metals (not considered here), it is very slow.

The quality of the basis sets adopted for a given system can be judged on the basis of general considerations and comparison with similar compounds. The size of the selected basis set is a compromise between the cost of the calculation, the accuracy of the results and the risks of numerical instabilities. As a first example, let us consider bulk $KMnF_3$ [57], described with the basis set reported in Table 14.

Since the system is fully ionic, the basis set for bulk calculations has been derived from basis sets optimized for the isolated K^+, Mn^{2+} and F^- ions. The exponents of the most diffuse single-Gaussian *sp* and *d* shells have been reoptimized in the bulk. The basis set of Table 14 is expected to be reasonably good: there are three valence *sp* shells on the anion and two on the cation (the 4*s* orbital of K and Mn is nearly totally empty); *d* electrons are described by two

Shell	Mn			K			F		
	exponents	coefficients		exponents	coefficients		exponents	coefficients	
		$s(d)$	p		$s(d)$	p		$s(d)$	p
s	292601	0.000227		172500.	0.000220		13770.	0.000877	
	42265	0.0019		24320.	0.00192		1590.0	0.00915	
	8947.29	0.0111		5140.	0.01109		326.5	0.0486	
	2330.32	0.0501		1343.9	0.04992		91.66	0.1691	
	702.047	0.1705		404.5	0.1702		30.46	0.3708	
	242.907	0.3691		139.4	0.3679		11.50	0.41649	
	94.955	0.4035		54.39	0.4036		4.76	0.1306	
	39.5777	0.1437		22.71	0.1459				
sp	732.14	-0.0053	0.0086	402.0	-0.00603	0.00841	19.000	-0.1094	0.1244
	175.551	-0.0673	0.0612	93.5	-0.0805	0.0602	4.530	-0.1289	0.5323
	58.5093	-0.1293	0.2135	30.75	-0.1094	0.2117	1.387	1.0	1.0
	23.129	0.2535	0.4018	11.92	0.258	0.3726			
	9.7536	0.6345	0.4012	5.167	0.684	0.4022			
	3.4545	0.2714	0.2222	1.582	0.399	0.186			
sp	38.389	0.0157	-0.0311	17.35	-0.0074	-0.0321	0.440	1.	1.
	15.4367	-0.2535	-0.0969	7.55	-0.129	-0.062			
	6.1781	-0.8648	0.2563	2.939	-0.6834	0.1691			
	2.8235	0.9337	1.6552	1.19	1.08	1.500			
				0.674	1.03	1.060			
sp	1.2086	1.0	1.0	0.389	1.	1.	0.179	1.	1.
sp	0.4986	1.0	1.0	0.216	1.	1.			
d	22.5929	0.0708							
	6.1674	0.3044							
	2.0638	0.5469							
	0.7401	0.5102							
d	0.249	1.0							
sp	=	=	=	0.4017	1.0	1.0	=	=	=
sp	=	=	=	0.2216	1.0	1.0			
sp	0.067	1.0	1.0	0.0281	1.0	1.0	0.150	1.0	1.0

Table 14. Exponents and coefficients of the Gaussian-type functions adopted for the study of $KMnF_3$. In the lower part of the table, the first two rows give the exponents and coefficients of the functions modified in the atom with respect to the ion in the bulk; the last row refers to the two outer *sp* functions, added to describe the atomic tails. The atomic basis set is used for the evaluation of the binding energies given in Table 15. The symbol "=" stands for "unmodified".

shells: a contraction of four Gaussians for the inner part and a single Gaussian for the outer part. The calculation with this basis set is cheap (about 30 minutes per energy point on a medium size workstation), because:

a) the unit cell contains 5 atoms only;
b) the system has high symmetry;
c) the external Gaussians of the two cations have large exponents (0.50 and 0.22 $bohr^{-2}$ for Mn and K, respectively) and that of the anion is not too diffuse (0.18 $bohr^{-2}$).

In fact, the cost of calculation, with the CRYSTAL code, increases dramatically with decreasing exponent of the most diffuse Gaussian. Note also that an *all-electron* basis set is used. Pseudopotential calculations are only marginally cheaper than all-electron calculations, for a system like the one considered here, because:

case	basis set	N	t	BE	a_0	B	ΔE	C_{44}	$C_{11}-C_{12}$
a)	Table 14	83	1.00	0.604	4.280	64.6	0.293	30.9	91.8
b)	5 – 1d instead of 4 – 1d	83	1.18	0.605	4.280	64.1	0.292	30.8	90.9
c)	b) +sp Mn (α=0.25)	87	1.55	0.606	4.284	63.3	0.334	30.3	91.3
d)	c) +d on F (α=0.7)	102	2.12	0.608	4.280	63.5	0.319	30.2	89.8
e)	d) +d on K (α=0.4)	107	2.54	0.609	4.276	63.9	0.325	29.6	94.8

Table 15. Basis set effects on bulk properties of $KMnF_3$. BE, a_0, ΔE, B, C_{44}, C_{11} and C_{12} are the binding energy (hartree), the lattice parameter (Å), the energy difference between the ferro- and the antiferromagnetic phases (millihartree), the bulk modulus and elastic constants (GPa). N is the number of atomic orbitals in the basis set; t is the CPU time with respect to case a).

a) Inner shells require the calculation of a very small number of bielectronic integrals, so that the cost of the INTEGRAL (see Chapter 8) step is nearly unchanged.
b) With a unit cell of this size, the cost of the SCF part of the calculation is not dominated by the diagonalization (about 50% of the time in SCF is spent in the reconstruction of the Fock matrix; INTEGRAL and SCF require about the same CPU time).

The basis set of Table 14 can be improved in many ways: polarization functions (d orbitals) can be added to K^+, and F^-; for Mn, an additional diffuse sp valence shell can be added; the 4-1G contraction, used for d electrons, can be substituted by a 5-1G one, which allows a better description of both the core and valence region. The results of this progressive improvement of the basis set are shown in rows 2-5 of Table 15. It turns out that all the quantities are very stable with respect to basis set improvements; only in the case of ΔE, the energy difference between the ferro- and the antiferromagnetic states, is the maximum variation of about 15% observed. This result is a consequence of the fully ionic nature of $KMnF_3$, and of its highly symmetric structure: the electron charge distribution of K^+ and F^- is essentially spherical, so that polarization functions are nearly useless; the $4sp$ shell of Mn is empty, so that the additional sp shell at 0.25 $bohr^{-2}$ is not used.

In order to document the relationship between basis set flexibility, polarizability of the ions and symmetry of the system, let us consider the influence of d polarization functions on another two ionic compounds, namely Na_2O and K_2O [48]. In Table 16 some bulk properties are reported, obtained with and without d functions on the cation. It turns out that d functions have negligible influence on all but one property of the two compounds, namely, the C_{44} elastic constant, where reductions of 10 and 42% are observed when d orbitals are introduced.

These results can be interpreted as follows. The cations are in a high symmetry position; when the unit cell is modified according to the deformations required for the evaluation of B, C_{11} and $C_{11}-C_{12}$, their local symmetry remains cubic, so that d orbitals are used essentially for describing the *breathing* of the ion, which, however, can also be described by the s and p functions of the valence shells. For these quantities, the difference between K (variation of the order of 3%) and Na (1%) is due to the larger polarizability of the former ion, or, in other words, to the smaller energy difference between valence s and

	Na_2O			K_2O		
	no *d*	*d*	$\Delta\%$	no *d*	*d*	$\Delta\%$
E_{tot}	−398.693	−398.695	−0.002	−1273.184	−1273.193	−0.010
a_0	5.498	5.487	−0.2	6.550	6.466	−1.3
B	58.7	58.7	0.0	33.3	34.6	+3.5
C_{11}	127.3	126.14	−0.9	71.8	74.1	+3.1
C_{12}	23.9	23.8	−0.4	14.2	14.8	+4.1
C_{44}	37.8	34.4	−10.2	19.7	13.9	-41.7

Table 16. Effect of *d* polarization functions, when added to the cations, on bulk properties of simple oxides. $\Delta\%$ is the percentage difference between the calculation with and without *d* functions (in the E_{tot} case, it is the absolute difference).

virtual *d* levels. The deformation required for the evaluation of the C_{44} elastic constant, however, drastically reduces the atomic point symmetry: the cation is no longer in a centrosymmetric position and the ion can undergo a dipolar relaxation, for which the combination of *p* and *d* orbitals is required. This effect is less dramatic for the Na compound, because the ion is less polarizable.

Type	Basis set			ΔE				δE_1	δE_2
	Definition	N(Mg)	N(O)	Pv	Il	St	Pe		
BS0		9	9	–	–	–	–	14.5	15.5
BS1	BS0+*d*(Mg)	14	9	5.7	6.1	–	3.8	14.9	13.6
BS2	BS0+d(O)	9	14	11.6	17.3	11.2	1.0	20.2	16.1
BS3	BS0+*Spt*(Mg)	13	9	4.3	3.2	–	2.0	13.4	13.2
BS4	BS0+*Spt*(O)	9	13	7.7	10.8	2.4	2.4	17.6	12.6
SUM				29.3	37.4	13.6	9.2	22.6	9.0
BS5	BS0+*Spt*+*d* on both	18	18	24.4	33.6	11.6	8.1	23.7	10.8

Table 17. Basis set effects on the total energy of $MgSiO_3$ perovskite(*Pv*), $MgSiO_3$ ilmenite(*Il*), MgO periclase(*Pe*) and SiO_2 stishovite(*St*); ΔE is the energy decrease (in millihartree) due to the indicated modification with respect to the reference BS0 basis set (8-5-1G contraction for oxygen and 8-6-1G for magnesium; see ref [46]); the BS0 reference energies (in Hartree per molecular units) are -713.5207(*Pv*), -713.5352(*Il*), -438.8721(*St*) and -274.6641(*Pe*). The SUM entry gives the sum of the energy decrease resulting from the BS1 to BS4 entries. The comparison with BS5 shows that basis set effects are approximatively additive. N(Mg) and N(O) are the number of atomic orbitals on the atoms. δE_1 and δE_2 are the energy differences (in millihartree) *Pv* minus *Il* and *Pv* minus (*St*+*Pe*) respectively.

The last example describes the relative stability of $MgSiO_3$ perovskite, $MgSiO_3$ ilmenite and the assemblage of the two oxides, MgO periclase and SiO_2 stishovite, at $P = 0$ [46]. The relative stability of the three phases, as a function of pressure, has been discussed in section 5. Table 17 shows the effect of basis set improvements of a given atom in different conditions of ionicity, point symmetry at the atomic site and coordination. The energy gain obtained for the four systems when a set of *d* functions is added to oxygen and magnesium and the double (single for Mg) valence *sp* shell is split in three (two) shells (8-411G basis instead of 8-51G for oxygen, and 8-5-11G instead of 8-6-1G for magnesium; the first shell is *s* type, the others are *sp*) is reported. It turns out that the effect of the improvements is relatively modest, of the order of a few millihartrees, but varies for the different systems, so that the relative stability of the three phases changes at the different levels of improvement of the reference basis set. The energy gain obtained with a further splitting of the valence *sp* shell or the inclusion of a second shell of *d* functions, is one order of magnitude smaller and has no influence on the relative stability of the three phases.

9. Numerical accuracy and numerical derivatives.

In this section, the influence of the computational parameters on total energy and energy related properties is documented with a few examples. In the CRYSTAL code [12, 11], numerical approximations are introduced in the treatment of the Coulomb and exchange, infinite, bielectronic series', which enter into the definition of the Fock matrix and the reciprocal space integration, which is required at each cycle of the SCF process, for reconstructing the density matrix. Input parameters (T1, T2, T3, T4, T5 and IS: see Chapter 8 and Users' Manual [12]) permit us to increase or reduce the accuracy of the calculation: high values of the Tx parameters correspond to evaluating, exactly, a large number of four-centre bielectronic integrals (otherwise evaluated through one- or two-centre multipole expansions), with a corresponding increase of cost. The IS parameter controls the number of reciprocal space points, κ, at which the Fock matrix is diagonalized, with a direct influence on the cost of the calculation.

Case	Tx	Molecule			Bulk		
		TE	KE	*t*	TE	KE	*t*
1	2 2 2 2 4	unstable	—	—	unstable	—	—
2	4 4 4 4 8	−223.78039	223.48430	96	-223.83341	223.49628	841
3	6 6 6 6 12	−223.78752	223.50083	132	-223.84090	223.50944	3501
4	7 7 7 7 14	−223.78755	223.50054	139	-223.84107	223.50854	5518
5	8 8 8 8 16	−223.78784	223.50157	162	-223.84142	223.50965	8043
6	9 9 9 9 18	−223.78833	223.50407	170	-223.84192	223.51210	11836
7	10 10 10 10 20	−223.78833	223.50413	178	-223.84311	223.51330	16845
8	20 20 20 20 26	−223.78840	223.50425	195			
	GAUSSIAN92	−223.78840	223.50426	- - -	- - -	- - -	- - -

Table 18. The influence of the computational parameters of the CRYSTAL program on total(TE) and kinetic(KE) energy (in hartree) of molecular and crystalline urea. Parameter *t* is the cost in CPU time on a medium-size workstation. A 6-21G basis set is used ([58]). The last entry refers to a calculation performed with the GAUSSIAN92 program [1].**

Roughly speaking, there are three aspects of the numerical problem that must be discussed in connection with the CRYSTAL code: a) the *catastrophic threshold*; b) the *exact* limit; and c) the problem of numerical derivatives of the energy, required for evaluating equilibrium geometry, pressure, bulk modulus, elastic constants and vibrational frequencies.

a) In some situations, the calculation can have a catastrophic behaviour and a totally unphysical wave function is obtained. There is always a lower limit for the Tx and IS tolerances, below which the catastrophic behaviour may be obtained (see Tables 18 and 19); this limit, however, depends on many factors, the most important of which is the basis set quality: very diffuse Gaussian functions generate numerical instability. As diffuse functions are required for the description of delocalized valence electrons, it turns out that numerical problems are more frequent in metals than in semiconductors, and in semiconductors than in ionic compounds; or, stated differently, while it is possible to use high-quality fully-optimized basis sets for ionic compounds, for metals, only low- or medium-quality basis sets are allowed. Many examples related to this problem can be found in [11]. In Table 18, the total energy of the isolated [58] urea molecule and of the urea molecular crystal are reported for different sets of computational parameters. Both for the molecule and the bulk, *Case1* tolerances are poor enough to generate an unphysical wavefunction and total energy. Catastrophic behaviour is also observed for $KMnF_3$ (Table 19) with very poor tolerances. Values of 7 7 7 7 14 for the Tx parameters will be considered in the following as *good* computational parameters for a CRYSTAL calculation.

b) The second question concerns the absolute error with respect to the *exact* total energy (within a given hamiltonian and basis set), and the speed of convergence towards this value. It is not easy to give qualitative answers, because, in general, we are not able to generate the *exact* total energy. Two possible partial answers are the following:
   i) In the case of molecular systems, a comparison with the energies produced by (very accurate) molecular codes is possible; if we suppose that the error for a given molecule and for the corresponding molecular crystal is roughly the same, then we can guess the order of magnitude of the numerical error for infinite systems also;
   ii) Convergence of the total energy can be explored in a relatively narrow range of values, from which extrapolation to *infinite* accuracy can, in some cases, be performed.

   In Table 18, convergence for bulk and molecular urea is shown; for the set of tolerances 7 7 7 7 14, the residual error with respect to the exact molecular result is of the order of one millihartree; taking into account the (vaguely) parallel trend for the molecule and the bulk, the absolute error for the latter is expected to be roughly the same. Table 19 shows similar data for $KMnF_3$, which is fully ionic. The extremely rapid convergence of the total energy with respect to IS is to be noted. The energy trend with respect to Tx parameters is very similar to that in Table 18, and the previous conclusions on the absolute error remain valid. When metallic systems are considered, the convergence, both with respect to Tx parameters (see the discussion in [11] concerning the exchange series) and with respect to the number of sampling points in reciprocal space, is much slower.

c) In most cases, the real problem is the *relative* numerical error, rather than the *absolute* error discussed above. This is the case when numerical differentiation of the total energy of the system with respect to some geometrical parameter is performed to obtain properties such as the pressure versus volume curve, the elastic tensor or when the relative stability of different phases is investigated. The problem is particularly delicate in a *direct space* approach, as used in CRYSTAL: all the truncation criteria work on the basis of overlaps between functions and distances between atoms, so that the number of mono- and bielectronic integrals, evaluated at each level of approximation, for two different geometrical configurations of the system, are completely different. Strategems have been implemented in the CRYSTAL code (see the GEOM option in Users' Manual [12]) that permit us to have, essentially, the same numerical error along the $E(\delta)$ curve, where δ represents the set of explored geometrical variables.

Case	Tx.	IS	Total energy		ΔE	t
			FM	AFM		
1	2 2 2 2 4	4	unstable	unstable	—	—
2	4 4 4 4 8	4	-2047.65452	-2047.65473	0.210	0.29
3	5 5 5 5 10	4	-2047.64004	-2047.64020	0.166	0.45
4	6 6 6 6 12	4	-2047.63984	-2047.64014	0.297	0.69
5	7 7 7 7 14	4	-2047.64309	-2047.64338	0.293	1
6	8 8 8 8 16	4	-2047.64316	-2047.64346	0.297	1.29
7	9 9 9 9 18	4	-2047.64312	-2047.64349	0.298	1.74
8	10 10 10 10 20	4	-2047.64473	-2047.64502	0.289	2.28
9	7 7 7 7 14	4	-2047.64309	-2047.64338	0.293	1
10	7 7 7 7 14	8	-2047.64308	-2047.64338	0.297	1.24
11	7 7 7 7 14	12	-2047.64308	-2047.64338	0.297	1.76

Table 19. The influence of computational parameters of the CRYSTAL program on total energy(in hartree) of the ferro- (FM) and antiferromagnetic phases (AFM) of $KMnF_3$. ΔE is the energy difference per molecular unit in millihartree; t is the CPU time relative to the reference case (7 7 7 7 14, 4). IS=4, 8, 12 correspond to 10, 35, 84 κ points respectively.

	Pol. degree	Number of points			
		11	9	7	5
C_{11}	2	128.00	127.00	126.17	**125.56**
	3	127.78	126.78	**126.00**	**125.39**
	4	**124.89**	**124.89**	**124.94**	
	5	**124.89**	**124.83**	**124.89**	
	6	**124.83**	**124.94**		
	7	**124.83**	**124.94**		
C_{44}	2	**31.07**	**30.99**	**30.93**	**30.94**
	3	**31.07**	**31.00**	**30.95**	**30.97**
	4	**30.84**	**30.99**	**31.03**	
	5	**30.84**	**30.89**	**31.04**	
	6	**30.99**	**31.14**		
	7	**30.98**	**31.16**		
$C_{11} - C_{12}$	2	92.75	92.36	91.89	**91.36**
	3	92.72	92.33	91.86	**91.33**
	4	**91.25**	**90.86**	**90.41**	
	5	**91.22**	**90.86**	**90.44**	
	6	**90.41**	**90.00**		
	7	**90.41**	**90.02**		
B	2	66.39	65.79	65.31	**64.95**
	3	66.62	65.91	65.35	**64.92**
	4	**64.58**	**64.58**	**64.56**	
	5	**64.56**	**64.57**	**64.56**	
	6	**64.58**	**64.55**		
	7	**64.57**	**64.56**		

Table 20. Bulk modulus and elastic constants of $KMnF_3$ (in GPa) obtained from a best fit of the indicated number of energy points to polynomial functions of various degree.

As a first example, let us consider the problem of the numerical evaluation of second-order derivatives of the total energy of the system. The results of fitting $E(\delta)$ are shown in Table 20, where δ indicates the deformation performed to obtain the bulk modulus B and the elastic constants C_{11}, $C_{11}-C_{12}$ and C_{44} of $KMnF_3$. The set of δ_i values have been selected according to the following criteria:

a) They are regularly spaced.
b) The interval $\delta_{max} - \delta_{min}$ must be small enough to explore the region where the $E(\delta)$ curve can be fitted with a simple function and the perturbation is small.
c) $\delta_{max} - \delta_{min}$ must be such that $|E(\delta_{max}) - E(\delta_{min})|$ is much larger than the numerical noise.

The compromise between points b) and c) requires that $|E(\delta_{max}) - E(\delta_{min})|$ is roughly between one and twenty millihartree (it has the values 1, 4, 2 and 22 millihartree for the four curves to which Table 20 refers).

Very often, the $E(\delta)$ curve is not parabolic in this interval, but includes important third- and fourth-order components, as is easily seen going down the first columns of the table (for $C_{11}-C_{12}$, there is also a sixth-order contribution of about 0.8 GPa). The columns with 9, 7 and 5 points have been obtained by dropping the external points, that is, by reducing the interval, so that the contribution from high-order components is less important. The best calculated values of the second-order derivatives in the minimum are easily identified in the table (all the values falling within a 1% interval of the selected value are indicated in boldface); the numerical noise can then be considered nearly constant along the curves. When lower-order derivatives are involved (the total energy in the minimum or equilibrium, geometry) the uncertainty in the calculated values, resulting from the fitting procedure, is much smaller, as expected.

As a last example, let us consider the energy difference between the ferro- and antiferromagnetic phases of $KMnF_3$, which is of the order of 0.3 millihartree; the variation with the computational parameters, reported in Table 19, is very small when computational conditions more severe than the standard ones are used (7 7 7 7 14, IS=8)). The ferro-/antiferromagnetic energy difference remains stable with respect to computational conditions, even at larger lattice parameter, where it may be reduced by an order of magnitute, that is, of the order of 10^{-5} hartree (see Figure 7.1 in ref [57], and Figure 2 in ref [31]).

10. Comparison with experimental data.

There is a last point that deserves some comment. The energy differences, referred to in Table 2 and the other energy related properties, discussed in previous sections cannot, in general, be compared *directly* with *corresponding* experimental quantities, because the latter refer to processes or reactions which are, in some respect, different or that take place in conditions which are different, from those the calculations try to mimic.

The most trivial example, which we will discuss in some detail, is the following: quantum mechanical calculations refer to the static limit ($T = 0$ and frozen nuclei), whereas experiments refer, obviously, to finite T. Manipulations of the calculated or experimental data are then required, in order to perform a proper comparison. In the following, we discuss how to correct a) binding energies, b) lattice parameters and c) elastic constants, for these two effects. The examples of these three corrections all refer to alkali halides because high-quality experimental data are available for these systems at various temperatures; this is not a common situation for other families of crystalline compounds.

As a first example, we report below the Born-Haber cycle which connects calculated and experimental binding energies of NaCl.

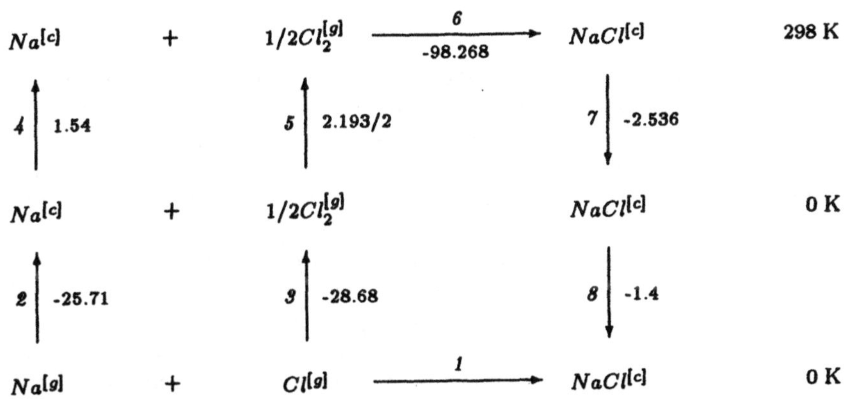

Figure 7. The Born-Haber cycle for NaCl

The calculated energy refers to step 1 (formation energy of the crystal at 0K from the isolated atoms, disregarding zero point effects), whereas the standard way experimental data are quoted refers to step 6 ($\Delta H_0^{298} = -98.268$ kcal/mol). However, all but one step of the cycle are available experimentally: the dissociation energy of Cl_2 (28.68 kcal/mol; the inverse of step 3) and the sublimation enthalpy of Na (25.71 kcal/mol; the inverse of step 2); the enthalpy difference between 0K and 298K of solid Na (1.54 kcal/mol; the inverse of step 4) and NaCl (2.536 kcal/mol; the inverse of step 7) are all known. The enthalpy difference between 0K and 298K of Cl_2 is evaluated in the hypothesis of ideal behaviour (2.193 kcal/mol; twice step 5). All the experimental data for reactions 2-7 are tabulated in ref. [59] (section 5, p. 16-59). Step 8 refers to the zero point energy (ε_0) which is usually not available experimentally. For its calculation, we used the Debye model (see for instance ref. [60], p. 100) in which ε_0 is related to the Debye's temperature Θ_D, through the equation:

$$\varepsilon_0 = 9/8 k_B \Theta_D \tag{5}$$

where k_B is the Boltzmann's constant. Debye temperatures for alkali halides can be found in ref. [47], p. 459. The resulting ε_0 value for NaCl is 1.4 kcal. The experimental *static* lattice energy, reported in Table 1, is the sum of steps 2-8:

$$BE_{static} = (2) + (3) + (4) + (5) + (6) + (7) + (8). \tag{6}$$

Syst.	a_{298}	a_0	Δ	δ
LiF	4.03[a]	3.99	0.04	0.018
NaF	4.62[a]	4.57	0.05	0.018
KF	5.34[a]	5.29	0.05	0.016
LiCl	5.13[a]	5.07	0.07	0.022
NaCl	5.64[b]	5.57	0.06	0.019
KCl	6.28[a]	6.20	0.08	0.020

Table 21. Lattice parameters at 298 K and at the static limit. δ is the zero point contribution, whereas Δ includes also temperature effects. All values in Å. [a] from reference [66]; [b] from reference [67].

The temperature dependence of lattice parameters and elastic constants is essentially linear, over a wide range of temperatures, but, at low temperature, anharmonic effects become negligible and the slope of the line becomes zero [61]. The extrapolation of the high temperature data to the static limit [61] requires the knowledge of the experimental data, at a given temperature and of their temperature dependence, in the high temperature (linear) region. It is not difficult to find quite accurate room temperature x-ray measurements of the lattice paramenter of alkali halides in the literature; the data reported by different authors are often in good agreement and no significant differences have been found between very early and more recent results. The room temperature experimental data we selected, for six compounds, are reported in the first column of Table 21.

However, only one paper has been found [62] that provides temperature-dependence data for the whole set of alkali halides we are discussing. In ref [62], the temperature dependence of the lattice constant, $a(\mathrm{T})$, is given in the form of a linear expansion coefficient, $\alpha(\mathrm{T})$:

$$\alpha(T) = \frac{da(\mathrm{T})}{d\mathrm{T}} = C_1 + C_2\mathrm{T} + \ldots + C_6\mathrm{T}^6, \tag{7}$$

from which (knowing the lattice constant at a given temperature) it is easy to obtain an explicit relationship of the form [62, 10]:

$$a(\mathrm{T}) = B_0 + B_1\mathrm{T} + B_2\mathrm{T}^2 + \ldots. \tag{8}$$

$a(0)$ values obtained from equation 8 are not static limit values, because, obviously, experimental C coefficients include zero point effects. Rather, the lattice constant at the static limit (a_0) can be obtained [61, 63, 10] as an intercept of the tangent to the $a(\mathrm{T})$ curve at some point, T_i, on the a axis in the *linear* high temperature region, that is:

$$a(\mathrm{T}_i) = a_0 + D\mathrm{T}_i, \tag{9}$$

where D is the first derivative of the $a(\mathrm{T})$ curve at T_i. The a_0 values obtained in this way are reported in the second column of Table 21; the third and the fourth columns of the same table provide the difference between the lattice parameter at room temperature and at the static limit (Δ) and between $a(0)$ and a_0 (δ), that is, the effect of zero point motion on the lattice parameter. It turns out that Δ can be as large as 1%, whereas δ is always smaller (0.5 %).

The manipulation of the elastic constants, in order to obtain the static limit data, is similar to that for the lattice parameter and will not be discussed here [61, 10, 63]. Several papers are available that provide room temperature elastic constants for the alkali halides, but only a few also provide temperature dependence [64]. By far the most complete set of data has been provided by Haussühl [65], who measured both the temperature dependence and the high temperature C_{ij}, for the whole set of alkali halides. Haussühl's room temperature data for C_{11}, C_{12}, C_{44} and B are reported in Table 22. Where available, other data from different authors [64] have been compared with those by Haussühl, and no significant differences emerged from this comparison. The static limit data can be quite different from room temperature data [10]. In particular, C_{11} can vary as much as 25%, as is shown in Table 22. Note also that the correction for C_{12} is positive in nearly all cases, whereas it is negative for C_{11} and C_{44}. The effect of zero point motion is small but not negligible: for example, Lewis *et al* [64] report 0.1, 1.4 and 0.8 GPa differences between the static limit and 4K elastic constants (in the order C_{11}, C_{12} and C_{44}) for NaCl. Although Table 22 indicates that it is important to refer to static limit data, in many cases the comparison between calculation and experiment is performed with reference to the room temperature or low temperature data only.

Note however that:

a) T and zero point effects are smaller for covalent systems or for ionic compounds with larger ionic charges, than for alkali halides; for example, the room temperature and the athermal limit values for the CaF_2 [49] lattice parameter are 5.463 and 5.445 Å respectively (0.3% difference).
b) In most cases, the T-dependence of the investigated quantity is not available, so that it is impossible to correct for a proper comparison with the calculated data.

	B		C_{11}		C_{12}		C_{44}	
	298	Δ	298	Δ	298	Δ	298	Δ
LiF	68.8	−8.1	111.3	−24.5	47.6	+0.2	63.5	−5.2
NaF	48.5	−5.3	97.0	−18.4	24.3	+1.3	28.1	−1.7
KF	30.2	−5.3	65.6	−14.1	14.6	+1.2	12.5	−0.8
LiCl	31.6	−5.3	49.4	−13.8	22.8	−1.0	24.6	−3.0
NaCl	25.1	−3.5	49.4	−11.7	12.9	+0.6	12.6	−1.0
KCl	18.2	−2.6	40.8	−10.1	6.9	+1.1	6.3	−0.4

Table 22. Experimental bulk modulus (B) and elastic constants (C_{ij}) in GPa at room temperature (298) and correction Δ to be applied for obtaining the static limit value. Data from ref. [65].

11. Conclusions

In Chapter 8, 9, and 10, three reliable public *ab initio* programs for the study of periodic systems are presented. The pioneering stage characterized by there being only a few, secret, home-made *ab initio* programs, with uncontrolled and unclear accuracy, seems overcome. In this Chapter, many results have been presented concerning the total electronic energy and related properties of crystalline compounds, obtained with the CRYSTAL code. Many tables have been reported, aimed at documenting the stability of the results with respect to computed features of the code and size of the basis set; many different properties have been explored, and large families of crystalline compounds have been considered, in order to give an overall indication of the quality of the results produced by the code. The result is that useful information on many chemical and physical properties of crystalline compounds can be obtained at a relatively low cost, with a medium-size workstation. Much additional work is obviously required (and is actually in progress) in order to overcome some of the many important limitations characterising the present *ab initio* codes, to obtain better agreement with experiment and to be able to tackle more complicated crystalline structures.

References

1. Gaussian 92, Revision C.4, M. J. Frisch, G. W. Trucks, M. Head-Gordon, P. M. W. Gill, M. W. Wong, J. B. Foresman, B. G. Johnson, H. B. Schlegel, M. A. Robb, E. S. Replogle, R. Gomperts, J. L. Andres, K. Raghavachari, J. S. Binkley, C. Gonzalez, R. L. Martin, D. J. Fox, D. J. Defrees, J. Baker, J. J. P. Stewart, and J. A. Pople, Gaussian, Inc., Pittsburgh PA, 1992.
2. B.G. Johnson, P.M.W. Gill and J.A. Pople, J. Chem. Phys. **98**, 5612 (1993).
3. A. García, C. Elsässer, Jing Zhu, S. G. Louie and M.L. Cohen, Phys. Rev. B **46**, 9829 (1992).
4. C. Filippi, D.J. Singh and C.J. Umrigar, Phys. Rev. B **50**, 14947 (1994).
5. F. Sim, A. St-Amant, I. Papai and D.R. Salahub, J. Am. Chem. Soc. **114**, 4391 (1992).
6. A. Szabó and N.S. Ostlund, *Modern Quantum Chemistry*, McGraw-Hill, New York (1993).
7. J.P. Perdew, Phys. Rev. B **33**, 8822 (1986). (errata: ibid. **34**, 7406 (1986). J.P. Perdew, J.A. Chevary, S.H. Vosko, K.A. Jackson, M.R. Pederson, D.J. Singh and C. Fiolhais, Phys. Rev. B **46**, 6671 (1992).
8. R. Dovesi, C. Roetti, C. Freyria Fava, E. Aprà, V.R. Saunders and N.M. Harrison, Phil. Trans. R. Soc. Lond. A **341**, 203 (1992).
9. M. Prencipe, A. Zupan, R. Dovesi, E. Aprà and V.R. Saunders, Phys. Rev B **51**, 3391 (1995).
10. M. Prencipe *Tesi di laurea*, Università di Torino, Torino(1990).
11. C. Pisani, R. Dovesi and C. Roetti, *Lecture Notes in Chemistry* , vol 48, Springer, Heidelberg (1988).
12. R. Dovesi, C. Pisani, C. Roetti, M. Causà and V.R. Saunders, CRYSTAL88, QCPE program N.577, Quantum Chemistry Program Exchange, Indiana University, Bloomington, Indiana (1989). R. Dovesi, V.R. Saunders and C. Roetti, CRYSTAL92 User's Manual, Gruppo di Chimica Teorica, Università di Torino, Torino (1992).
13. S. Froyen and M.L. Cohen, J. Phys. C: Solid State Phys. **19**, 2623 (1986).
14. H.J.F. Jansen and A.J. Freeman, Phys. Rev. B **33**, 8629 (1986).
15. J.L. Feldman, M.J. Mehl and H. Krakauer, Phys. Rev. B **35**, 6395 (1987).
16. G. te Velde and E.J. Baerends, Phys. Rev. B **44**, 7888 (1991).
17. P. Cortona, Phys. Rev. B **44**, 8454 (1991).
18. C.O. Rodriguez and M. Methfessel, Phys. Rev. B **45**, 90 (1992).
19. L. Pueyo, V. Luaña, M. Florez and E. Francisco *Structure, Interactions and Reactivity* ed. S. Fraga (Amsterdam: Elsevier), vol. B, p.504 (1992).
20. E. Aprà, M. Causà, M. Prencipe, R. Dovesi, V.R. Saunders, J. Phys. Cond. Matter **5**, 2969 (1993).
21. H. Spetzler, C.G. Sammis and R.I. O' Connell, J. Phys. Chem. Solids **33**, 1727 (1972).
22. M. Causà, R.Dovesi and C.Roetti, Phys. Rev. B **43**, 11937 (1991).
23. R. Dovesi, M. Causà, R. Orlando, C. Roetti, and V.R. Saunders, J. Chem. Phys. **92**, 7402 (1990).
24. L. Ojamäe, K. Hermansson, R. Dovesi, C. Roetti, and V.R. Saunders, J. Chem. Phys. **100**, 2128 (1994).
25. M. Causa', R. Dovesi, C. Pisani and C. Roetti, Surface Sci. **175**, 551 (1986).

26. M. Causà, R. Dovesi, and F. Ricca, Surface Science **280**, 1 (1993).
27. R. Orlando, R. Dovesi, P. Azavant, N.M. Harrison and V.R. Saunders, J. Phys. Cond. Matter **6**, 8573 (1994).
28. W.C. Mackrodt, N.M. Harrison, V.R. Saunders, N.L. Allan, M.D. Towler, E. Aprà and R. Dovesi, Phil. Mag. A **68**, 653 (1993).
29. M. Catti, G. Valerio, R. Dovesi and M. Causà, Phys. Rev. B, **49**, 14179 (1994).
30. R. Dovesi and R. Orlando Phase Trans. **52**, 151 (1994).
31. J.M. Ricart, R. Dovesi, C. Roetti and V.R. Saunders Phys. Rev. B **52**, 2381 (1995).
32. M. Catti, A. Pavese, E. Aprà and C. Roetti, Phys. Chem. Minerals **20**, 104 (1993).
33. M. Catti, A. Pavese, R. Dovesi and V. R. Saunders, Phys. Rev. B **47** 9189, (1993).
34. M. Catti, G. Valerio and R. Dovesi, Phys. Rev. B **51**, 7741 (1995).
35. Ph. D'Arco, G. Sandrone, R. Dovesi, R. Orlando and V.R. Saunders Phys. Chem. Minerals **20**, 407 (1993).
36. L. W. Finger and R. M. Hazen, Acta Cryst. B **47**, 561 (1991).
37. A.M. Dziewonski and D.L. Anderson, Phys. Earth Planet. Int. **25**, 297 (1981).
38. E. Ito and E. Takahashi, J. Geophys. Res. **94**, 10637 (1991).
39. N. L. Ross and R. M. Hazen Phys. Chem. Minerals **17**, 228 (1990).
40. M. A. Glazer, Acta Cryst. B **28**, 3384 (1972).
41. Y. Wang, F. Guyot, A. Yaganeh-Haeri and R. C. Liebermann, Sciences, **248**, 468 (1990).
42. T. Yagi, H.K. Mao, P.M. Bell Phys. Chem. Minerals, **3**, 97 (1978).
43. M. O' Keeffe, H. T. Mao and J. O. Bell, Phys. Chem. Minerals **4**, 299 (1979).
44. R. J. Hemley, D. J. Jackson and R. G. Gordon, Phys. Chem. Minerals **14**, 2 (1987).
45. Y. Kudoh, E. Ito and H. Takeda, Phys. Chem. Minerals **14**, 350 (1987).
46. Ph. D'Arco, G. Sandrone, R. Dovesi, E. Aprà and V.R. Saunders Phys. Chem. Minerals **21**, 285 (1994).
47. N. W. Ashcroft, and N. d. Mermin, *Solid State Physics*, (Holt-Saunders, 1976), p. 459.
48. R. Dovesi, C. Roetti, C. Freyria Fava, M. Prencipe and V.R. Saunders, Chem. Phys. **156**, 11 (1991).
49. M. Catti, R. Dovesi, A. Pavese and V.R. Saunders, J. Phys. Cond. Matter **3**, 4151 (1991).
50. M. Catti, A. Pavese, R. Dovesi, C. Roetti and C. Causà, Phys. Rev. B **44**, 3509 (1991).
51. S. Hull, T. W. D. Farley, W. Hayes and M. T. Hutchings, J. Nucl Mater. **160**, 125 (1988).
52. Y. Sumino, L. O. Anderson and I. Susuki, Phys. Chem. Minerals **9**, 38 (1983).
53. R. A. Bartels and V. H. Vettro, J. Phys. Chem Solids **33**, 1991 (1972).
54. R. Colle and O. Salvetti, J. Chem. Phys. **93**, 534 (1990).
55. Ph. D'Arco, M. Causà, C. Roetti ans B. Silvi, Phys. Rev. B **47**, 3522 (1993).
56. N. M. Harrison, V. R. Saunders, E. Aprà, M. Causà and R. Dovesi, J. Phys. Cond. Matter **4**, L261 (1992).
57. F. Freyria Fava, *Tesi di Laurea*, Università di Torino, Torino (1995).
58. R. Dovesi, M. Causà, R. Orlando, C. Roetti and V.R. Saunders J. Chem. Phys. **92**, 7404 (1990).
59. *The Handbook of chemistry and Physics*, 72nd ed., edited by David R. Lide (CRC Press, Boca Raton, FL, 1991-92).
60. F. L. Hill, *An introduction to statistical Thermodinamics*, (Addison Wesley, 1960), p 100
61. J. A. Gaber, and A. V. Granato, Phys. Rev. B **11**, 3990 (1975).
62. J. E. Rapp, and H. D. Merchant, J. Appl. Phys. **44**, 3919 (1973).
63. G. Liebfried and W. Ludwig, Solid State Physics, **12**, 275 (1961).
64. J. T. Lewis, A. Lehoczky, and C. V. Briscoe, Phys. Rev. B **161**, 877 (1967); L. S. Ching, J. P. Day, and A. L. Ruoff, J. Appl. Phys. **44**, 1017 (1973); W. C. Overton, and R. T. Swim, Phys. Rev B **84**, 758 (1951).
65. S. Haussühl, Z. Phys. **159**, 223 (1960).
66. W. G. Wyckoff, *Crystal Structures*, 2nd ed. (Wiley, New York, 1968), vol 1.
67. V. S. Urusov, and V. U. Blinov, Izves. Ak. Nauk SSSR, Ser. Khimichestraya **12**, 278 (1970).

Lattice Dynamics and Thermodynamic Properties

Michele Catti

Dipartimento di Chimica Fisica ed Elettrochimica, Università di Milano,
via Golgi 19, 20133 Milano, Italy

May 29, 1996

Summary. The dynamical properties of atoms in crystalline solids are reviewed, in the framework of the quasi-harmonic approximation for the Born-Oppenheimer potential energy hypersurface. Model interatomic potential functions, fitted to either theoretical or experimental data, are discussed. Methods and results of calculations of phonon spectra and thermodynamic functions, both by use of model potentials and of *ab initio* approaches, are presented.

Key words: Interatomic potentials – Born-Oppenheimer surface – Quasi-harmonic approximation – Phonon spectra – Thermodynamics of crystals

1. Introduction

A major part of the interest, in this series of contributions, is focused on all possible outcomes of a quantum-mechanical calculation of the ground-state total energy for a system of electrons in the field of atomic nuclei arranged periodically, according to the rules of space group symmetry. As is well known for molecular systems also, the information provided concerns not only the strictly electronic and chemical bonding features but also the mechanics of nuclear motion. This is due essentially to the validity of the Born–Oppenheimer (BO) approximation. We recall that, owing to the much higher velocity of the electron with respect to nuclear motion, the dynamical state of the electronic system can be thought of as conforming instantaneously to a change in the relative positions of the nuclei, so that the motions of the two subsystems (electrons and nuclei) are uncoupled. Therefore, the total energy, Φ, of the system depends on the positions of the nuclei (which are denoted collectively as $\mathbf{R}$) in a parametric way:

$$\Phi = \Phi_{ee} + \Phi_{en} + \Phi_{nn} \equiv \Phi(\mathbf{R}) \qquad (1)$$

This function, which was denoted in Chapter 3 as $E_0(\mathbf{R})$, represents a hypersurface in the multidimensional space of the geometrical configurations of the system of atomic nuclei and can be determined, in practice, by sampling a number of points in that space and performing a full quantum-mechanical calculation of the total energy for each point. The function, $\Phi(\mathbf{R})$, has, thus, the meaning of a classical potential energy, which is able to describe the statics and the dynamics of the system of atomic nuclei, in the framework of the formalism of Newtonian mechanics. The application of classical mechanics is, of course, limited by some conditions, which will be touched upon below, but which are not very severe for heavy particles, such as nuclei.

It should be emphasized that $\Phi(\mathbf{R})$ has been derived by intrinsically static methods, corresponding to conditions at the athermal limit $T = 0$ K. Therefore, following this line of development leads one to introduce thermal effects on the nuclear motion, by some kind of *a posteriori* assumption, which, at least, overcomes the problem of a temperature-independent static potential energy. Further, the BO function of type $\Phi(\mathbf{R})$ could have been obtained by quite different means with respect to that outlined above: typically, by the use of euristic formulae of the *model potential* kind, where parameters optimized on empirical properties appear. A mixed method is very often used, based on euristic model potentials whose parameters are optimized not on empirical quantities but on quantum-mechanically computed energy values. In that case, the model potential simply takes the meaning of a numerical interpolation formula; the *ab initio* calculations are usually carried out, not directly on the solid phase of interest but on one or more molecular fragments which simulate reasonably well, the most important chemical interactions within the unit cell. No matter by what method the $\Phi(\mathbf{R})$ function was obtained, from now on, the considerations concerning its use apply on the same footing.

The energy potential function can be utilized from a purely static point of view, in order to determine the structural configuration, $\mathbf{R}_0$, which is stable at 0 K: this means searching for the minimum of $\Phi(\mathbf{R})$. Such an operation may seem to be very simple and to require the knowledge of $\Phi(\mathbf{R})$ only in a small range about the minimum point. This is often not the case, however, for at least two reasons. Firstly, to be certain that the point concerned is really a minimum and not a saddle or pseudo-equilibrium point, it is necessary that not only the equation $\nabla\Phi(\mathbf{R}_0) = 0$ holds, but also that the Hessian matrix of second derivatives of E is positive and definite. Secondly, while finding a local minimum point is relatively easy, finding the absolute minimum in the presence of several secondary minima may be a very complicated process. On the other hand, the systems with many energy minima (i.e., many possible metastable structure configurations, in addition to the stable one) are generally the most interesting to study. Apart from the problem of the most appropriate numerical method (among others: steepest descent, conjugated gradients, Newton-Raphson in the local minimum case, simplexus or simulated annealing for the absolute minimum), searching for an absolute minimum may require the exploration of a large domain in the $\mathbf{R}$ space. However, if, for instance, an *ab initio* method of Hartree-Fock type with a basis set of localized Gaussian functions is used, the same optimized parameters of the basis set will surely not be satisfactory in all points of the domain. It should be added that a very important by-product of the static problem is the determination of the stability conditions of the solid phase at variable pressure (at 0 K) and also of the athermal equation of state

$p = p(V)$. Further, the elastic constants of the phase can also be derived from a knowledge of suitable components of the Hessian matrix of the energy, $\Phi(\mathbf{R})$.

The really challenging problem is, however, twofold: modelling the local dynamics of the periodic atomic system by introducing the variable, time and representing its average thermodynamic behaviour by use of the variable, temperature. This can be done in two different ways, starting in both cases from the knowledge of the interatomic potential energy function, $\Phi(\mathbf{R})$. The first, more direct way is to employ the method of Molecular Dynamics (MD). That means applying, straightforwardly, the Newtonian equations of motion of classical dynamics to the atoms of the system, in the potential energy field, $\Phi(\mathbf{R})$. Actually a "box" of limited size (say, 100-10,000 atoms) is considered, removing the translational symmetry inside and applying periodic boundary conditions outside. The equations of motion are integrated numerically with respect to time, by the method of finite differences, choosing appropriately the finite time step and the starting dynamical conditions. As a result, the time evolution of positions and velocities of all atoms in the box are obtained, for as many time steps as allowed by computational resources. In principle, therefore, all local information on the microscopic dynamics of the single atomic particles is provided by this kind of calculation. Moreover, the atomistic dynamics are linked in a natural way to macroscopic thermodynamics by equating the total time-averaged kinetic energy, $1/2\sum_i m_i\langle v_i^2\rangle$, to $3/2Nk_B\langle T\rangle$, where $\langle T\rangle$ is the mean temperature. Similarly, the total average potential energy can be computed by summing the contributions from all particles and thus, the total energy (equal to the standard internal energy, E) is obtained as the sum of the kinetic and potential terms. A straightforward connection is therefore established between temperature and the velocities of atomic motion, so that the initial temperature of the MD simulation can be changed by simply rescaling the starting velocities of all particles. According to standard concepts of statistical thermodynamics, the calculation can be performed for a microcanonical ensemble (called NVE, because the number of particles N, the volume and the energy are kept constant), for a constant pressure ensemble, (NpH) [1], for canonical (NVT) or constant pressure canonical (NpT) ensembles [2]. The cases at constant pressure allow the unit-cell geometry and the symmetry of the crystal to change during the simulation, so as to study, for instance, structural phase transitions in the solid state.

The MD simulation technique is very powerful and turns out to be particularly appropriate for dealing with heavily disordered systems, which may be characterized by high atomic mobility (e.g. ionic conductors) or with systems at very high temperature, where the anharmonic effects become important. Excluding such conditions, however, a second method for simulating the dynamics and thermal effects in solids can be used with equal, or sometimes greater, success [3]. This is called historically *lattice dynamics* (but we should specify, in the quasi-harmonic approximation) and is based on approximating the potential energy, $\Phi(\mathbf{R})$, by the first non-zero quadratic term of the Taylor expansion about the equilibrium configuration, $\mathbf{R}_0$. Thus, the classical dynamical problem can be solved, straightforwardly, in an analytical way, in the form of normal modes of vibration oscillating sinusoidally. Furthermore, as the problem of the harmonic oscillator is solvable exactly by quantum-mechanical methods, the total energy of the normal vibration modes can be dealt with on a quantum basis. This allows us to reproduce correctly the crystal dynamics at low temperature, where the classical formula of the energy equipartition is insufficient and thus, methods like molecular dynamics are quite unsatisfactory. The temperature is introduced simply through Boltzmann's statistical mechanical formulae, so that analytical expressions can be derived immediately for all thermodynamic functions. Summarizing, the attractive features of lattice dynamics are the following: simple, analytical solutions of the dynamical problem, instead of tables of numerical results; a correct quantum formulation of the total energy working in the low temperature regime. It should be added that the quadratic coefficients of the potential energy Taylor expansion (force constants) can be calculated, not only by numerical differentiation of $\Phi(\mathbf{R})$ but also directly by first-principles quantum-mechanical methods, as will be discussed in the sections below.

Before entering into the details of mainstream lattice dynamics and in order to conclude this short overview of theoretical methods for studying nuclear motions in solids, the Car-Parrinello *ab initio* molecular dynamics [4] should be mentioned, too. This method differs from all those previously discussed for a fundamental reason: no knowledge of the potential energy function, $\Phi(\mathbf{R})$, is required as a starting point, but rather, the problems of electronic and nuclear motions are tackled and solved at the same time. Briefly, the classical newtonian equations of motion for the nuclei are written together with fictitious newtonian equations, where the role of position coordinates is played by the orbital-like monoelectronic functions, which are the unknowns of the self-consistent-field electron problem. The latter is typically formulated within the scheme of the Kohn-Sham equations of the density functional theory. In the pseudo-newtonian equations, the variable, time, has no real physical meaning, but rather labels all possible orbital functions, among which the variational method has to determine those which minimize the energy, considered as their functional. Thus, at each time step, the electron problem is solved and the time evolution of nuclear positions and momenta (as in classical MD) is computed contemporaneously.

2. Model potential functions

In the most general case, the potential energy function, $\Phi(\mathbf{R})$, can be represented as the sum of terms depending on interatomic distances, r_{ij} (a two-body potential), on bond angles φ_{ijk} (a three-body potential), on torsional angles θ_{ijkl} (a four-body potential) and so on:

$$\Phi(\mathbf{R}) = \sum_{ij} E_{ij}(r_{ij}) + \sum_{ijk} E_{ijk}(\varphi_{ijk}) + \sum_{ijkl} E_{ijkl}(\theta_{ijkl}) + \cdots \qquad (2)$$

Three- and four-body potentials express interactions of an essentially covalent type, while the two-body ones may have either covalent, ionic or dispersive character. The fundamental question is to choose an approximate but realistic model, including in (2) only the most significant terms on physico-chemical grounds, in relation to the nature of the crystalline phase studied. The next problem is to determine explicit values for the parameters, contained in the chosen potential formulae [5] and this will be discussed below. Typical expressions used for two-body potentials follow [6].

The Born-Mayer (or Buckingham) potential, despite its simplicity, works well in a variety of situations where the bonding may be partially ionic, dispersive and covalent:

$$E_{ij} = e^2 \frac{z_i z_j}{r_{ij}} + B_{ij} \exp\left(-\frac{r_{ij}}{\rho_{ij}}\right) - \frac{c_{ij}}{r_{ij}^6} \tag{3}$$

In particular, the partially covalent character of the two-body interaction may be accounted for by letting the atomic charges z_i take empirical values smaller than the ideal fully ionic ones. The dispersive coefficients can be derived from experimental refractivity data, through the London formula, or may be considered as empirical parameters. The repulsive coefficients are also often written as

$$B_{ij} = \exp\left(-\frac{r_i + r_j}{\rho_{ij}}\right) \tag{4}$$

where the r_i lengths are called repulsive radii and their relative values are roughly similar to ionic radii. Repulsive radii and hardness parameters, ρ_{ij}, have to be determined by a fitting procedure.

The harmonic (quadratic) and Morse potentials, effective for covalent bonds, have the following forms, respectively:

$$E_{ij} = \frac{1}{2} k_r (r_{ij} - r_{ij}^0)^2 \tag{5}$$

and:

$$E_{ij} = D_{ij}\{1 - \exp[-a(r_{ij} - r_{ij}^0)]\}^2 - D_{ij} \tag{6}$$

In the first, the parameters r_{ij}^0 and k_r represent, respectively, the equilibrium bond distance and the harmonic force constant for the bond length deformation. In the Morse potential, D_{ij} corresponds to the two-body bond energy and a is, again, related to the force constant of the bond. Thus, these parameters may either be taken directly from experimental data (bond lengths, bond stretching frequencies and bond energies) or they may be fitted to experimental or theoretical energy-related properties.

The three-body angular potentials are usually of harmonic type:

$$E_{ijk} = \frac{1}{2} k_\varphi (\varphi_{ijk} - \varphi_{ijk}^0)^2 \tag{7}$$

The two parameters appearing in this expression are related to the bond-bending force constant and to the equilibrium bond angle.

A further way to define the potential energy function, using a fully empirical approach, is the following: for each two-body atom-atom interaction, two parameters are given, corresponding to the bond force constant ($A_{ij} = \partial^2\Phi(\mathbf{R})/\partial r_{ij}^2$) and to the residual bond force ($F_{ij} = \partial\Phi(\mathbf{R})/\partial r_{ij}$). The energy derivatives are calculated at the equilibrium interatomic distances and the net forces on every atom must vanish.

The interatomic potentials examined so far correspond to a rigid ion or rigid atom model, since the ability of atoms to polarize their electron density is not taken into account. Therefore, the dielectric and optical properties of the crystal cannot be reproduced at all. One way to overcome such limitations is to use the dipole shell model [7], according to which the i-th atom is considered to be made up of a core (the nucleus plus the core electrons) and a shell (the valence electrons). Core and shell are characterized by their own charges X_i and Y_i, respectively, with the obvious condition that $X_i + Y_i = z_i$. Further, they are located at a distance d_i from each other and interact elastically by a harmonic-type law, with energy: $1/2k_id_i^2$. The atomic polarizability is given by the formula $\alpha_i = Y_i^2/k_i$. The model parameters Y_i, k_i and d_i have to be determined by fitting to experimental dielectric properties. Further refinements and improvements are given by the breathing shell [8] and quadrupolar shell [9] models.

The undetermined parameters, contained in the above potential expressions, are optimized, in some cases, on empirical properties, in other cases, on BO energy functions calculated by quantum-mechanics. Mixed empirical-theoretical schemes have also been proposed. The empirical properties are usually the experimental equilibrium crystal structure and the elastic constants [10][11]; in the case of the shell model, dielectric properties also have to be included. The structure depends on the first derivatives of the potential energy with respect to unit-cell constants and atomic coordinates, which must be vanishingly small. The elastic constants are related to the energy second derivatives according to $C_{hk} = (1/V)\partial^2\Phi(\mathbf{R})/\partial\eta_h\partial\eta_k$, where the strain components $\eta_h(h = 1, \cdots, 6)$ are related to changes in the six unit-cell constants, due to lattice strain. The structural relaxation induced by the unit-cell deformation must be taken into account, too, so that second derivatives with respect to atomic coordinates are needed as well [11]. Examples of parameter optimization for potentials of the type given in equations 3-7 are given in references [12] and [13]. In more accurate work, the vibrational frequencies (see section 3), obtained from spectroscopic measurements, can also be included in experimental data, used for the parameter fitting [14].

A number of studies have been carried out by fitting the parameters of model potentials to theoretical $\Phi(\mathbf{R})$ hypersurfaces, in the case of α-quartz and other polymorphs of SiO_2. Molecular clusters such as H_4SiO_2, $H_6Si_2O_7$ or $H_{12}Si_5O_{16}$ have been used, employing both HF (Hartree-Fock) [15, 16, 17] and LDF (local density functional) [18] quantum-mechanical approaches. The potential formulae applied range from the Born-Mayer (equation 3) to force-field-type expressions, such as those in equations 5-7 and combinations of these, some including the shell model. Different fitting strategies are used: the "observed" quantities are, sometimes, the numerical theoretical profiles of the energy against a single generalized coordinate of the molecular cluster, while, in other cases, they are simply the theoretical second derivatives of $\Phi(\mathbf{R})$ with respect to two atomic displacements or two internal coordinates. The mixed theoretical-empirical approach of reference [17], where experimental information, as well as quantum-mechanical energy results, on H_4SiO_4 are included in the fitting, has proved particularly successful. Some insufficiency of theoretical results coming from molecular clusters has been claimed by some authors, so that using calculations on periodic systems instead might improve substantially the quality of the fitted potentials.

3. Lattice dynamics

3.1. Harmonic approximation

The motion of atoms around their equilibrium positions in the crystal is approximated as harmonic vibrational motion [3, 19], depending on time with sinusoidal-like behaviour. This result follows necessarily from the form which the potential energy, $\Phi(\mathbf{R})$, is given, by truncating, to second-order, its Taylor expansion around the equilibrium configuration, $\mathbf{R}_0$. Let every atom in the crystal be labelled by the s index ($s = 1, \cdots, p$), identifying it among the p atoms in the unit cell and by the lattice vector, $\mathbf{T}$, identifying the unit cell containing the atom. The translational symmetry which rules the equilibrium atomic positions is expressed by the relation: $\mathbf{R}^0_{s\mathbf{T}} = \mathbf{R}^0_s + \mathbf{T}$, where, by omitting the $\mathbf{T}$ index, the reference unit cell is understood. The dynamical variables are the vectors $\mathbf{u}_{s\mathbf{T}}(t)$, expressing the shifts of atoms from their equilibrium positions as functions of time. Clearly, the relation: $\mathbf{R}_{s\mathbf{T}}(t) = \mathbf{R}^0_{s\mathbf{T}} + \mathbf{u}_{s\mathbf{T}}(t)$ holds. Thus, the harmonic approximation gives $\Phi(\mathbf{R})$ the following form:

$$\Phi(\mathbf{R}) = \Phi(\mathbf{R}_0) + \frac{1}{2} \sum_{jj'ss'\mathbf{T}} \left(\frac{\partial^2 \Phi}{\partial u_{js} \partial u_{j's'\mathbf{T}}} \right)_{\mathbf{u}=0} u_{js} u_{j's'\mathbf{T}} \tag{8}$$

where the j index means the Cartesian component ($j = 1, 2, 3$). $\Phi(\mathbf{R})$ is the potential energy referred to a unit cell of the crystal and $\Phi(\mathbf{R}_0)$ is its static component ($\mathbf{u} = 0$). The harmonic truncation of the Φ function is valid in so far as the atomic shifts $\mathbf{u}$ are very small with respect to interatomic distances. This approximation is, of course, used to represent the atomic vibrational motion in molecules also. The second derivatives of the potential energy, $(\partial^2\Phi/\partial\, u_{js}\partial\, u_{j's'\mathbf{T}})_{\mathbf{u}=0} \equiv V_{jj'}(s, s', \mathbf{T})$, take the meaning of force constants, which couple the shifts for pairs of different atoms in the crystal and can be calculated, either analytically or numerically from the knowledge of the $\Phi(\mathbf{R})$ function. By consideration of the newtonian equations of motion for atoms:

$$m_s \ddot{u}_{js} = -\partial\Phi/\partial u_{js} = -\sum_{j's'\mathbf{T}} V_{jj'}(s, s', \mathbf{T}) u_{j's'\mathbf{T}} \tag{9}$$

a set of second-order differential equations is obtained, which have solutions in the form of travelling waves of the type:

$$u_{js\mathbf{T}}(t) = U_{js} \exp[\imath(\mathbf{k} \cdot \mathbf{T} - \omega t)] \tag{10}$$

These are called vibration normal modes and are characterized by the following features: all atoms oscillate harmonically with the same angular frequency, ω and with amplitudes, U_{js}, which depend on the type of atom in the unit cell but not on the unit cell itself; atoms related by a lattice vector, $\mathbf{T}$, undergo a phase shift equal to $\mathbf{k} \cdot \mathbf{T}$.

The wave vector $\mathbf{k}$, identifying the travelling direction of the wave and controlling the phase shifts of atoms belonging to different unit cells, plays a central role in all the formalism of lattice dynamics. It arises only from the translational symmetry of the system and not from the specific assumptions of the physical model. In fact, the phase factor $\exp(\imath\mathbf{k} \cdot \mathbf{T})$ appearing in the oscillating shift of an atom when it is translated by the lattice vector $\mathbf{T}$, is identical to the phase factor induced in the electronic wave function when the electron position is subjected to the same translation (Bloch's theorem: see Chapter 2). This is simply related to the group-theoretical result that the irreducible representation of the group of translations contains elements of type $\exp(\imath\mathbf{k} \cdot \mathbf{T})$. For the same reason, the independent wave vectors $\mathbf{k}$ are only those contained in the unit cell of the reciprocal lattice (or, equivalently, in the first Brillouin zone). The vibration modes characterized by $\mathbf{k} \approx \mathbf{0}$ ($\lambda = 2\pi/|\mathbf{k}| \to \infty$) correspond to very large portions of the crystal where all atoms oscillate in-phase; this resembles totally symmetrical solutions in which the translational symmetry holds not only for equilibrium $\mathbf{R}_0$ but also for instantaneous atomic positions, $\mathbf{R}$. On the other hand, when $\mathbf{k}$ is close to the boundary of the first Brillouin zone, the wavelength, λ, takes its minimum values (of the order of a few unit cells), so that phase shifts operate at very short range. It should be observed finally that the U_{js} quantities are actually complex amplitudes, comprising a phase factor which is relative to the motion of atoms belonging to the same unit cell: $U_{js} = |U_{js}| \exp(\imath\varphi_{js})$.

3.2. Born matrix and dispersion relation

If the travelling wave (equation 10) is substituted into the Newtonian equations (given in equation 9), it turns out to be an acceptable solution only on condition that the following homogeneous linear equations, in the complex amplitudes, U_{js}, are satisfied:

$$\sum_{j'=1}^{3} \sum_{s'=1}^{p} [B_{jj'}(s, s', \mathbf{k}) - m_s \omega^2 \, \delta_{jj'} \, \delta_{ss'}] U_{j's'} = 0 \tag{11}$$

where the quantities $B_{jj'}(s, s', \mathbf{k}) = \sum_{\mathbf{T}} V_{jj'}(s, s', \mathbf{T}) \exp(\imath \mathbf{k} \cdot \mathbf{T})$ are Fourier transforms of the force constants. It is convenient to contract the pair of indexes js into a single one; thus, the amplitudes can be considered to be components of a $3p \times 1$ column matrix $\mathbf{U}$ and the $B_{jj'}(s, s', \mathbf{k})$ quantities to be components of the $3p \times 3p$ square matrix $\mathbf{B}(\mathbf{k})$. The latter is called dynamical or Born matrix, and can be shown to be hermitian. Thus, equation 11 can be rewritten in matrix form as $(\mathbf{B} - \mathbf{m}\omega^2)\mathbf{U} = 0$ and the condition to obtain a non-zero $\mathbf{U}$ solution is that the generalized secular equation holds:

$$\det[\mathbf{B}(\mathbf{k}) - \mathbf{m}\omega^2] = 0 \tag{12}$$

The $3p \times 3p$ mass diagonal matrix, $\mathbf{m}$, has components $m_{jsj's'} = m_s \delta_{jj'} \delta_{ss'}$. Therefore, a travelling wave is not an acceptable solution to the equations of motion, unless its frequency ω is related to the wave vector $\mathbf{k}$ by equation 12. The latter is an algebraic equation of order $3p$, in the squared angular frequency ω, which then gives solutions of type $\omega_1(\mathbf{k}), \cdots, \omega_n(\mathbf{k}), \cdots, \omega_{3p}(\mathbf{k})$. These $3p$ functional relationships between ω and $\mathbf{k}$ are called branches of the dispersion relation, $\omega(\mathbf{k})$. In other words, in the normal modes of vibration the frequency and the wavevector are not independent quantities, but are related by a multi-valued ($3p$-valued) function called the dispersion relation.

Three branches of the dispersion relation share a peculiar behaviour: the frequency goes to zero linearly as $|\mathbf{k}|$ approaches zero, along any direction in the Brillouin space. These are called *acoustic* branches, because, for small $|\mathbf{k}|$ values, the corresponding vibration modes behave as long-wavelength acoustic waves (frequency inversely proportional to wavelength). All other branches of the dispersion relation show finite non-zero frequencies at the Brillouin zone centre and are known as *optic* modes. These correspond to vibrations with large phase shifts between atoms in the same unit cell, so that the unit-cell dipole moment may oscillate, too and give rise to interaction with electromagnetic radiation.

By using periodic boundary conditions, the finite dimensions of the crystal (N unit cells) can be accounted for, without introducing the symmetry-breaking defect of surface. As a consequence, the only values inside the Brillouin zone allowed, for the wave vector $\mathbf{k}$, are those satisfying the relation $\mathbf{k} = \sum_j (n_j/N_j)\mathbf{b}_j$, where $N_1N_2N_3 = N$, n_j is any integer number and $\mathbf{b}_j$ is a vector of the reciprocal lattice basis (see Chapter 2). Therefore, the total number of vibration modes is $3p$ (the number of eigenfrequencies, ω_n) $\times N$ (the number of allowed wave vectors, $\mathbf{k}$) and is then equal to the number of degrees of freedom of the crystal. Every normal mode of vibration is identified by means of the index n (running from 1 to $3p$) of the dispersion relation branch and of the wave vector $\mathbf{k}$ (among the N values allowed by the periodic boundary conditions).

By substitution of the dispersion relation into equation 11, the homogeneous linear equations in the complex amplitudes U_{js} can be solved and thus, the eigenvectors, $\mathbf{U}$, of the dynamical matrix are determined except for a scale factor. For each eigenvalue, $\omega_n(\mathbf{k})$, there is one normalized eigenvector, $\mathbf{e}_n(\mathbf{k})$ and so their number is equal to that of vibration modes. The conditions of orthonormality hold:

$$\sum_j \sum_s e^*_{js,n}(\mathbf{k}) e_{js,n'}(\mathbf{k}') = \delta_{nn'}\delta_{\mathbf{k}\mathbf{k}'} \tag{13}$$

The most general solution of the classical harmonic vibrational dynamics of atoms in the crystal is given by a linear combination of all normal modes, weighted by appropriate coefficients, $A_n(\mathbf{k})$:

$$u_{js\mathbf{T}}(t) = \sum_n \sum_{\mathbf{k}} A_n(\mathbf{k}) e_{js,n}(\mathbf{k}) \exp\{\imath[\mathbf{k}\cdot\mathbf{T} - \omega_n(\mathbf{k})t]\} \tag{14}$$

Of course, the absolute complex amplitudes are given by:

$$U_{js,n}(\mathbf{k}) = A_n(\mathbf{k}) e_{js,n}(\mathbf{k}) \tag{15}$$

3.3. Experimental methods and empirical fitting

The dispersion relation, $\omega(\mathbf{k})$, contains the most important information concerning vibrational motions of atoms in the crystal. It is, therefore, worthwhile to present briefly some experimental methods to obtain the spectrum of oscillation frequencies of a solid substance. First, the classical vibrational spectroscopies (infrared and Raman), which are also used to study molecular systems, in the gaseous or liquid phase, should be recalled. These techniques, however, when applied to solids present a fundamental limitation: only vibration modes with $\mathbf{k} \approx 0$ may be revealed, so that just a small part of the eigenfrequencies spectrum (that at the centre of the Brillouin zone) can be explored. In fact, the interaction between radiation and matter produces a detectable intensity only if a sufficiently large number of unit cells vibrate in-phase; that means very large wavelengths and thus, a vanishing wave vector. For $\mathbf{k}$ appreciably far from the zone centre, on the other hand, a destructive interference arises from out-of-phase motions of small clusters of unit cells and the spectroscopic signal is destroyed. Infrared and Raman spectra of crystals are, however, very useful, because of the importance of frequency data close to zone centre and because the relative experimental equipment is easily accessible.

In order to measure the frequencies corresponding to non-zero wave vectors, inelastic neutron scattering should be used [19]. Neutrons interact with the normal modes of vibration of the lattice (or phonons, see section 4.1), according to two fundamental conservation laws. In the very important case of one-phonon scattering (absorption), these laws take the following form:

$$E' - E = \hbar\omega_n(\mathbf{k}) \qquad \text{(energy conservation)} \tag{16}$$
$$\mathbf{p}' - \mathbf{p} = \hbar\mathbf{k} + \mathbf{K} \qquad \text{(crystal momentum conservation)} \tag{17}$$

where $\mathbf{K}$ is any vector of the reciprocal lattice.

The initial (incident beam), final (scattered beam) energy and the momentum of the neutrons are easily measured from their velocity and direction. Thus, the energy and momentum transferred to the vibrating crystal lattice can be derived; for different directions and velocities of the scattered beam, several wave vectors, $\mathbf{k}$, are explored and the corresponding frequencies, $\omega(\mathbf{k})$, are obtained. It should be observed, however, that such experiments need a very powerful neutron source (nuclear reactor or spallation source), expensive equipment and large single crystals (about 1 cm^3 of volume) as samples, so that measuring a full phonon spectrum along a few $\boldsymbol{\kappa}$ directions may be a hard task to accomplish.

An example of phonon dispersion relations determined by neutron scattering for CaF_2 (fluorite) [20] is shown in Figure 1.

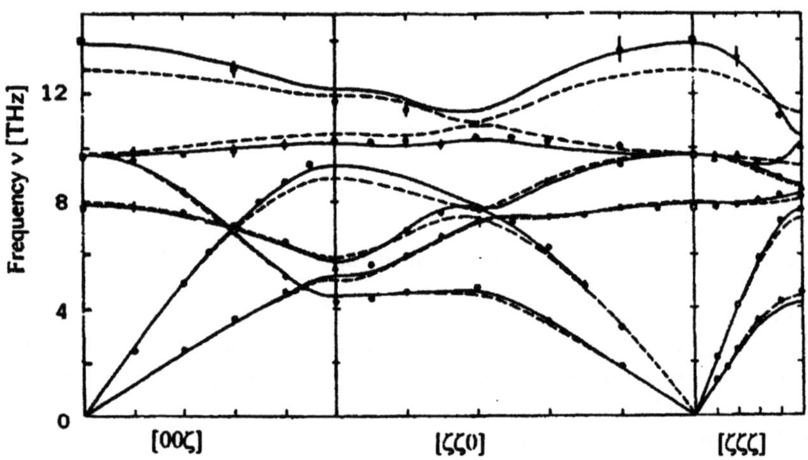

Figure 1. Phonon dispersion relations for CaF_2 in the [001], [110] and [111] directions (see ref. 20).

Three directions are explored in the wave vector reciprocal space: [001], [110] and [111]; the diagram is divided in three parts correspondingly. Frequency values from neutron scattering measurements are marked as circles and triangles; squares correspond to infrared and Raman zone-centre data. The broken and full curves denote calculated dispersion curves from a rigid-ion and a shell-model potential, respectively. Parameters of the potentials included bond force constants A_{ij} and residual forces F_{ij} for Ca-F, Ca-Ca, F1-F1 and F1-F2 interactions. One single and two degenerate acoustic branches are observed, together with four optic branches, in the dispersion relation. The shell model potential appears to account for optic frequencies better than the rigid-ion model does.

3.4. Quantum-mechanical derivation of force constants

The lattice dynamics in the harmonic approximation can be coupled with *ab initio* treatments of the electronic properties of the solid in two ways. The simplest one is to derive the Fourier transforms of force constants, $B_{jj'}(s, s', \mathbf{k})$, by second-order differentiation of the $\Phi(\mathbf{R})$ function, obtained as a ground-state total energy for different structural configurations $\mathbf{R}$. For $\mathbf{k} \neq 0$, it is necessary to devise suitable supercells, in order to account for the lowering of translational symmetry and this may raise the computational cost very much. For zone-centre vibrational modes, on the other hand, structural deformations preserve the full lattice periodicity and calculations are much simpler. Further, in some favourable cases, the problem can be formulated, straightforwardly, in "normal coordinates", so that at least part of the Born matrix at $\mathbf{k} = 0$ is obtained in diagonal form, yielding the eigenfrequencies directly. This is the case for two zone-centre optic modes of fluorite, CaF_2 [21]. In the Raman active mode, two centrosymmetrical fluorines vibrate along the cubic cell diagonal against the fixed Ca atom at the origin. The normal coordinate, u, is the change in the F-F distance and the reduced mass is $M = m_F/2$. In the infrared active mode, the two F atoms vibrate in-phase, with the same displacement; the normal coordinate is the change in the F- Ca distance and the reduced mass is $M = 2m_F m_{Ca}/(2m_F + m_{Ca})$. By polynomial fitting of the total energy Φ against u, the force constants B can be derived as $d^2\Phi/du^2$ and the corresponding frequencies are given as $\nu = (B/M)^{1/2}$ for each case. In the quoted study, where the total energy, Φ, was computed by the periodic Hartree-Fock approach, the results ν_{Raman}=10.1 THz and ν_{infrared}=8.1 THz were obtained, in very good agreement with the experimental values of 9.9 and 8.3 THz, respectively.

The other way to obtain the harmonic force constants, quantum-mechanically, is based on the analytical theory of linear response [22]. By straightforward application of the Hellmann- Feynman theorem, the force associated with an atomic displacement, $u_{js\mathbf{T}}$, is:

$$\frac{\partial \Phi}{\partial\, u_{js\mathbf{T}}} = \int n(\mathbf{r}) \frac{\partial\, V(\mathbf{r})}{\partial\, u_{js\mathbf{T}}} d\mathbf{r} \tag{18}$$

where $n(\mathbf{r})$ and $V(\mathbf{r})$ are the electron density and the bare external potential acting on the electrons, respectively, at the position $\mathbf{r}$. If $n(\mathbf{r})$ is expanded to linear order against the atomic displacement and the force is integrated, an expression of Φ to second-order in the atomic displacements is obtained, whose bilinear coefficients are the force constants requested:

$$\begin{aligned} \left(\frac{\partial^2 \Phi}{\partial\, u_{js} \partial\, u_{j's'\mathbf{T}}} \right)_{\mathbf{u}=0} &\equiv V_{jj'}(s, s', \mathbf{T}) = \\ &= \int \left[\frac{\partial n(\mathbf{r})}{\partial\, u_{js}} \frac{\partial V(\mathbf{r})}{\partial\, u_{js\mathbf{T}}} + n_0(\mathbf{r}) \frac{\partial^2 V(\mathbf{r})}{\partial u_{js} \partial u_{j's'\mathbf{T}}} \right] d\mathbf{r} \end{aligned} \tag{19}$$

The Fourier transforms of force constants (components of the Born matrix) can be derived by introducing oscillating displacements (of the type given in equation 10) into the above expression. It is, therefore, clear that, if the response of the electron density to oscillating atomic shifts is known, the dynamical matrix can be calculated directly. The linear response theory has been developed, thoroughly, in the framework of the density functional (DFT) approach, by application of first-order perturbation theory to the change in the electron density, induced by changes of the SCF-DFT electron potential energy, which, in their turn, are caused by atomic displacements. The matrix elements with respect to a plane-wave basis set are expressed by analytical formulae.

The linear response theory formalism has been applied to the calculation of phonon dispersion curves in semiconductors, i.e. Si, Ge and several members of the III-V family[22]. In Figure 2, results are shown for Si and GaAs, over a number of directions in the first Brillouin zone. Comparison with experimental data (diamonds) shows that this *ab initio* calculation within the harmonic model is very successful.

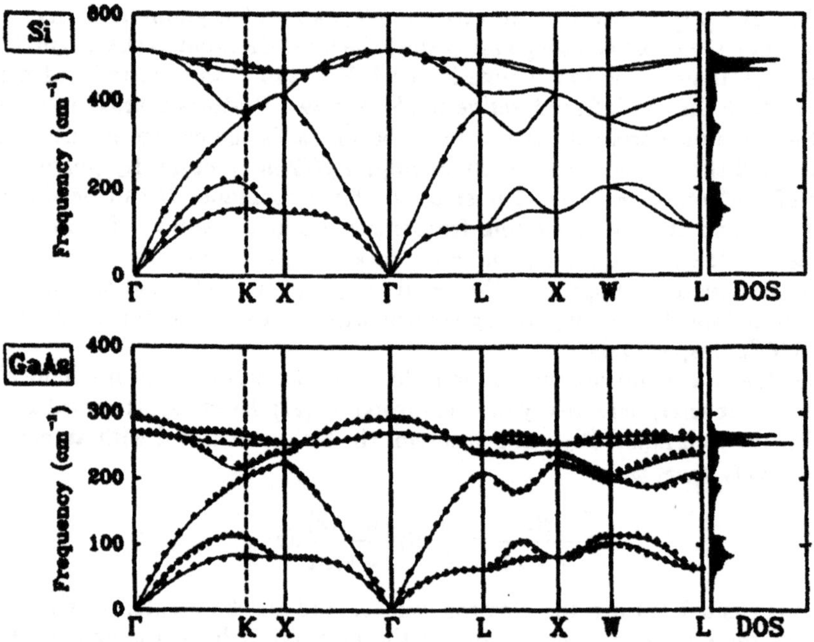

Figure 2. Phonon dispersions for Si (above) and GaAs (below) from *ab initio* calculations (see ref. 22).

4. Thermodynamic properties

4.1. *Energy quantization and phonons*

Within the framework of the purely classical theory of harmonic lattice dynamics summarized above, the total (kinetic + potential) vibrational energy per unit cell, corresponding to a given $n\mathbf{k}$ normal mode, is equal to:

$$E_{vib}(n\mathbf{k}) = \frac{1}{2}\omega_{n\mathbf{k}}^2 \sum_s m_s \sum_j |U_{js}|^2 = \frac{1}{2}\omega_{n\mathbf{k}}^2 A^2(n\mathbf{k}) \sum_s m_s \sum_j |e_{js}(n\mathbf{k})|^2. \quad (20)$$

In the most general case, a superposition of appropriately weighted normal modes is present and the total energy depends only on the squares of these weights or excitation degrees of the vibration modes. The average values of the $A(n\mathbf{k})$ coefficients are related to temperature by the formulae of classical statistical mechanics and by means of these, one may obtain the classical value of the average total vibrational energy $\langle E_{vib}(n\mathbf{k})\rangle = 3pNk_BT$. This expression reproduces the experimental data only for temperatures very far from absolute zero. On the other hand, the harmonic approximation leads to a very simple dynamic problem (the harmonic oscillator), which can be tackled successfully even on a quantum-mechanical basis, as is well known. All of the theory concerning the decomposition of motion into normal modes of vibration holds quite well in the quantum formulation. What changes is simply, the formula for the total energy, which becomes, for the single $n\mathbf{k}$ mode:

$$E_{vib}(n\mathbf{k}) = \left(\frac{1}{2} + \nu_{n\mathbf{k}}\right)\hbar\omega_{n\mathbf{k}} \quad (21)$$

The degree of excitation of the normal mode is now expressed by the vibrational quantum number $n\mathbf{k}$, instead of by the continuous variable $A(n\mathbf{k})$. Boltzmann statistics can be applied, deriving the average value of the vibrational quantum number for each mode and then summing over all modes to obtain the average total vibrational energy:

$$\langle \nu_{n\mathbf{k}} \rangle = \frac{1}{\exp\left(\frac{\hbar\omega_{n\mathbf{k}}}{k_BT}\right) - 1} \quad (22)$$

$$\langle E_{vib} \rangle = \sum_{n\mathbf{k}} \left[\frac{1}{2} + \frac{1}{\exp\left(\frac{\hbar\omega_{n\mathbf{k}}}{k_BT}\right) - 1}\right]\hbar\omega_{n\mathbf{k}} \quad (23)$$

It can easily be shown that the latter expression is equivalent to its classical counterpart in the limit of very large T, including the quantum zero-point term.

Equation 22 is very important because it leads in a very natural way to a particle interpretation (phonons) of lattice vibrations, which parallels the wave-like view of normal modes (travelling waves). Thus, a phonon of energy $\hbar\omega_{n\mathbf{k}}$ is associated with the normal mode of angular frequency $\omega_{n\mathbf{k}}$; the quantum number $\nu_{n\mathbf{k}}$ represents the number of phonons with energy $\hbar\omega_{n\mathbf{k}}$ and the formula, in equation 22, gives the average number of such phonons. This particle interpretation shows that equation 23 represents the statistical distribution function of phonons with respect to their energy $\hbar\omega_{n\mathbf{k}}$, according to the well known Bose-Einstein formulation. Phonons are, therefore, particles belonging, like photons, to the statistical species of bosons.

4.2. Thermodynamic functions and density of states

The quantity $\langle E_{vib}\rangle$ given in equation 23 represents the vibrational contribution to the thermodynamic, macroscopic internal energy of the solid system.

Analogously, the vibrational terms of all other thermodynamic functions can be calculated from a knowledge of the full $\omega_{n\mathbf{k}}$ frequency spectrum. The most important formulae concern F (Helmholtz free energy), S (entropy) and C_V (constant-volume heat capacity):

$$\langle F_{vib}\rangle = -k_B T \ln Z = \sum_{n\mathbf{k}} \left\{ \frac{1}{2}\hbar\omega_{n\mathbf{k}} + k_B T \ln\left[1 - \exp\left(\frac{-\hbar\omega_{n\mathbf{k}}}{k_B T}\right)\right]\right\} \tag{24}$$

$$\langle S_{vib}\rangle = \sum_{n\mathbf{k}} \left\{ -k_B \ln\left[1 - \exp\left(\frac{-\hbar\omega_{n\mathbf{k}}}{k_B T}\right)\right] + \frac{1}{T}\frac{\hbar\omega_{n\mathbf{k}}}{\exp\left(\frac{\hbar\omega_{n\mathbf{k}}}{k_B T}\right) - 1}\right\} \tag{25}$$

$$\langle C_V\rangle = \sum_{n\mathbf{k}} \frac{\hbar^2\omega_{n\mathbf{k}}^2}{k_B T^2} \frac{\exp\left(\frac{\hbar\omega_{n\mathbf{k}}}{k_B T}\right)}{\left[\exp\left(\frac{\hbar\omega_{n\mathbf{k}}}{k_B T}\right) - 1\right]^2} \tag{26}$$

In practice, the sums over all normal modes, $n\mathbf{k}$, are calculated by taking into account all branches of the dispersion relation, but only a very limited number of the N allowed wave vectors. There are suitable ways to select a few tens or hundreds of $\mathbf{k}$ points in the Brillouin zone, so as to obtain quite satisfactory results.

There is perhaps a more convenient way to calculate the thermodynamic functions, which is based on use of the density of vibrational (phonon) states $g(\omega) = \sum_n g_n(\omega)$. This function gives the number of normal modes (phonons) with angular frequency (energy) in the range between ω and $\omega + d\omega$ ($\hbar\omega + d(\hbar\omega)$) and is computed numerically by a simple count, within the set of wave vectors explored for each branch term $g_n(\omega)$. By knowledge of the density of states, any thermodynamic function, f, can be calculated through the formula:

$$f = \int f(\omega) g(\omega) d\omega \tag{27}$$

where $f(\omega)$ is the contribution to f from the single mode of frequency ω and the integral is approximated by a sum over the eigenfrequencies, obtained as solutions of equation 12. The density of phonon states is very important, because it provides an overall picture of the whole phonon energy spectrum and contains implicitly all information on the thermodynamic behaviour of the solid. A quite similar role is played by the density of electron states in the theory of electron energy bands. The density of phonon states can be determined by an inelastic neutron scattering experiment more simply than the full phonon spectrum, resolved in the $\mathbf{k}$ space, can be: in fact, a polycrystalline sample is sufficient, instead of a very large single crystal which needs to be oriented according to the different $\mathbf{k}$ directions.

Table 1. Parameters of the rigid ion (RIM, first section) and shell model (SM, second section) potentials optimized for $CaCO_3$ (calcite) (see ref. 14).

	O – O	Ca – O	C – O
B_{ij}(eV)	14683.5	1870.3	54129×10^8
	20431.8	1605.4	13258×10^8
ρ_{ij}(Å)	0.2107	0.2893	0.0402
	0.2127	0.2965	0.0415
c_{ij}(eVÅ^6)	3.47	0	0
	3.47		
z_O(e)	−0.995		
	−1.095	$z_{O_s} = -2.662$	$z_{O_C} = 1.567$
k_ϕ(eVrad^{-2})	2.550		
	2.881		
k_s(eVÅ^{-2})	0		
	177.4		

4.3. Quasi-harmonic model

From the expression of the vibrational Helmholtz free energy (equation 24), by differentiation with respect to volume, a formula can be derived for the vibrational pressure:

$$p_{vib} = -\left(\frac{\partial F_{vib}}{\partial V}\right)_T = \frac{1}{V}\sum_{n\mathbf{k}} \gamma_{n\mathbf{k}} E_{n\mathbf{k}}^{vib} \tag{28}$$

where:

$$\gamma_{n\mathbf{k}} = -\left(\frac{V}{\omega_{n\mathbf{k}}}\right)\frac{\partial \omega_{n\mathbf{k}}}{\partial V} \tag{29}$$

is called the Grueneisen parameter of the $n\mathbf{k}$ normal mode. Hence, the thermal expansion coefficient is also obtained (K is the elastic bulk modulus):

$$\alpha = \frac{1}{V}\left(\frac{\partial V}{\partial T}\right)_p = \frac{1}{K}\left(\frac{\partial p}{\partial T}\right)_V = \left(\frac{1}{KV}\right)\sum_{n\mathbf{k}} \gamma_{n\mathbf{k}} C_{V,n\mathbf{k}} \tag{30}$$

However, rigorously speaking, the harmonic model would not allow the frequencies to depend on volume. The force constants, $V_{jj'}(s, s', \mathbf{T})$, are defined as second derivatives of the potential energy at the equilibrium structure configuration and thus, cannot have a volume dependence; neither can the eigenfrequencies, which are related to force constants through equation 11. Therefore, all Grueneisen parameters, the vibrational pressure and the thermal expansion would be vanishingly small in a strictly harmonic approximation, which would then fail to account for some fundamental properties of solids. This difficulty is overcome by simply assuming that Taylor expansions of the potential energy, of the type given in equation 8, can be performed in the proximity, not only of the equilibrium structure but also of deformed configurations (corresponding to different values of the volume). The quasi-harmonic model is, thus, obtained [23], in

Table 2. Some thermodynamic properties of calcite (heat capacities at constant volume and pressure, bulk elastic modulus, coefficient of thermal expansion and Grueneisen parameter) calculated by the RIM and SM potentials (first and second lines, respectively) and compared to experimental values (third line) at three different temperatures (in K; see ref. 24).

T(K)	0	300	500
C_v(JK^{-1}mol^{-1})	0.0	16.03	19.91
	0.0	15.89	19.70
		16.69	20.56
C_p(JK^{-1}mol^{-1})	0.0	16.13	20.17
	0.0	16.05	19.91
		16.69	20.56
K(GPa)	72.07	71.59	69.92
	73.52	72.91	71.29
		73.1	67.9
α(10^{-5}K^{-1})	0.0	2.51	3.19
	0.0	2.34	2.86
		1.93	2.08
γ	0.830	0.841	0.845
	0.725	0.798	0.776
		0.624	0.509

which a central role is played by the Grueneisen parameters, $\gamma_{n\mathbf{k}}$, of vibrational modes; these can be computed by performing Born matrix diagonalizations at several volumes (relaxing the corresponding structural configurations), and by differentiating, numerically, the eigenfrequencies according to equation 29.

At T values much smaller than the melting temperature, the evolution of the crystal structure as a function of T and p can be simulated successfully by the quasi-harmonic approximation. The most straightforward way is to set temperature and pressure to given values and to minimize the Gibbs free energy $G = F + pV$ with respect to changes in the structural variables (lattice constants and atomic ccordinates). The procedure is actually not simple: for a number of values of the volume, the atomic coordinates have first to be relaxed and a full lattice-dynamical calculation has to provide the set of eigenfrequencies. Then the pressure can be calculated and compared to the external one; in order to balance the difference, the volume has to be changed and the whole cycle repeated, until the difference between internal and external pressure is vanishingly small. This scheme has been implemented in the computer code PARAPOCS [23], where the anisotropy of lattice strain is also taken into account.

As an example of application of the quasi-harmonic model, the study of calcite $CaCO_3$ (space group $R\bar{3}c$) can be referred to [14, 24]. A potential including terms given in equations 3 and 7 (for the O-C-O bond angle) (RIM) and another one supplemented by the shell model for oxygen atoms (SM) were fitted to the equilibrium structure, the elastic constants and the zone-centre infrared and Raman vibrational frequencies. The parameters obtained in each case are reported in Table 1. By use of these two potentials, full quasi-harmonic structural equilibrations of calcite were carried out at three different temperatures (0, 300 and 500 K). All relevant thermodynamic quantities were obtained as functions of temperature; a few (heat capacities per atom in the formula unit, elastic bulk modulus, thermal expansion coefficient and Grueneisen parameter) are reported in Table 2 and compared to the corresponding experimental values. It turns out that the agreement is very good for properties related to the purely harmonic model (heat capacity and bulk modulus), while it appears less satisfactory for α and γ, which depend on the quasi-harmonic regime. In the latter case, the SM potential performs better than the RIM one. The simulated properties depend on the whole phonon dispersion spectrum, while only the zone-centre frequencies were used for the fitting of potential parameters.

It should, therefore, be stressed that it is critical to simulate the correct value for the thermal expansion behaviour of solids, even at low temperature where intrinsic anharmonicity should be quite negligible. In this respect, a considerable improvement can be expected, in the future, by use of potential functions reproducing the theoretical BO energy hypersurface, rather than a small set of selected empirical data.

References

1. M. Parrinello and A. Rahman, Phys. Rev. Lett. **45**, 1196 (1980).
2. S. Nose, Mol. Phys. **52**, 255 (1984).
3. M. Born and K. Huang, *Dynamical Theory of Crystal Lattice*, Clarendon Press, Oxford (1954).
4. R. Car and M. Parrinello, Phys. Rev. Lett. **55**, 2471 (1985).
5. M.P. Tosi, Solid State Phys. **16**, 1 (1964).
6. C.R.A. Catlow and A.N. Cormack, Inter. Rev. Phys. Chem. **6**, 227 (1987).
7. B.G. Dick and A.W. Overhauser, Phys. Rev. **112**, 90 (1958).
8. V. Nusslein and U. Schroeder, Phys. Status Solidi **21**, 309 (1967).
9. K. Fischer, Phys. Status Solidi b **66**, 295 (1974).
10. C.R.A. Catlow and W.C. Mackrodt, *Computer Simulations of Solids. Lecture Notes in Physics*, Vol. 166. Springer, Berlin (1982).
11. M. Catti, Acta Crystallogr. A **45**, 20 (1989).
12. G.D. Price and S.C. Parker, Phys. Chem. Minerals **10**, 209 (1984).
13. M. Catti, Phys.Chem. Minerals **16**, 582 (1989).
14. A. Pavese, M. Catti, G.D. Price and R.A. Jackson, Phys. Chem. Minerals **19**, 80 (1992).
15. A.C. Lasaga and G.V. Gibbs, Phys. Chem. Minerals **17**, 485 (1991).
16. S. Tsuneyuki, M. Tsukada, H. Aoki, Y. Matsui. Phys. Rev. Lett. **7**, 869 (1988).
17. B.W.H. van Beest, G.J. Kramer, R.A. van Santen. Phys. Rev. Lett. **16**, 1955 (1990).
18. J. Purton, R. Jones, C.R.A. Catlow, M. Leslie. Phys. Chem. Minerals **19**, 392 (1993).
19. M.T. Dove, *Introduction to Lattice Dynamics*, Cambridge University Press (1993).
20. M.M. Elcombe and A.W. Pryor, J. Phys. C: Solid State Phys. **3**, 492 (1970).
21. M. Catti, R. Dovesi, A. Pavese and V.R. Saunders, J. Phys.: Condens. Matter **3**, 4151 (1991).
22. P. Giannozzi, S. de Gironcoli, P. Pavone and S. Baroni, Phys. Rev. B **43**, 7231 (1991).
23. S.C. Parker and G.D. Price, Adv. Solid State Chem. **1**, 295 (1989).
24. M. Catti, A. Pavese and G.D. Price, Phys.Chem. Minerals **19**, 472 (1993).

Loss of Symmetry in Crystals: Surfaces and Local Defects

Cesare Pisani

Department of Inorganic, Physical and Materials Chemistry, University of Torino, via Giuria 5, I-10125 Torino, Italy

May 30, 1996

Summary. The use of periodic techniques for the study of crystalline surfaces (slab model) and of local defects in crystals (supercell model) is discussed. Special attention is given to projected band structures and to the possible appearance of surface or defect states. The (111) surface of silicon and substitutional impurities in silicon are used as examples.

Key words: Supercell – Crystalline Surface – Slab – Surface State – Surface Formation Energy – Projected Band-structure – Regular Adsorption – Defect – Defect Formation Energy – Defect State – Green Function

1. Introduction

Three-dimensional (3-D) translational symmetry has been exploited in the previous chapters to calculate the structure and properties of ordinary crystals. This is obviously a mathematical artifact: real crystals are limited by surfaces and contain various kinds of defects in their interior. As long as the number of crystalline atoms in the vicinity of such imperfections is a vanishingly small fraction of the total, the average crystalline properties should not differ appreciably from those that can be determined on the basis of the *perfect crystal* assumption. On the other hand, these special features are a subject of great theoretical and practical interest on their own. Surface science is concerned with phenomena and processes that occur at the interface between a condensed phase and the vacuum; heterogeneous catalysis is an example of the importance of this discipline. A recent, special issue of the journal Surface Science, provides an informative overview of topical problems and research tools in this field [1]. As concerns defects, their nature and distribution is often critically important to determine the structural and electronic properties of crystals [2]; the controlled doping of semiconductors is an obvious example. The study of the reference perfect crystal, to be denoted in the following as *host crystal* (HC), is closely related to that of its imperfections. The HC electronic structure determines the boundary conditions of the defect zone; in turn, the quantum-mechanical (QM) study of defects permits us to better qualify our initial statement about the ratio between the number of atoms in the vicinity of the defect and that of "proper" crystal atoms. For the study of the thermodynamic properties of surfaces and defects, the knowledge of the energetics of the HC is an essential reference. This chapter considers a few selected aspects of this vast subject.

In reviewing techniques used, we will not consider in detail the use of the *cluster model*, an extremely popular model because of its simplicity and flexibility [3]. In this approach, a piece of the material is ideally cut out from its surroundings to simulate a portion of the surface or a defect zone within a crystal. This cluster of a few atoms is treated with the standard techniques of molecular quantum chemistry. Simple devices are sometimes adopted for eliminating spurious effects related to the limited cluster size and to the presence of the boundary. For instance, dangling bonds at the frontier are saturated with hydrogen atoms or the whole cluster is immersed in an electrostatic non-homogeneous field to simulate crystalline field effects. The cluster model can provide useful information if the cluster is properly designed and if the results are critically interpreted: in particular, it allows the use of high quality basis sets and permits us to investigate electron correlation effects. On the other hand, the results are often critically dependent on cluster size and geometry.

Two instructive studies can be cited for this purpose. In the first one, the importance of a proper design of the cluster in adsorption studies on metals is documented by Zonneyville *et al* with reference to the adsorption of carbon monoxide at different sites of the (111) face of cobalt [4]. When the cluster was simply "cut out" from the metal, preserving its geometry, both the absolute value of the adsorption energy and the relative order of stability on the different sites were found to depend very much on the cluster symmetry, size and shape. More sensible and more stable results were obtained by adopting a model where the cluster was a fully optimized droplet of 12+1 cobalt atoms, whose central atom exhibited the same coordination as in bulk cobalt. Adsorption at the sites of interest (top, bridge and hollow) was simulated by letting CO approach the droplet in correspondence to a surface Co atom, to the midpoint of an atom-atom bond or to the centre of a triangle of Co atoms. In the second study of interest, Teunissen *et al* calculated the protonation energy of ammonia in a cage of acidic chabazite [5]. Four different clusters were cut out of the crystal, with dangling bonds saturated; the largest one comprised 60 atoms. The results were compared with those of periodic calculations, where the full three-dimensional structure of the crystal was taken into account; the interaction between different ammonia molecules in the periodic structure was found to be negligible. This comparison showed that convergence of cluster results with increasing cluster size, to the limit provided by the periodic calculation, is slow and somewhat erratic. This is especially true for the data concerning the ammonium ion, where the effect of the Madelung field is very important.

In accordance with the scope of this book, we shall treat here, in some detail, only QM *periodic* approaches which adopt the same computational tools as used for the treatment of perfect crystals. With such approaches, size and shape problems which plague cluster techniques are either totally absent or are much less critical; on the other hand, the model is less flexible and computationally more demanding.

Regarding the kind of problems that can be the object of investigation, the present discussion will be limited to *ideal*, or *quasi-ideal*, surfaces and to *local* defects. Let us clarify these points.

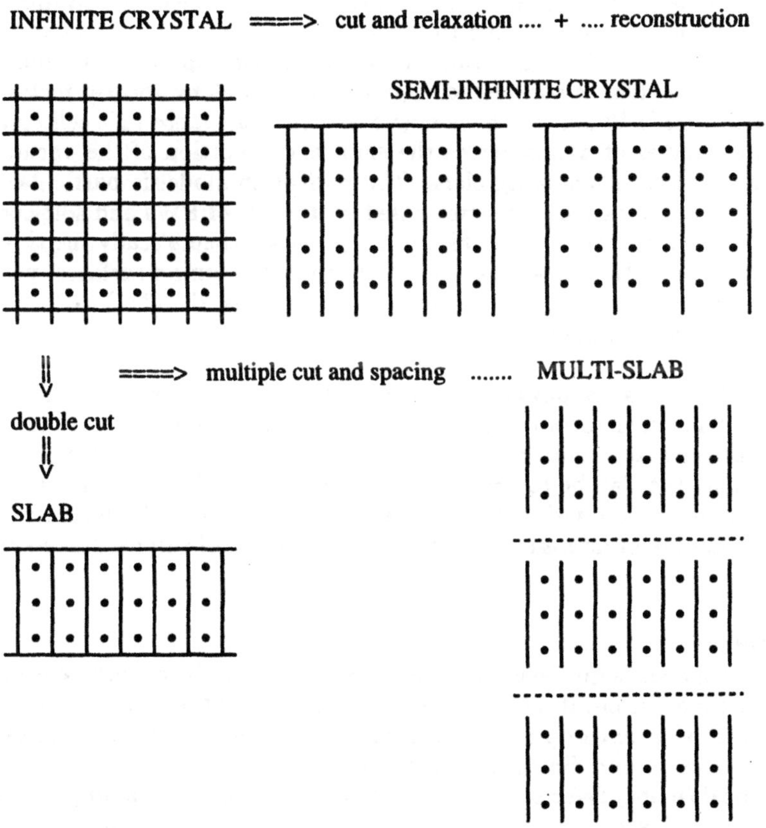

Figure 1. Models for simulating surfaces starting from a perfect 3-D crystal.

Surfaces:
By cutting a 3-D crystal through a crystalline plane (hkl), we generate two ideal semi-crystals, each limited by an ideal surface. Each semi-crystal preserves 2-D periodicity parallel to the selected face but loses all symmetry elements which involve displacements in a perpendicular direction (conventionally, the z-direction). The ideal surface may undergo relaxation, without loss of translational symmetry, or exhibit partial reconstruction, whereby the 2-D unit cell becomes larger. This is a typical *finite+infinite* problem: a subsystem consisting of a few layers close to the surface, whose properties we are interested in, is connected to the rest of the semi-crystal, an infinite system whose electronic structure is known. Green function (GF) techniques are ideal for studying prob-

lems like these (see Appendix A) and have been applied, in fact, to the study of semi-crystals, both with reference to approximate [6] and *ab initio* [7] Hamiltonians. A different technique, the ***slab model***, is considered here: a slab (or thin film) is created, formed by a few atomic layers, parallel to the (hkl) surface; relaxation or partial reconstruction can be taken into account. The unit cell of this periodic 2-D structure comprises a finite number of atoms and can, therefore, be studied using the same techniques as for the perfect crystal. Note that, at variance with the semi-crystal, the slab may possess symmetry elements (a mirror plane, a glide plane, a 2-axis parallel to the surface, an inversion center) which involve displacements in the z-direction. In the ***multi-slab*** approach, an infinite number of identical slabs is considered, regularly spaced along the z-direction: a typical separation is 10 a.u. We have now a 3-D crystal, whose unit cell comprises a finite number of atoms across the slab, and a portion of the vacuum region separating two neighboring slabs. The multi-slab model is particularly advantgeous for computational schemes based on the use of PWs and soft-core PPs (see Chapter 3, section 3.2) [8]. Figure 1 represents schematically the three different models In section 2, we discuss a few topics concerning surfaces, with reference to the slab technique [for some of these questions, the (111) surface of silicon is taken as an example]:

- What kind of surfaces are physically possible and worthy of investigation?
- How are the band-structures of the slab and of the HC related to each other?
- What are surface states and how can they be recognized and characterized?
- How thick must the slab be to provide a satisfactory model of the surface?
- How can the interaction with a regular overlayer be studied and interpreted?
- What quantities are of interest in studying the energetics of surface processes?

Local defects:
A great variety of defects may exist in crystals. Reference can be made to specialized literature for a classification and a general discussion of their characteristics [2]. Defects can be macroscopically extended in two dimensions (for example, stacking faults) or in one dimension (dislocation lines). We limit ourselves to considering local (point) defects, such as substitutional or interstitial impurities, self-interstitials and vacancies: in this case, the geometry and chemical composition of the HC are changed only in a small region about the defect. An important class of point defects are local imperfections at surfaces; the HC here is the semi-infinite crystal or a 2-D slab. As an example, we can cite the case of a catalytic process taking place near a vacancy at an otherwise perfect surface. We consider explicitly, in the following, only local defects in 3-D crystals but all considerations can be applied to the 2-D case. The presence of the point defect completely destroys the translational periodicity of the HC; only some point operations may be left. We are, again, in the presence of a finite+infinite problem and again, GF techniques can be adopted (Appendix A). For the application of these techniques, the HC solution is usually an essential pre-requisite. This is the reason why ***ab initio embedding*** codes for the study of point defects have become operational only recently, both in the framework of the density functional (DF) [9], and of the Hartree-Fock (HF) [10] approximation. The discussion of these methods is outside the scope of this book. Here, we limit ourselves to recalling

that among the non-standard methods adopted for the study of perfect crystals, there is one, the recursion method, [11] which is ideally suited to point defects because it does not exploit translational periodicity (see Chapter 3, section 5). For the application of standard crystalline methods to the solution of the point-defect problem, it is essential to restore 3-D periodicity. For this purpose, the *supercell* (SC) *model* can be adopted. This rather artificial trick consists of introducing super-unit-vectors, $\mathbf{A}_1$,$\mathbf{A}_2$ and $\mathbf{A}_3$, which are linear combinations of integer coefficients, j_1, j_2 and j_3, with the original unit vectors of the HC, $\mathbf{a}_1$, $\mathbf{a}_2$ and $\mathbf{a}_3$. The point defect is then reproduced identically within each SC. The system can now be treated as a perfect crystal using standard crystalline methods. Of course, the model is more acceptable the larger the distance between defects, that is, the longer the super-unit-vectors, but the computational cost increases rapidly with SC size, that is, approximately proportional to $(j_1 j_2 j_3)^3$ (see Chapter 3, section 5). In section 3, we address briefly two important questions concerning the local defect problem:
- How can charged defects be treated with the SC model?
- How rapidly do calculated defect properties converge with increasing SC size?

2. Crystalline surfaces

2.1. The stability of ionic crystal surfaces

Not all crystalline surfaces are physically stable or worthy of investigation. This is specially true for ionic or semi-ionic crystals. We summarize, in this section, the analysis carried out by Tasker concerning this point [12]. Consider a certain (hkl) plane, and construct a slab parallel to this plane. The slab will be composed of a certain number of *repeat units* which are, in turn, composed of atomic layers parallel to the selected face. The resulting structure can be classified in one of three categories (*Tasker types*): type 1 consists of neutral layers, each with the same stoichiometry as the host crystal; type 2 consists of charged layers, arranged symmetrically so that the repeat unit presents no net dipole perpendicular to the surface; type 3 consists of charged layers alternating in such a way that the repeat unit has a net dipole per unit area, d, normal to the surface. While surfaces of type 1 and 2 may exist, those of type 3 are unstable and can only be prepared with substantial reconstruction or with the adsorption of charged species. In fact, if we consider a slab of type 3, consisting of n repeat units, a net potential difference builds up across the slab: $\delta V = 2\pi d n$, which would correspond to a huge surface energy for macroscopically large n.

Following Tasker, we reproduce, in Figure 2, the schemes of a few low-index faces of the rocksalt, zincblende and fluorite structures, which show the three types just discussed. Note that, in some cases, the surface may be "possible" or "impossible" according to the chosen termination.

2.2. Two-dimensional Band Structure and Surface States

The electronic structure of a thick slab will be very similar to that of the HC but will contain, in addition, characteristic surface features which become particularly evident in the band-structure. A 3-D perfect crystal can be viewed as an

Rocksalt (NaCl)

```
 (100){1}      (110){1}      (111){3}

[-X-M-X-M-    [-X-M-X-M-    [-M-M-M-M-
 -M-X-M-X-     -M-X-M-X-    [-X-X-X-X-
 -X-M-X-M-     -X-M-X-M-     -M-M-M-M-
 -M-X-M-X-     -M-X-M-X-     -X-X-X-X-
 -X-M-X-M-     -X-M-X-M-     -M-M-M-M-
 -M-X-M-X-     -M-X-M-X-     -X-X-X-X-
```

Zincblende(ZnS)

```
 (100){3}      (110){1}      (111){3}

[-M---M---    [-X-M---X-    [-M---M---
[-X---X---    [---X-M---    [---X---X-
 ---M---M-     -X-M---X-    [---------
 ---X---X-     ---X-M---     ---M---M-
 -M---M---     -X-M---X-     -X---X---
 -X---X---     ---X-M---     ---------
```

Fluorite (CaF_2)

```
 (100){3}      (110){1}      (111){2or3}

[-M---M---    [-X-M-X-X-    [-X-X-X-X-
[-X-X-X-X-    [-X-X-M-X-    [-M-M-M-M-
 ---M---M-     -X-M-X-X-    [-X-X-X-X- ]
 -X-X-X-X-     -X-X-M-X-    [--------- ]
 -M---M---     -X-M-X-X-     -X-X-X-X- ]
 -X-X-X-X-     -X-X-M-X-     -M-M-M-M- ]
 -X-X-X-X-     -X-X-M-X-     -X-X-X-X-
 -M---M---     -X-M-X-X-     ---------
```

Figure 2. Representation of stacking sequences corresponding to different ionic structures and different crystal planes. The *Tasker type* of the sequence is indicated in braces (see text); the set of square brackets indicates the basic repeat unit (the proper repeat unit, as defined in Appendix B, may contain a number of basic repeat units, each associated with a fractionary translation parallel to the selected face). In the fluorite structure, the M species has twice the charge of the X species.

infinitely extended slab, with no limiting surfaces. In this picture, the general κ vector of the Brillouin Zone (see Chapter 2) is conveniently decomposed into a *parallel component*, parallel to the selected face and into a *perpendicular component*: $\kappa = \kappa_{\|} + \kappa_{\perp}$. Appendix B shows how to obtain a *projected band-structure*, that is one where all eigenvalues $\varepsilon_i(\kappa_{\|} + \kappa_{\perp})$ are attributed to $\kappa_{\|}$, independently of $\kappa_{\perp}$. For a given $\kappa_{\|}$, allowed (black) energy intervals appear, corresponding to eigenvalues associated with extended crystalline states, alternating with forbidden (white) energy intervals. Overall, the projected band-structure looks like a mixture of white and black lenses and layers. When comparing the HC projected structure with that of a slab of finite thickness, interesting similarities and discrepancies appear (for this discussion, reference is made to Figure 3):

- The first impression is one of strict correspondence between the projected band-structure and those of the slabs, after performing a rigid shift along the energy axis by a finite amount, $\delta\varepsilon$, which is small in the present case ($\delta\varepsilon \approx$ -0.08 Hartree) but can be much larger. What does this mean? The potential zero can be arbitrarily chosen in an infinite, perfect 3-D crystal. In a semicrystal or a slab, it is customary to set to zero the potential at an infinite distance from the surface (in a slab, two different zeros may be chosen, if the slab itself has non-zero electric dipole moment in the z-direction). This choice determines the position of all one-electron levels. Within a one-electron model, the general eigenvalue of the slab will correspond to the energy required to extract an electron from the associated eigenstate and bring it outside the crystal over the selected face. In the case of a metal, the Fermi energy, e_F, determined with this convention, corresponds to the workfunction relative to the selected face. The band-structure of the HC can be adjusted to that of the slab, by shifting all HC eigenvalues by an amount, $\delta\varepsilon$, equal to the difference between a core level of an atom at the center of the slab and the corresponding one in the bulk.
- Instead of black and white lenses and layers, the slab band-structure exhibits a discrete set of bands, which may run very close to each other in the neighbourhood of $\kappa_{\|}$ points where the HC projected band structure exhibits short black segments, separated by large white intervals. This is an obvious consequence of the finite slab thickness: for each $\kappa_{\|}$ point, the number of slab eigenvalues equals the size of the basis set (the number of basis functions in the unit 2-D cell) and is proportional to the slab thickness.
- With increasing slab thickness, the similarity with the HC becomes clearer, and the black-white structure makes itself evident. At the same time, there are some peculiar features which remain practically constant and have no counterpart in the HC. There may be, in particular, some bands which cross the white lenses, corresponding to energy eigenvalues which are forbidden in the HC for functions of that symmetry and are clearly associated with the presence of the surface. The corresponding eigenstates are called *surface states*. They are localized in proximity to the surface and their amplitude decreases exponentially with increasing distance from the surface. A surface resonance is a state which is also predominantly localized at the surface but its eigenvalue lies in an allowed interval of the HC band-structure for the same $\kappa_{\|}$ and, therefore, has an oscillatory tail which extends to infinite distance inside the crystal.

- Surface states and resonances are particularly sensitive to surface relaxation. Occupied surface states are often considerably stabilized by relaxation and eventually, transform into resonances; the opposite is true for virtual surface states. The fact that the position of surface states does not change with increasing slab thickness, is a good indicator that the slab is thick enough to represent a good model of the selected crystalline face. We also note that in thin slabs limited by two equivalent surfaces, the bands corresponding to surface states are doubled, indicating a bonding-antibonding interaction between states at the two faces; the extent of the splitting is reduced rapidly with increasing slab thickness.
- The chemical characterization of surface states is often obvious from an examination of the projected band-structure and of the associated density of states [13].

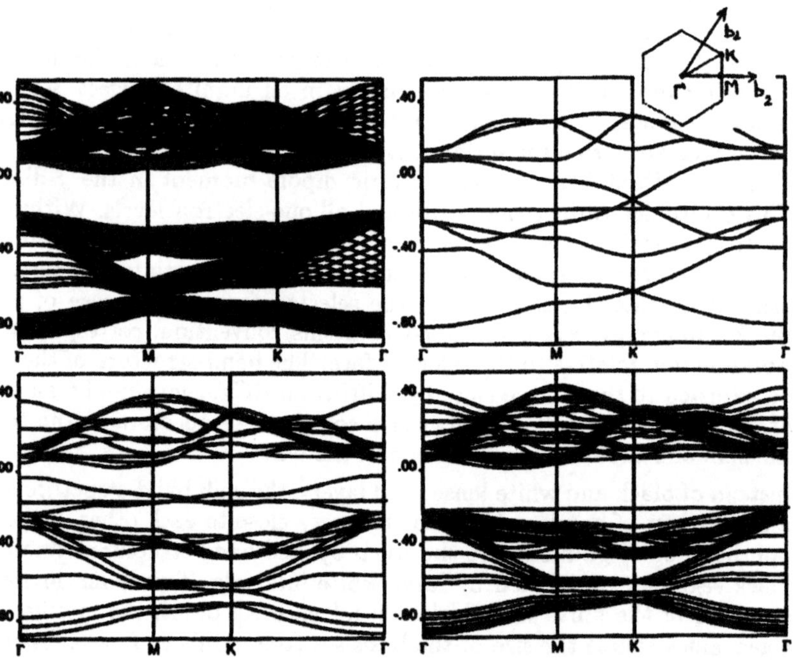

Figure 3. Band-structures of silicon with reference to (111) surfaces. Top left: (111)-projected band-structures for bulk Si; top right: band-structure for a 2-layer slab; bottom left: same as in preceding plot for a 6-layer slab; bottom right: same for a 12-layer slab, with H saturation of surface dangling bonds. The thin horizontal line in the last three plots is the Fermi level. The inset shows the path followed in the 2-D Brillouin zone (BZ): see Appendix B. Energies are in Hartree. The calculations have been performed with CRYSTAL (Chapter 8), using pseudopotentials and a double-z basis set for valence electrons.

2.3. Regular Chemisorption at Surfaces

Perhaps the most common application of the slab model is the study of the regular adsorption of molecular or atomic species at crystalline surfaces. In the simplest case, a monolayer (M) of the adsorbed species is considered, which has the same 2-D periodicity as the slab (S). Lower coverages can be studied by using a SC model, where the two unit vectors of M are multiples of those of S. M can be positioned either over one of the two surfaces of S (M-S) or symmetrically over both faces (M-S-M). The new 2-D crystal can be treated with the same standard tools as used for S alone. By changing the distance of M from the surface, the internal geometry of M and possibly, that of the crystalline atoms closer to the surface, relatively complicated adsorptive or catalytic processes can be studied, such as the interaction of water with zirconia [14]. Rather than documenting this kind of studies, we discuss briefly how to obtain useful information from an analysis of the band-structures of the two interacting systems, M and S. The excellent booklet by Hoffmann can be very useful to go deeper in this subject, starting from concepts of current use in molecular organic chemistry [15].

Consider first the two non-interacting systems, M and S. While S has a physical meaning, the isolated M system is usually an unphysical structure, since the periodic arrangement of the constituent species takes place only on adsorption. Nonetheless, M may be viewed as resulting from a rigid displacement of the adsorbate to a large distance away from the surface. From another point of view, it can be useful to have an estimate of the energetic cost of ordering the overlayer, which must be compensated by the formation of bonds with the surface. Since M and S are characterized by the same translational symmetry, the two BZs coincide and the two band-structures can be legitimately super-imposed: the resulting band-structure will be referred to as M+S. In this case, there is no arbitrariness in the absolute location of the bands, since the two potential zeros coincide (at a large distance above S and below M). Only one-electron states with the same symmetry (in particular, with the same $\boldsymbol{\kappa}_{\|}$) can interact. Also, as is known from molecular chemistry, the two functions $\psi_{iM}(\boldsymbol{\kappa}_{\|})$ and $\psi_{jS}(\boldsymbol{\kappa}_{\|})$ must be close in energy, in order to interact appreciably. In summary, we can expect that the band-structure of the M-S system will be similar to M+S, except where there is crossing or near-crossing of bands of the two sub-structures. A large modification of the independent energy levels will be an index of strong chemical interaction. There is still an important point to be taken into account, however: namely, the relative geometric position of the two subsystems, both concerning vertical distance and horizontal displacement. The M+S structure will be independent of these parameters but, for a bonding interaction to take place, it is also essential that the two interacting wave-functions [$\psi_{iM}(\boldsymbol{\kappa}_{\|})$ and $\psi_{jS}(\boldsymbol{\kappa}_{\|})$] are located with respect to each other, in such a way that appreciable overlap takes place between them with the correct phase. To ascertain this point, it is important to know the chemical composition of the two orbitals and it can be useful to sketch a simplified scheme of their periodic structure, by taking into account the phase factor $exp(\imath\boldsymbol{\kappa}_{\|} \cdot \boldsymbol{T}_{\|})$ associated with a displacement by a 2-D lattice vector $\boldsymbol{T}_{\|}$ (see Chapter 2).

Figure 4 represents the simple case of H on (111) silicon. The valence band structure of the isolated H monolayer (M) essentially consists of a semi-occupied band of moderate width because the distance between neighbouring hydrogens is large (3.86 Å). When super-imposed on the band-structure of the 6-layer silicon slab, the M band results in the same energy range as the bands associated with dangling bonds (*db*) of silicon. For each $\boldsymbol{\kappa}_{\|}$, the two wave-functions, $\psi_M(\boldsymbol{\kappa}_{\|})$ and $\psi_{db}(\boldsymbol{\kappa}_{\|})$, have exactly the same structure. The best bonding interaction is then found when hydrogens are located above surface silicons, at a distance such that there is best overlap between s H orbitals and surface sp_z hybrid orbitals. This result confirms elementary chemical considerations. More complicated cases may occur when M interacts primarily with surface states or resonances, resulting from a combination of $sp_{x,y}$ hybrid orbitals: see, for instance, references [15] and [16].

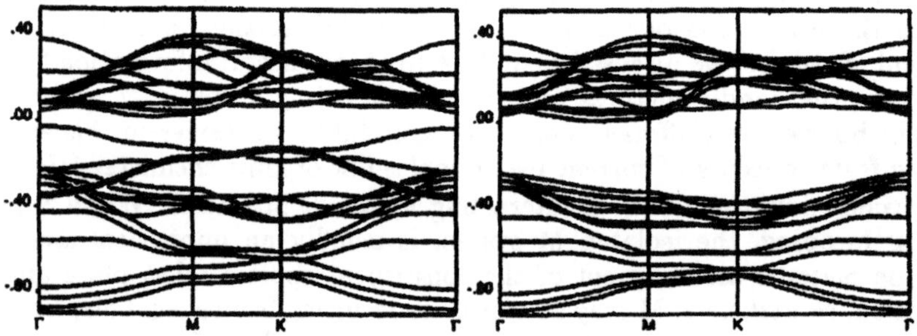

Figure 4. Band-structures concerning the symmetric adsorption of atomic H on a 6-layer (111) slab of silicon. Left: Superposition of the two non-interacting structures; the H band is the heavy line. Right: band-structure for the interacting systems, with H on top of the silicon surface. The calculations have been performed with CRYSTAL, using a pseudopotential, double-z basis set; both band-structures correspond to a closed-shell solution. Symbols and units are as in Figure 3.

2.4. Energy parameters in Surface Studies

Table 1 illustrates, schematically, important energy data which can be obtained from periodic slab investigations. The table is self-explanatory. Some care must, however, be taken when applying these formulae, because they often correspond to small differences between large quantities. If the computations do not adopt exactly the same computational technique, the same cut-off parameters and the same numerical accuracy, the final result may be affected by substantial relative errors. This may happen, in particular, when calculating the surface formation energy, by combining information coming from a 3-D calculation (α) and a 2-D calculation (γ). Boettger has recently discussed the risks involved in using such formulae and he has suggested, instead, the use of a procedure whereby both the α parameter and the surface formation energy are determined from a series of 2-D calculations, corresponding to different film thicknesses [17].

3. Local defects

3.1. The Supercell Model and the Problem of Charged Defects

A major problem for the application of the SC model to the study of local defects consists of the spurious interaction between defects in neighbouring SCs. Each

Table 1. Reference systems and energy parameters of interest in surface studies. HC denotes the perfect host crystal; S the $(hkl)n$-slab; M the ad-layer.

E	Reference system	Conventional unit	Chemical composition
α	Bulk HC	3-D unit cell	$\sum \nu X_\nu$
β	S (unrelaxed)	2-D unit cell, area A	$n \sum \nu X_\nu$
γ	S (relaxed)	2-D unit cell, area A	$n \sum \nu X_\nu$
δ	S (relaxed,reconstr.)	2-D supercell, area $A' = tA$	$tn \sum \nu X_\nu$
ϵ_μ	isolated ad-species	atom or molecule	Y_μ
ζ	M(isolated)	2-D supercell, area $A'' = vA$	$\sum \mu Y_\mu$
η	M-S (optimized)	2-D supercell, area A''	$vn \sum \nu X_\nu + \sum \mu Y_\mu$
θ	M-S-M (optimized)	2-D supercell, area A''	$vn \sum \nu X_\nu + 2 \sum \mu Y_\mu$

Energy expression	Quantity
$(\gamma - n\alpha)/(2nA)$	surface formation E per unit area (no reconstruction)
$(\beta - \gamma)/(2A)$	surface relaxation E per unit area (no reconstruction)
$(\delta - t\gamma)/(2A')$	surface reconstruction E per unit area
$\sum \mu\epsilon_\mu - \eta$	adsorption E with respect to isolated atoms or molecules
$(2\sum \mu\epsilon_\mu - \theta)/2$	adsorption E with respect to isolated atoms or molecules
$\zeta - \eta$	adsorption E with respect to the isolated M monolayer
$(2\zeta - \theta)/2$	adsorption E with respect to the isolated M monolayer

defect causes a disturbance in its proximity, consisting of a rearrangement of the electron distribution and a displacement of nuclei. In many important cases, *defect states* may be present as well, analogous to the surface states described in section 2.2: however, while surface states may appear within "white lenses" with specific $\kappa_{\|}$ values, no translational symmetry is left in a crystal with a local defect, so defect states may be present only in band gaps of the HC. The size of the SC must be sufficiently large that the perturbed zone created by each defect is contained within its SC (this statement must be interpreted in a broad sense; for instance, one-electron wavefunctions associated with defect states should be essentially localized within the cell). In the next section, we shall show through an example that accurate *ab initio* SC defect studies are feasible with present-day computational tools, in the case of simple *neutral* defects.

With *charged* defects, the problem is much more difficult. In this case, it is not possible to use the SC technique directly, because the infinite repetition of the charged unit leads to unphysical divergent contributions to the Hamiltonian and to the total energy. The usual trick for solving this problem consists of compensating the charged defect with a uniform background of opposite charge. Operationally, this requires no modification of the computer code if the Ewald technique is adopted for the treatment of the long range Coulomb terms (Chapter 8): in this method, each fraction of the unit cell charge is implicitly compensated with a background charge, to insure charge neutrality. The description of the system is now as follows: in each SC, we have a local defect (for instance, a cation vacancy) plus the uniform background of opposite charge. However, the electrostatic field created in the reference SC by the compensated defects in all other cells is far from negligible and can influence the electronic distribution and the ionic relaxation. Leslie and Gillan [18] proposed a correction E'' for the SC formation energy of a charged defect in a cubic lattice, that cancels, to some extent, the interaction between the periodic array of compensated point defect charges:

$$E'' = -\alpha Q^2/\varepsilon_0 L$$

where L is the lattice parameter of the SC, α is the appropriate Madelung constant, Q is the charge of the point defect and ε_0 is the static dielectric constant of the HC. The proposed correction seems to work reasonably well for ionic systems: it has been used in conjunction with DF [19] and HF [20] *ab initio* studies of vacancies in MgO. However, it is far from satisfactory from a general point of view:

i) the correction is specific to cubic lattices;
ii) it is, in a way, semi-empirical, because the experimental dielectric constant is used;
iii) the correction is performed *a posteriori*: it is then implicitly assumed that ionic and electronic relaxation effects due to the presence of the other defects are unimportant.

In a sense, the vast field of charged defect investigations is still in its infancy, as far as *ab initio* approaches are concerned. This is probably one of the cases where useful information can be gathered through a critical comparison between the results of SC and embedded cluster studies [20].

3.2. Convergence of Calculated Defect Properties with SC size: Substitutional Carbon in Silicon

In this sub-section, we present, in summary, the results of a recent SC study by Orlando *et al* [21] concerning a carbon substitutional impurity in silicon. The calculations have been performed with the HF CRYSTAL code (Chapter 8). Pseudopotentials have been adopted for core electrons of both C and Si; for valence electrons, a good split-valence plus polarization set was used. The cases S8, S16, S32 and S64 were considered, where the number after "S" indicates the number of HC unit cells in each SC. Stars of atoms near the substitutional impurity, and within the SC were allowed to relax; the number of such stars is 1, 1, 3 and 4 in the four cases. In all cases, the cubic symmetry is maintained: S8 and S64 are simple cubic, S16 is face-centered, S32 is body-centered. Substitutional defect energies are calculated with respect to energies per atom of perfect silicon and diamond. Correlation contributions to defect formation energies have been estimated *a posteriori* using the DF formula by Perdew (see Chapter 11).

Table 2 summarizes the most important results of this study. The following conclusions can be drawn. The relaxation of silicon atoms around the defect is expected to depend on their electrostatic interaction with the negatively charged defect (C is appreciably more electronegative than Si) and on the propagation of the displacement of the four first neighbours towards the C atom (C is noticeably smaller than Si). The results show that the latter, "covalent" effects dominate the relaxation process, while electrostatic effects are quite short-ranged and are completely screened by first neighbour Si atoms. The C-Si bondlength reduces from 2.36 to 2.12 Å; second- and third-neighbour silicon atoms relax, in turn, to minimize the deviation of the Si-Si bond-length from the perfect crystal value. The relaxation energy is very high (about 2eV). SCs containing at least 32 atoms are necessary in order to describe relaxation properly.

Table 2. Defect formation energies and net Mulliken charges for a carbon substitutional impurity in silicon, as a function of SC size. Energies in parentheses include correlation corrections (see text). Si^{I} and Si^{II} are first and second neighbours of the C impurity, respectively. Net charges are referred to fully relaxed structures.

	Substitution energy (eV)		Net Mulliken charges		
SC	Unrelaxed	Relaxed	C	Si^{I}	Si^{II}
S8	3.93 (4.98)	2.86 (3.74)	-1.085	0.341	-0.094
S16	3.94 (5.00)	2.81 (3.68)	-1.101	0.367	-0.049
S32	3.93 (4.98)	2.06 (2.87)	-1.201	0.375	-0.022
S64	3.92 (4.98)	2.01 (2.83)	-1.202	0.376	-0.025

4. Conclusions

The use of periodic models for the study of surfaces and local defects has gained importance in recent years. The periodic approach is, in a sense, in an intermediate position with respect to molecular cluster models and the GF approach. With respect to both techniques, the periodic approach will remain a reliable, clearly defined reference of fundamental importance for calibrating methods and for testing approximations.

Acknowledgments

This work was supported in part by the Italian MURST (fondi 40%), and by the Italian CNR. The help of Dr. R. Orlando, in calculating silicon band-structures, is gratefully acknowledged.

A. Green Function Techniques for the Study of Imperfections in Crystals

The purpose of this Appendix is to give an idea of how GF techniques can be applied to the study of surfaces and point defects in crystals. For a general introduction to GFs and their application in physics, excellent textbooks exist [22].

For a given crystal, let us denote a set of localized basis functions by $\{\phi_\mu\}$ and the corresponding Hamiltonian and overlap matrix, by H and S. For a complex energy value $z = \varepsilon + \imath s$, the *Green matrix*:

$$G(z) = (zS - H)^{-1} \tag{1}$$

is defined everywhere except at those z values where the secular equation is satisfied: $det(zS-H)=0$ and the matrix inversion in equation 1 is not possible. Since S and H are both Hermitian, the secular equation is satisfied only for real $z=\varepsilon_j$ values. These are, of course, the one-electron eigenvalues corresponding to eigenfunctions $\psi_j(\mathbf{r})$, that is:

$$\begin{aligned} \sum_\nu H_{\mu\nu} c_{\nu j} &= \varepsilon_j \sum_\lambda S_{\mu\lambda} c_{\lambda j} \\ \psi_j(\mathbf{r}) &= \sum_\mu c_{\mu j} \phi_\mu(\mathbf{r}) \end{aligned} \tag{2}$$

The ε_j values may be either isolated or they may fill energy intervals (bands). In the former case, they are called ***poles*** of $G(z)$; in the latter case, they are said to belong to an ***analiticity cut***. For a λ value within the cuts, the eigenfunctions are denoted as: $\psi(\mathbf{r};\lambda)=\sum_\mu c_\mu(\lambda)\phi_\mu(\mathbf{r})$. The Green matrix elements $G_{\mu\nu}(z)$ are polynomials in z and are, therefore, analytic functions of z everywhere except at poles or along cuts. A number of important properties derive from this simple fact. In particular, a strict relationship is easily shown to exist between the Green matrix and the density of states (DOS) $N(e)$ (see Chapter 3, section 4.2):

$$\begin{aligned} N_{\mu\nu}(e) &= \sum_j^{(poles)} c^*_{\mu j} c_{\nu j} \delta(e-\varepsilon_j) + \int^{(cuts)} d\lambda\, c^*_\mu(\lambda) c_\nu(\lambda)\, \delta(e-\lambda) = \\ &= (2\imath\pi)^{-1} \lim_{s\to+0} [G_{\mu\nu}(e-\imath s) - G_{\mu\nu}(e+\imath s)] \end{aligned} \tag{3}$$

$$\text{where} \quad G_{\mu\nu}(z) = \int_{-\infty}^{\infty} de N_{\mu\nu}(e)/(z-e) \tag{4}$$

Relation 3 is used in the recursion method (Chapter 3, section 5) for obtaining the crystalline DOS, after calculating, approximately, the inverse of $(zS-H)$ by a continued fraction technique, which exploits the quasi-diagonal matrix of H and S. Vice versa, after solving the periodic problem by a standard technique, we can obtain from the DOS the crystalline GF according to equation 4.

The power of GF techniques becomes evident when considering the effect of perturbations on a reference system whose solution is available. In the present case, the reference system is the HC: all corresponding quantities will be denoted with a zero subscript. For simplicity, we will consider only the case where the basis set is the same for the perfect and the imperfect crystal (for more general cases, see, for instance, references [23]). Let us denote by V the perturbation matrix:
$V = H - H^0$, whence: $[G(z)]^{-1} = [G^0(z)]^{-1} - V$. From elementary matrix algebra, the identity is obtained:

$$G(z) = [I - G^0(z)V]^{-1} G^0(z) \tag{5}$$

If $V_{\mu\nu}$ is different from zero only when both μ and ν belong to a finite subset (C) of the representative basis set (in other words, when the perturbation is local), the useful relationship (embedding equation) is immediately obtained:

$$G_C(z) = \{[G^0_C(z)]^{-1} - V_C\}^{-1} \tag{6}$$

which permits us to determine the electronic structure in the defect region, by limiting the computation to that region, while taking into account exactly the boundary conditions. The above equation was derived and applied to the study of defects in crystals for the first time by Koster and Slater [24]. Among the enormous number of applications possible, we can cite its use for the study of surfaces by Kalkstein and Soven [6]: the HC is cut in two by setting to zero all matrix elements of H between functions belonging to the two halves, that is: $V_{\mu\nu} = -H^0_{\mu\nu}$, if μ and ν belong to two different halves of the crystal. V is different from zero in a region C, comprising few layers near the cut. Equation 6 can then be used; due to translational 2-D symmetry, the problem can be factorized into a separate one for each $\kappa_{\|}$ value. In summary, the sequence of steps to be taken for a standard GF approach to the problem of imperfections in crystals is as follows:

a) Use a standard procedure for solving the HC problem;
b) From the HC DOS, obtain $G^0(z)$ (equation 4);
c) Individualize the C zone;
d) Construct the V_C matrix;
e) Solve equation 6;
f) Reconstruct the local DOS N_C, according to 3.

Steps d) to f) must be solved to self-consistency

B. Projected Band-Structures

The cartesian components of the three vectors $\boldsymbol{a}_1, \boldsymbol{a}_2$ and $\boldsymbol{a}_3$, which define the primitive cell of the HC may be arranged columnwise in a 3×3 matrix A ($\boldsymbol{a}_1 = a_{11}\boldsymbol{i} + a_{21}\boldsymbol{j} + a_{31}\boldsymbol{k}$, etc.) ; the volume of the cell is $V = |\det(A)|$. Consider now a new cell defined by three translation vectors $\boldsymbol{a}'_1, \boldsymbol{a}'_2$ and $\boldsymbol{a}'_3$; the first two belong to a plane Π, parallel to the selected face, the third one points in a perpendicular direction (this is the most common case, though not the general one). We can write:

$$A' = AL \tag{7}$$

where the matrix A' contains columnwise the cartesian coordinates of the vectors $\boldsymbol{a}'_1, \boldsymbol{a}'_2$ and $\boldsymbol{a}'_3$ and L is a matrix of integers: $\boldsymbol{a}'_j = l_{1j}\boldsymbol{a}'_1 + l_{2j}\boldsymbol{a}'_2 + l_{3j}\boldsymbol{a}'_3$. The volume of the new cell (let us call it the P-cell) is $V' = nV$, with $n = |\det(L)|$. We shall be specially interested in a triplet $\boldsymbol{a}'_1, \boldsymbol{a}'_2$ and $\boldsymbol{a}^0_3$, corresponding to the lowest value of n, $n^0 = |\det(L^0)|$; if $n^0 = 1$, then the new cell is equivalent to the original one. If we choose $\boldsymbol{a}'_3 = m\boldsymbol{a}^0_3$, the corresponding cell contains $n = mn^0$ crystalline units, arranged in a sequence of m repeat units stacked in a pile perpendicular to Π (see Figure 5).

From the definition of the fundamental vectors of the reciprocal lattice [$\boldsymbol{b}_1 = (2\pi/V)\boldsymbol{a}_2 \times \boldsymbol{a}_3$, etc.: see Chapter 2] we have easily:

$$B' = BM \tag{8}$$

with

$$M = \frac{1}{n}\begin{pmatrix} l_{22}l_{33} - l_{32}l_{23} & l_{23}l_{31} - l_{33}l_{21} & l_{21}l_{32} - l_{31}l_{22} \\ l_{32}l_{13} - l_{12}l_{33} & l_{33}l_{11} - l_{13}l_{31} & l_{31}l_{12} - l_{11}l_{32} \\ l_{12}l_{23} - l_{22}l_{13} & l_{13}l_{21} - l_{23}l_{11} & l_{11}l_{22} - l_{21}l_{12} \end{pmatrix}, \tag{9}$$

Vectors b'_1 and b'_2 lie in a plane parallel to P; b'_3 equals b_3^0/m and is perpendicular to Π (Figure 5). The general vector k of reciprocal space can be expressed as follows:

$$k = \sum x_j b_j = \sum x'_j b'_j \quad \rightarrow \quad |k\rangle = B|x\rangle = B'|x'\rangle \tag{10}$$

where $|k\rangle$ is the column vector of the cartesian co-ordinates of k and the column vectors $|x\rangle$ and $|x'\rangle$ contain its fractionary co-ordinates with respect to the two basis sets. From equations 8 and 10 the relation is obtained:

$$|x\rangle = M|x'\rangle \tag{11}$$

which allows us to express the "old" fractionary co-ordinates in terms of the "new" ones. Two k points are equivalent with respect to the conventional cell (or to the Π-cell), if the corresponding $|x\rangle$ (and respectively, $|x'\rangle$) differ by a vector $|i\rangle$ with integer components. Following the definition of a parallel and a perpendicular component of k, introduced in section 2.2 ($k = k_{\|} + k_{\perp}$), it is clear that k points, characterized by the same $k_{\|}$ and $k_{\perp} = jb'_3 (j = 0, 1, ..., n-1)$, correspond to the same point in the Π-BZ but are inequivalent with respect to the original BZ.

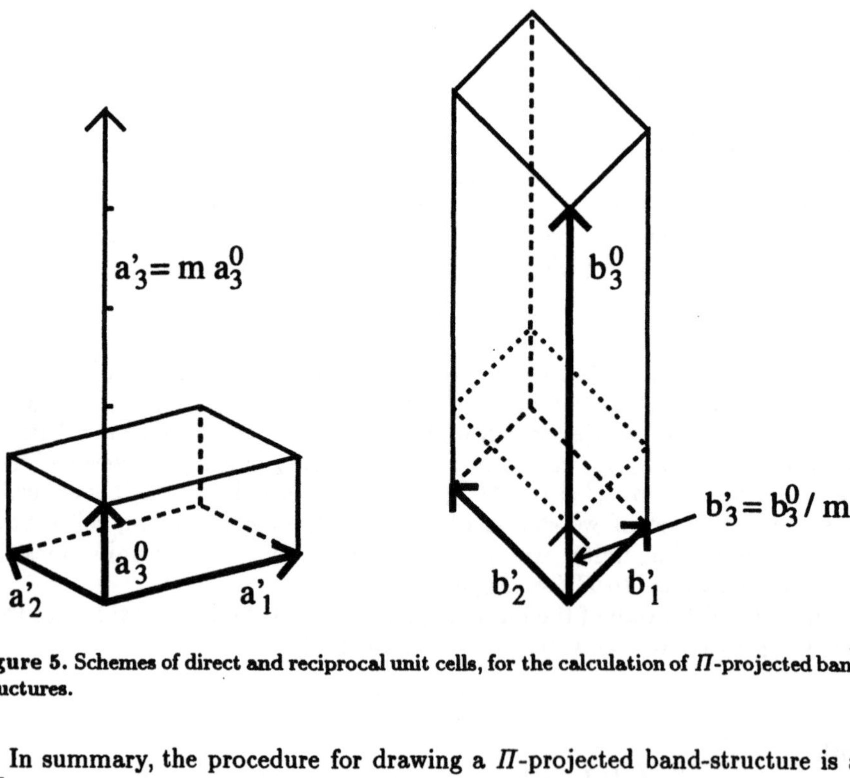

Figure 5. Schemes of direct and reciprocal unit cells, for the calculation of Π-projected band-structures.

In summary, the procedure for drawing a Π-projected band-structure is as follows:

a) Select the triplet a'_1, a'_2 and a_3^0, and construct the corresponding L_0 matrix;

b) Choose an appropriate integer $n = mn_0$ and construct the M matrix (equation 9);
c) Choose a path in the plane $\boldsymbol{k}_{\parallel} = \boldsymbol{0}$, formed by segments connecting some special points $\underline{S}$, each characterized by fractionary coordinates $|x'^{s0}\rangle = (x_1'^{s}, x_2'^{s}, 0)$;
d) For each $\underline{S}$, construct n displaced points $\underline{S}^j \leftrightarrow |x'^{sj}\rangle = (x_1'^{s}, x_2'^{s}, j)$, with $j = 0, 1, ..., n-1$ and obtain the corresponding $|x^{sj}\rangle$ using equation 11;
e) For each j, calculate the HC band structures along the *displaced path* which joins in a sequence the $\underline{S}^j$ points and superimpose all of them: the black/white structure, described in section 2.2, will appear for m large enough.

As an example, consider the case of the (111) surface of silicon, an *fcc* lattice. From the standard definition $[\boldsymbol{a}_1 = (a/2)(\boldsymbol{j}+\boldsymbol{k}); \boldsymbol{a}_2 = (a/2)(\boldsymbol{i}+\boldsymbol{k}); \boldsymbol{a}_3 = (a/2)(\boldsymbol{i}+\boldsymbol{j})]$ we have:

$$A = \frac{a}{2}\begin{pmatrix} 0 & 1 & 1 \\ 1 & 0 & 1 \\ 1 & 1 & 0 \end{pmatrix}, \quad \rightarrow \quad V = \frac{a^3}{4} \tag{12}$$

The triplet $\boldsymbol{a}'_1, \boldsymbol{a}'_2$ and $\boldsymbol{a}_3^0$ can be chosen, as follows: $\boldsymbol{a}'_1 = -\boldsymbol{a}_2 + \boldsymbol{a}3$, $\boldsymbol{a}'_2 = \boldsymbol{a}_1 - \boldsymbol{a}_3$, $\boldsymbol{a}_3^0 = \boldsymbol{a}_1 + \boldsymbol{a}_2 + \boldsymbol{a}_3$ (and $\boldsymbol{a}'_3 = m\boldsymbol{a}_3^0$), that is:

$$L^0 = \begin{pmatrix} 0 & 1 & 1 \\ -1 & 0 & 1 \\ 1 & -1 & 1 \end{pmatrix}, \; ; \; V = \begin{pmatrix} 0 & 1 & m \\ -1 & 0 & m \\ 1 & -1 & m \end{pmatrix}, \quad \rightarrow \quad n^0 = 3 \; ; \; n = 3m \tag{13}$$

$$A' = AL = \frac{a}{2}\begin{pmatrix} 0 & 1 & 2m \\ 1 & 0 & 2m \\ 1 & -1 & 2m \end{pmatrix}, \; ; \; M = \frac{1}{3}\begin{pmatrix} 1 & 2 & 1/m \\ -2 & -1 & 1/m \\ 1 & -1 & 1/m \end{pmatrix}, \tag{14}$$

The vectors $\boldsymbol{b}'_1 = (2\pi/3a)(-2\boldsymbol{i} + 4\boldsymbol{j} - 2\boldsymbol{k})$ and $\boldsymbol{b}'_2 = (2\pi/3a)(-4\boldsymbol{i} + 2\boldsymbol{j} + 2\boldsymbol{k})$ which define the projected unit cell in reciprocal space, are immediately seen to form an angle $\alpha = \arccos[(\boldsymbol{b}'_1 \cdot \boldsymbol{b}'_2)/(|\boldsymbol{b}'_1||\boldsymbol{b}'_2|)] = \arccos(0.5) = \pi/3$. The BZ is an hexagon. A suitable path for constructing the projected band structure is $\underline{\Gamma}$ - $\underline{M}$ - $\underline{K}$ (see Figure 3), where the three points are defined by: $\underline{\Gamma} = 0\,\boldsymbol{b}'_1 + 0\,\boldsymbol{b}'_2$; $\underline{M} = 0\,\boldsymbol{b}'_1 + (1/2)\,\boldsymbol{b}'_2$; $\underline{K} = (1/3)\,\boldsymbol{b}'_1 + (1/3)\,\boldsymbol{b}'_2$. The fractionary co-ordinates of the displaced points are:

$$\begin{array}{lclclcl} \underline{\Gamma}^j & \leftrightarrow & |x'^{\Gamma_j}\rangle & = & (0,0,j) & \leftrightarrow & |x^{\Gamma_j}\rangle & = & (J,J,J) \\ \underline{M}^j & \leftrightarrow & |x'^{M_j}\rangle & = & (0,\frac{1}{2},j) & \leftrightarrow & |x^{M_j}\rangle & = & (\frac{1}{3}+J, -\frac{1}{6}+J, -\frac{1}{6}+J) \\ \underline{K}^j & \leftrightarrow & |x'^{K_j}\rangle & = & (\frac{1}{3},\frac{1}{3},j) & \leftrightarrow & |x^{K_j}\rangle & = & (\frac{1}{3}+J, -\frac{1}{3}+J, J) \end{array}$$

where $j = 0, ..., 3m-1$ and $J = j/(3m)$.
With CRYSTAL, after choosing a shrinking factor IS=$6m$, the integers which define the three line extremes are: $(2j, 2j, 2j)$; $(2m+2j, -m+2j, -m+2j)$; $(2m+2j, 2m+2j, 2j)$, for each j. The P-projected band-structure is shown in Figure 3, top left plot, for $m = 6$.

References

1. C.B. Duke (ed.), *Surface Science. The First Thirty Years*, North-Holland, Amsterdam (1994).
2. F. Agullo-Lopez, C.R.A. Catlow and P.D. Townsend, *Point Defects in Materials*, Academic, London (1988); G. Benedek, A. Cavallini and W. Schröter (eds.), *Point and Extended Defects in Semiconductors*, NATO ASI Series B: Physics Vol. 202, Plenum, New York (1989).
3. G. Pacchioni, P.S. Bagus and F. Parmigiani (eds.), *Cluster Models of Surface and Bulk Phenomena*, NATO ASI Series B: Physics Vol. 283, Plenum, New York (1992); R.W. Grimes, C.R.A. Catlow and A.L. Shluger, *Quantum Mechanical Cluster Calculations in Solid State Studies*, World Scientific, Singapore (1992).
4. M.C. Zonneyville, J.J.C. Geerlings and R.A. van Santen, J. Catal. **148**, 417 (1994).
5. E.H. Teunissen, C. Roetti, C. Pisani, A.J.M. de Man, A.P.J. Jansen, R. Orlando, R.A. van Santen and R. Dovesi, Model. Simul. Mater. Sci. Eng. **2**, 921 (1994).
6. D. Kalkstein and P. Soven, Surf. Sci. **26**, 85 (1971).
7. A. Gonis, X.-G. Zhang, J.M. MacLaren and S. Crampin, Phys. Rev. B **42**, 3798 (1990).
8. J.E. Northrup and M.L. Cohen, Phys. Rev. B **29**, 1966 (1984).
9. P.J. Feibelman, Phys. Rev. B **43**, 9452 (1991).
10. C. Pisani, F. Corà, R. Nada and R. Orlando, Comp. Phys. Commun. **82**, 139 (1994); C. Pisani, U. Birkenheuer, F. Corà, R. Nada and S. Casassa, *EMBED96 User Documentation*, University of Torino, Torino (1996).
11. V. Heine, R. Haydock and M. Kelly, in *Solid State Physics*, Vol. 35 (H. Ehrenreich, F. Seitz and D. Turnbull, Eds.), Academic, New York (1980).
12. P.W. Tasker, J. Phys. C: Solid St. Phys. **12**, 4977 (1979).
13. M. Causà, R. Dovesi, C. Pisani and C. Roetti, Surf. Sci. **215**, 259 (1989).
14. R. Orlando, C. Pisani, E. Ruiz and P. Sautet, Surf. Sci. **275**, 482 (1992).
15. R. Hoffmann, *Solids and Surfaces*, VCH, New York (1988).
16. G. Angonoa, J. Koutecky, A.N. Ermoshkin and C. Pisani, Surf. Sci. **138**, 51 (1974).
17. J.C. Boettger, Phys. Rev. B **49**, 16798 (1994).
18. M. Leslie and M.J. Gillan, J. Phys. C: Solid St. Phys. **18**, 973 (1985).
19. A. De Vita, I. Manassidis, J.S. Lin and M.J. Gillan, Europhys. Lett. **19**, 605 (1992).
20. C. Pisani, F. Corà, R. Dovesi and R. Orlando, J. Elec. Spect.**82**, 139 (1994).
21. R. Orlando, R. Dovesi, P. Azavant, N.M. Harrison and V.R. Saunders, J. Phys.: Cond. Matter **6**, 1 (1994).
22. E.N. Economou, *Green Functions in Quantum Physics*, Springer, Berlin (1979).
23. C. Pisani, R. Dovesi and P. Ugliengo, Phys. Stat. Sol. (b) **116**, 249 (1983); P.J. Feibelman, Phys. Rev. B **43**, 9452 (1991).
24. G.F. Koster and J.C. Slater, Phys. Rev. **95**, 1167 (1954).

One-Electron Density Matrices and Related Observables

Wolf Weyrich

Lehrstuhl für Physikalische Chemie I, Fakultät für Chemie, Universität Konstanz, D-78434 Konstanz, Germany

May 30, 1996

Summary. In this Chapter the concept of density matrices and especially of the reduced one-electron density matrix is reviewed with their various properties, among which phase space, chemical bonding, the relation to observables and the experimental access to density matrices are particularly emphasised. Among the observables, electron position and momentum densities play the most important part, whence the state of the art of experimental determination of those densities is sketched with the aim to give some feeling for the potential and the reliability of modern experimental data. A number of entrance points to literature are given.

Key words: Density Matrix – Reduced Density Matrices – Phase Space – Wigner Function – Moyal Function – Electron–Electron Correlation – Natural Spin Orbitals – Hartree–Fock Orbitals – Kohn–Sham Orbitals – Localisation – Delocalisation – Translational Symmetry – Chemical Bonding – Electron Position Density – Electron Momentum Density – Form Factor – Structure Factor – Compton Profile – Reciprocal Form Factor – Reconstruction of Density Matrices

1. Introduction

The Hartree–Fock program CRYSTAL explores the energy eigenstate of a solid variationally. The resulting expectation value forms an upper bound for the true

Correspondence to: Wolf Weyrich – e-mail: Weyrich@chclu.chemie.uni-konstanz.de energy of the solid and is invariant against small variations of the state, i.e. of the quality of the wave function. The same is true for the eigenvalues and eigenfunctions of the one-electron Hartree–Fock equation.

For an optimum theoretical description of the solid it is therefore desirable not only to fulfil the necessary criterion of energy minimisation, but also to assess the quality of the wave function by comparison with experimental data for observables that do not commute with energy and hence are sensitive to variations of the state already in first order. In properly designed experiments it is even possible to determine probability (density) distributions for those experimental data and therefore to obtain some information about the orientation of the state vector of the solid in the appropriate Hilbert space.

With these ideas in mind, the aim of this Chapter is to present a brief overview of the theoretical background as well as of the state of the art of experiments that determine electronic density distributions, both in order to give the reader some insight and feeling for the power and the limitations of the comparison and the analysis of such experimental and theoretical data. The analysis particularly involves concepts of chemical bonding.

Because of the brevity of the overview, the Chapter can only facilitate the access to the pertinent literature, not replace it. The citations are for the same reason far from being complete. The author hopes that the attempt to summarise some insights he has gained over the years will nevertheless be useful for the reader.

2. Density matrices

Solids are systems with a very large number N of electrons. The goal of obtaining information about the orientation of the state vector in Hilbert space would therefore require experiments that determine N-electron density distributions. We do not have such experiments and have rather to be content with single-electron or at most two-electron density distributions. It is therefore not only Hartree–Fock (as well as density functional) theory that necessitates to consider the relationships between the N-particle and the single-particle picture.

The most transparent connection between quantum states and probability distributions at the N-particle and the few-particle level is provided by density matrix theory ([1.2]–[1.6], [1.8], [1.9]), which in addition offers the bonus of a joint description of position and momentum density distributions.

Single-electron density matrices are used in CRYSTAL in order

– to save the computing time for multiplying and adding the expansion coefficients of the crystal orbitals again and again,

$$P_{ik} = \sum_j c_{ij}\, n_j\, c^*_{kj}$$

with i, k = basis-function indices, n_j = occupation number of orbital j,

– and to exploit the range of interatomic interactions for the truncation of infinite series.

Apart from those two features, density matrices have many more important properties. Some of them are indicated in the following subsections.

2.1. Density matrices as general information carriers

Density matrices Γ are an equivalent alternative to wave functions Ψ for the description of pure quantum states,

$$\Gamma(z,z') = \Psi(z)\Psi^*(z')$$

with z as a (set of) position or momentum coordinate(s). The prime indicates that z' can take any value independently from the value of z. In contrast to wave functions, density matrices possess the additional feature to be superimposable,

$$\Gamma = \sum_i X_i \Gamma_i$$

with i = number of a quantum state with mole fraction X_i, for the quantum-mechanical description of microcanonical ensembles (ideal gas, spin gas etc. with thermal excitation and/or molecular mixture etc.). From their definition it follows that density matrices are Hermitian,

$$\Gamma^\dagger = \Psi^*(z')\Psi(z) = \Psi(z)\Psi^*(z') = \Gamma \ ,$$

that their diagonal is positive definite and represents the probability density $\rho(z)$ to find the system at the point z,

$$\rho(z) = \Gamma(z,z) = \Psi(z)\Psi^*(z)$$

(hence the name 'density matrix'), and that their trace is their norm,

$$\mathrm{Tr}\,\Gamma = \int \Gamma(z,z)\,\mathrm{d}z = \int \Psi(z)\Psi^*(z)\,\mathrm{d}z \ .$$

In the following we will first stay with the more familiar position representation. When considering phase space, the momentum representation will come into play again, and the second set of primed coordinates will gain a much deeper meaning. For the sake of simplicity in the present context, we furthermore specialise to N-electron systems in the Born–Oppenheimer approximation right from the beginning.

2.2. Reduction of density matrices

Density matrices can be rigorously reduced from the N-particle level,

$$\Gamma(\mathbf{x}_1, \mathbf{x}_2, \mathbf{x}_3, \ldots, \mathbf{x}_N; \mathbf{x}_1', \mathbf{x}_2', \mathbf{x}_3', \ldots, \mathbf{x}_N')$$

with $\mathbf{x}_i \equiv (\mathbf{r}_i, \sigma_i)$, $\mathbf{r}_i$ = position coordinate, σ_i = spin coordinate, to the two-particle level

$$\Gamma^{(2)}(\mathbf{x}_1, \mathbf{x}_2; \mathbf{x}_1', \mathbf{x}_2')$$

as well as to the one-particle level

$$\Gamma^{(1)}(\mathbf{x}_1; \mathbf{x}_1') \equiv \gamma(\mathbf{x}, \mathbf{x}')$$

by equating $\mathbf{x}_i = \mathbf{x}_i'$ and integrating over $\mathbf{x}_i$ for all $i > \nu$ for $\Gamma^{(\nu)}$ to yield a simpler description of quantum systems (pure state or ensemble) for all physical two- and one-particle observables.

Since we are considering electrons, which are indistinguishable, the $\binom{N}{\nu}$ different possibilities of selecting ν electrons out of N to reduce Γ yield the same *partial* $\Gamma^{(\nu)}_{\text{partial}} \equiv D^{(\nu)}$. If Γ is normalised to $\operatorname{Tr}\Gamma = 1$, the reduction process to a particular set of ν electrons does not change that norm, i.e. $\operatorname{Tr} D^{(\nu)} = 1$ (McWeeny normalisation [1.5]). All these partial $D^{(\nu)}$ are combined to the *total* $\Gamma^{(\nu)}$ given above, which therefore has the properties

$$\Gamma^{(\nu)} = \binom{N}{\nu} D^{(\nu)}, \qquad \operatorname{Tr}\Gamma^{(\nu)} = \binom{N}{\nu}.$$

(Löwdin normalisation [1.2]). Both normalisations, $D^{(\nu)}$ and $\Gamma^{(\nu)}$, are found in literature; we adhere to the more widely used latter one.

2.3. Eigenvalues and eigenfunctions of density matrices

The eigenfunctions of the reduced density matrices provide a rigorous definition of two-particle wave functions = natural spin geminals $\xi_j(\mathbf{x}_1, \mathbf{x}_2)$ (NSGs) with

$$\Gamma^{(2)}(\mathbf{x}_1, \mathbf{x}_2; \mathbf{x}_1', \mathbf{x}_2') = \sum_j \xi_j(\mathbf{x}_1, \mathbf{x}_2)\, n_j^{(2)} \xi_j^*(\mathbf{x}_1', \mathbf{x}_2')$$

and one-particle wave functions = natural spin orbitals $\psi_j(\mathbf{x})$ (NSOs) with

$$\gamma(\mathbf{x}, \mathbf{x}') = \sum_j \psi_j(\mathbf{x})\, n_j\, \psi_j^*(\mathbf{x}')$$

in an N-electron system. The corresponding eigenvalues $0 \le n_j^{(2)} \le [N/2]$ and $0 \le n_j \le 1$ (all real because of the Hermiticity of density matrices) are the occupation numbers of the NSGs and NSOs, respectively, and fulfil the trace relationships

$$\sum_{j=1}^{\infty} n_j^{(2)} = \binom{N}{2} = \frac{N(N-1)}{2}$$

and

$$\sum_{j=1}^{\infty} n_j = N \; ,$$

since traces are invariant against the unitary transformation of the density matrix to its diagonal eigenvalue matrix.

2.4. Orbitals and electron–electron correlation

Hartree–Fock orbitals are a special case of NSOs with

$$n_j = \begin{cases} 0 & \text{for unoccupied (virtual) spin orbitals} \\ 1 & \text{for occupied spin orbitals} \end{cases}$$

and an 'idempotent' one-electron density matrix

$$\gamma^2 = \gamma$$

(independent-particle model).

In real electronic systems, however, electronic motion is correlated. Using the Hartree–Fock orbitals as a starting point, the (in the Hartree–Fock sense) 'occupied' NSOs have occupation numbers

$$n_j < 1$$

and the 'virtual' NSOs have

$$n_j > 0 \; .$$

As a perturbationally justified rule of thumb, the respective deviation from 1 and 0 is the larger the closer the orbital energy ε_j of the related Hartree–Fock orbital lies to the Fermi energy ε_F.

In atoms and molecules, the Fermi energy is $\varepsilon_F = (\varepsilon_{HOMO} + \varepsilon_{LUMO})/2$ or $\varepsilon_F = \varepsilon_{SOMO}$. Correspondingly it is located in the middle of the band gap of insulators and semiconductors. Hence, as another rule, correlation effects are the larger the smaller the band gap or $\varepsilon_{LUMO} - \varepsilon_{HOMO}$ is.

In perturbationally first order only the n_j differ from the Hartree–Fock approximation, whereas in second and higher order also the NSOs are no longer Hartree–Fock orbitals. There are infinitely many NSOs with n_j close to 0 in a rigorous sense, and the one-electron density matrix of real electronic systems is *not* idempotent,

$$\gamma^2 < \gamma \; .$$

Another (additionally) possible reason for non-idempotency of γ is that its parent Γ belongs to an ensemble. Given a non-idempotent γ, there is still no way of differentiating whether it originates from a pure quantum state with electron-electron correlation ('N-representable') or from an ensemble of quantum states ('ensemble-representable'). Non-idempotency with fulfilled bounds for the eigenvalues thus ensures ensemble representability, but not N-representability. Only idempotent density matrices are necessarily N-representable (Hartree–Fock-representable, e. g.).

2.5. Phase space

Density matrices provide the quantum-mechanical phase-space density, as it becomes clear when considering then transition from the position to the momentum representation (see, e. g., Refs. [3.5] and [3.6]).

For a transparent notation of the phase-space relationships it is useful to change the notation of the coordinates slightly. For each particle i, the row coordinate (argument of Ψ) is now denoted by $\mathbf{x}_i'$, the column coordinate (argument of Ψ^*) by $\mathbf{x}_i''$. For electrons in matter, each coordinate is a four-dimensinal vector (position plus spin), and the algebraic operations with the vectors are defined in that 4D space.

Since the density matrices contain two such independent vectors per particle, the matrices are defined in a space with eight dimensions per particle. In other words, that matrix space is spanned by [4 (4D) × 2 (′,″) × ν (particles)] mutually perpendicular (unit) basis vectors. The special property of the diagonal (a 4ν-dimensional subspace) to contain the probability density distribution suggests a coordinate transformation such that half of the new basis vectors span the diagonal and the other half the subspace perpendicular to the diagonal. That separation is achieved by forming the average and the difference for each pair of single-primed and double-primed coordinates, i. e. vectorially [3.5]

$$\begin{aligned} \mathbf{x}_i &= \frac{\mathbf{x}_i' + \mathbf{x}_i''}{2} = (\mathbf{r}_i, \sigma_i) \quad \text{('extracule coordinate')}\,, \\ \mathbf{u}_i &= \mathbf{x}_i'' - \mathbf{x}_i' = (\mathbf{s}_i, \sigma_i) \quad \text{('intracule coordinate')}\,. \end{aligned}$$

As emphasised, the vector addition and subtraction are done in the 4D space of particle i, and the components of the resulting vectors provide then the lengths for the components in the new $(\mathbf{x}_i, \mathbf{u}_i, i = 1, \ldots, \nu)$ coordinate system. Geometrically the transformation is a rotation by 45° with subsequent compression of $\mathbf{x}_i$ and dilation of $\mathbf{u}_i$ by a factor of $\sqrt{2}$ (cf. Fig. 1). The diagonal $\mathbf{x}_i' = \mathbf{x}_i'' = \mathbf{x}_i$ $(i = 1, \ldots \nu)$ of $\Gamma^{(\nu)}$ is the spin-plus-position density

$$\rho^{(\nu)}(\mathbf{x}_1, \ldots, \mathbf{x}_\nu) = \Gamma^{(\nu)}(\mathbf{x}_1, \ldots, \mathbf{x}_\nu; \mathbf{x}_1, \ldots, \mathbf{x}_\nu)$$

and the off-diagonal part $\mathbf{x}_i' \neq \mathbf{x}_i''$ contains the Fourier–Dirac-transformed spin-plus-momentum density $\varpi^{(\nu)}(\mathbf{y}_1, \ldots, \mathbf{y}_\nu)$ with $\mathbf{y}_i \equiv (\mathbf{p}_i, \sigma_i)$, $\mathbf{p}_i$ = momentum coordinate, in the direction $\mathbf{u}_i$, spin-plus-position resolved parallel to the diagonal along $\mathbf{x}_i$.

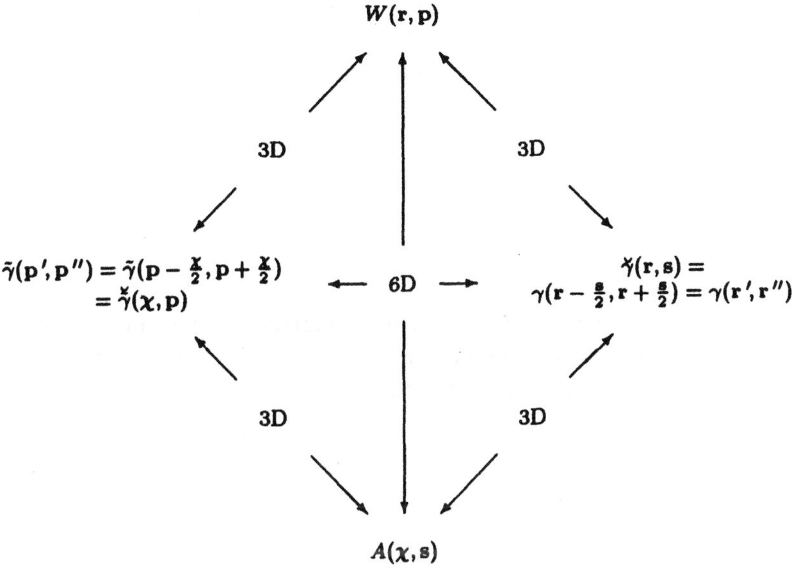

Figure 1. Interrelationships between the continuous representations of density matrices, explicitly given for the single-particle (= one-electron) case. 3D and 6D stand for the dimensionality of the Fourier–Dirac transformations. $\check{\gamma}$ and $\check{\tilde{\gamma}}$ are expressed in the coordinate systems $\mathbf{r},\mathbf{s}$ and $\mathbf{p},\boldsymbol{\chi}$ rotated relative to $\mathbf{r}',\mathbf{r}''$ and $\mathbf{p}',\mathbf{p}''$, respectively. By multiplying the number of coordinates and the dimensionality of the Fourier–Dirac transformations by ν, the interrelationships hold at any ν-particle level.

Fourier–Dirac transformation along each $\mathbf{r}_i$ of all $\mathbf{u}_i$ thus yields a true phase-space density, the Wigner function[1]

$$W^{(\nu)}(\mathbf{x}_1,\ldots,\mathbf{x}_\nu;\mathbf{y}_1,\ldots,\mathbf{y}_\nu) = (2\pi\hbar)^{-3\nu}\int\cdots\int \Gamma^{(\nu)}(\mathbf{x}_1,\ldots,\mathbf{x}_\nu;\mathbf{u}_1,\ldots,\mathbf{u}_\nu)\exp\left(\mathrm{i}\sum_{i=1}^{\nu}\mathbf{p}_i\cdot\mathbf{s}_i/\hbar\right)\mathrm{d}\tau_{s_1}\cdots\mathrm{d}\tau_{s_\nu}$$

in a mixed position–momentum representation. There are also a pure momentum representation $\tilde{\Gamma}^{(\nu)}(\mathbf{y}_1',\ldots,\mathbf{y}_\nu';\mathbf{y}_1'',\ldots,\mathbf{y}_\nu'')$ and a mixed form-factor representation $A^{(\nu)}(\mathbf{u}_1,\ldots,\mathbf{u}_\nu;\mathbf{v}_1,\ldots,\mathbf{v}_\nu)$ (Moyal function, with $\mathbf{v}_i = \mathbf{y}_i'' - \mathbf{y}_i' = (\boldsymbol{\chi}_i,\sigma_i)$ as the second intracule, viz. the momentum intracule coordinate) of density matrices.

[1] Here and in the remaining part of the Chapter, $\hbar$, m_e and e are retained for pedagogical reasons and in adherence to recommended rules. For the same reason, lengths are given in multiples of a_0, linear momenta in multiples of p_0, etc., rather than using the indifferent notion 'atomic units'. The particular benefit of retaining $\hbar$ is the clear distinction between wave vectors $\mathbf{k}$, $\boldsymbol{\kappa}$ and linear momenta $\mathbf{p} = \hbar\mathbf{k}$, $\boldsymbol{\chi} = \hbar\boldsymbol{\kappa}$ later in the text. Wave vectors are reciprocal to position with respect to Fourier transformation, whereas linear momenta are reciprocal to position with respect to Fourier–Dirac transformation.

Only the diagonals of $\Gamma^{(\nu)}$ and $\tilde{\Gamma}^{(\nu)}$ are positive-definite; the otherwise possible negative regions of density matrices in those four representations are a puzzling feature and a result of the Pauli principle, which forces electrons into oscillatory single-particle wave functions with nodes.

2.6. Density operators

As a matter of fact, the concept of a density matrix is independent of any representation in the same way as a quantum-state vector $|\mu\rangle$. The representation-free density matrix is also called 'density operator' $\hat{\Gamma}$. In Dirac notation one can write

$$\hat{\Gamma}_\mu = |\mu\rangle\langle\mu|$$

for pure quantum states and

$$\hat{\Gamma} = \sum_\lambda X_\lambda |\lambda\rangle\langle\lambda| = \sum_\lambda |\lambda\rangle\langle\lambda| X_\lambda = \sum_\lambda |\lambda\rangle X_\lambda \langle\lambda|$$

for ensembles of systems in states λ.

2.7. Expectation values of observables and reduction

The expectation values of observables with operator $\hat{\Omega}$ are calculated as

$$\langle\Omega\rangle = \mathrm{Tr}\,\hat{\Omega}\hat{\Gamma} = \mathrm{Tr}\,\hat{\Gamma}\hat{\Omega}\ ,$$

in analogy to $\langle\Omega\rangle = \langle\mu|\hat{\Omega}|\mu\rangle$.

The reduction of density matrices with respect to the particle number means forming the trace beforehand over those particles or particle coordinates that are not acted on by a class of operators $\hat{\Omega}$. Hence, $\Gamma^{(2)}$ contains the complete information about the quantum-state (or ensemble) for two-electron operators (Coulomb repulsion, e.g.), as does γ for one-electron operators (kinetic energy, nuclear attraction, dipole moment, polarisability, susceptibility, chemical shift etc. etc.). For spin-independent operators the spin coordinates can also be traced out, leading to spin-free density matrices such as $\Gamma^{(2)}(\mathbf{r}_1', \mathbf{r}_2'; \mathbf{r}_1'', \mathbf{r}_2'')$, $\gamma(\mathbf{r}', \mathbf{r}'')$, etc.

2.8. Continuous and discrete representations

So far four different *continuous* representations have been considered, the counterpart to wave functions $\Psi(x) = \langle x|\mu\rangle$ (position representation) and $\tilde{\Psi}(y) = \langle y|\mu\rangle$ (momentum representation). Discrete basis sets $\{|i\rangle\}$, $\{\Phi_i(z) = \langle z|i\rangle\}$, $z = x$ or y, lead to *discrete* representations:

$$|j\rangle = \sum_i |i\rangle\langle i|j\rangle = \sum_i |i\rangle\, c_{ij}\ , \qquad c_{ij} = \langle i|j\rangle\ ,$$

with the transformation relation

$$\Psi_j(z) = \sum_i \Phi_i(z)\, c_{ij}$$

into a continuous representation for vectors $|j\rangle$, and

$$\hat{\Gamma} = \sum_{i,k} |i\rangle\langle i|\hat{\Gamma}|k\rangle\langle k| = \sum_{i,k} |i\rangle \Gamma_{ik}\langle k|\ , \qquad \Gamma_{ik} = \langle i|\hat{\Gamma}|k\rangle\ ,$$

with the transformation relation

$$\Gamma(z', z'') = \sum_{i,k} \langle z'|i\rangle\langle i|\hat{\Gamma}|k\rangle\langle k|z''\rangle = \sum_{i,k} \Phi_i(z')\, \Gamma_{ik}\, \Phi_k^*(z'')$$

for density operators/matrices.

2.9. Density matrices in CRYSTAL

The density matrix $\mathbf{P} = [P_{ik}]$ in CRYSTAL is such a discrete representation of the one-electron density matrix $\gamma(\mathbf{r}', \mathbf{r}'')$. Only the row block with row numbers i belonging to the reference unit cell $\mathbf{m}' = 0$ is used because of the so-called cyclicity of $\mathbf{P}$ that originates from the translational symmetry (only $\mathbf{m}'' - \mathbf{m}'$ matters, not the absolute positions $\mathbf{m}', \mathbf{m}''$ of the unit cells). Furthermore, the spin coordinates σ', σ'' are traced out (hence $\gamma(\mathbf{r}', \mathbf{r}'')$) in the restricted (closed or open-shell) Hartree–Fock of CRYSTAL92, whereas in the unrestricted Hartree–Fock (UHF) scheme of CRYSTAL95 two different $\mathbf{P}$ are maintained: $\mathbf{P}_\alpha$ with $\sigma' = \sigma'' = \frac{1}{2}$ and $\mathbf{P}_\beta$ with $\sigma' = \sigma'' = -\frac{1}{2}$; off-diagonal $\sigma' \neq \sigma''$ (giving rise to $\mathbf{P}_{\alpha\beta} = \mathbf{P}_{\beta\alpha}^*$) is neglected in common UHF with the idea of a spin quantisation direction common to the whole crystal, thus limiting the possible patterns of magnetic structure.

2.10. Two-electron density matrix in Hartree–Fock

In the Hartree–Fock approximation as an independent-particle model, the two-electron density matrix can be expressed in terms of the one-electron density matrix,

$$\Gamma^{(2)}_{\mathrm{HF}}(\mathbf{x}_1',\mathbf{x}_2';\mathbf{x}_1'',\mathbf{x}_2'') = \frac{1}{2}\left[\gamma(\mathbf{x}_1',\mathbf{x}_1'')\,\gamma(\mathbf{x}_2',\mathbf{x}_2'') - \gamma(\mathbf{x}_1',\mathbf{x}_2'')\,\gamma(\mathbf{x}_2',\mathbf{x}_1'')\right] .$$

The pair distribution function $P(r_{12})$ calculated from that two-electron density matrix reflects the Fermi hole owing to the Pauli principle ('exchange hole'), but does not contain any Coulomb hole.

2.11. Localisation and delocalisation in Hartree–Fock

The degeneracy of the eigenvalues n_j of the one-electron density matrix in the Hartree–Fock approximation leads to the fact that the density matrices of any order ν and hence all measurable quantities are invariant against unitary transformations of the set of occupied spin orbitals (all $n_j = 1$).

The most important type of unitary transformation of the set of occupied orbitals is localisation of the completely delocalised canonical orbitals of the Hartree–Fock eigenvalue equation (canonical molecular orbitals = CMOs, Bloch functions) to localised molecular orbitals (LMOs) or Wannier functions and vice versa. The program CRYSTAL makes ample use of such transformations. The same holds formally for the set of unoccupied orbitals (all $n_j = 0$), which do not contribute to density matrices and measurable quantities at all.

Mixing of occupied and unoccupied orbitals, however, *changes* density matrices and observables, as it does in the case of the non-degenerate true (spin) orbitals of the real, correlated electronic systems.

2.12. Kohn–Sham orbitals

In the approach of the Kohn–Sham orbitals [2.2], which are the basis of any nowadays' implementation of density-functional theory [2.1]–[2.7], there is an important inherent restriction that cannot be emphasised enough.

The Kohn–Sham equation with its exactly N orbitals $\psi_j^{\mathrm{KS}}(\mathbf{r})$ for N electrons are constructed such that the true one-electron density $\rho(\mathbf{r})$ and hence the diagonal $\gamma(\mathbf{r},\mathbf{r})$ of γ is perfectly reproduced as $\sum_{j=1}^{N} |\psi_j^{\mathrm{KS}}(\mathbf{r})|^2$ and that simultaneously the energy functional $E[\rho(\mathbf{r})]$ is minimised.

This yields correct results for any observable that depends solely on the position coordinate $\mathbf{r}$, but gives systematically incorrect results for observables that contain the momentum coordinate $\mathbf{p}$ such as momentum densities $\varpi(\mathbf{p})$ or angular momentum $\mathbf{r} \times \mathbf{p}$.

The reason is the *contradiction* between an *exact* position density $\rho(\mathbf{r})$ and $n_j^{\mathrm{KS}} = 1$ for $j = 1, 2, \ldots, N$, $n_j^{\mathrm{KS}} = 0$ for $j > N$, since the exactly described electronic system has infinitely many NSOs with $n_j > 0$. Another way of stating this contradiction is the finding that *any* $\rho(\mathbf{r})$ can be generated with an *idempotent* density matrix $\gamma(\mathbf{r}', \mathbf{r}'')$ [2.8]–[2.10], whereas the true density matrix is *not* idempotent.

Thus, the Kohn–Sham approach is even unable to describe the interacting homogeneous electron gas with its occupation function $1 > n(\mathbf{k}) > 0$ (= occupation numbers of NSOs $\psi_{\mathbf{k}}$). Lam and Platzman [2.3] have introduced a first-order correction for this situation, later treated more generally and rigorously by Bauer [2.4].

In summary, the Kohn–Sham approach is correct for $\rho(\mathbf{r})$ and the form factor $F(\boldsymbol{\chi})$ equal to the corresponding three-dimensional subspaces $\gamma(\mathbf{r}, \mathbf{r})$ of $\gamma(\mathbf{r}', \mathbf{r}'')$ and $A(\boldsymbol{\chi}, 0)$ of $A(\boldsymbol{\chi}, \mathbf{s})$, but incorrect for the other parts of the density matrix in these two representations. In the Wigner and in the pure momentum representation it provides only correct projections of the density matrix into the $\{\mathbf{r}\}$ and $\{\boldsymbol{\chi}\}$ subspaces, respectively.

2.13. Range of interactions in molecules and solids

Because of the immunity of density matrices against localising or delocalising transformations of the orbitals of independent-particle models, they are an ideal tool to determine the *true* range of electronic interactions and hence chemical bonding in a molecule or solid. This feature becomes particularly useful when considering $\gamma(\mathbf{x}', \mathbf{x}'')$ rather than any basis-set dependent P. The assignment to atoms and pairs of atoms can be done simply by analysing the volume of γ at or around $\mathbf{r}', \mathbf{r}''$ equal to the atomic positions [1.10], [1.11].

It is this feature, together with the interest to investigate electronic phase space, that makes it so worthwile to calculate density matrices theoretically or even to obtain them from experiments.

To my knowledge the first position-representation density matrix of a solid, viz. of LiH, has been presented and discussed in Ref. [1.10], based on results from the program CRYSTAL. That density matrix reveals that the electronic structure shows interionic coupling only between direct neighbours in any of the directions $\langle 100 \rangle$, $\langle 110 \rangle$ and $\langle 111 \rangle$. The coupling, however, can involve more than a pair of ions. A more detailed analysis with corrected normalisation and an extension to LiF and LiFHF is about to appear [1.12].

2.14. Density matrices and scattering experiments

Density matrices are closely related to scattering experiments that probe the electronic structure of a material [3.1]–[3.6].

X-ray and electron *diffraction* yield the position density $\rho(\mathbf{r})$, i. e. the *diagonal* $\mathbf{r}' = \mathbf{r}'' = \mathbf{r}$ of $\gamma(\mathbf{r}', \mathbf{r}'')$ (in conjunction with circularly polarised photons or spin-polarised electrons even the spin-position densities $\rho_\alpha(\mathbf{r})$ and $\rho_\beta(\mathbf{r})$, i. e. $\gamma(\mathbf{x}', \mathbf{x}'')$ with $\mathbf{x}' = \mathbf{x}'' = \mathbf{x}$).

Neutron diffraction enables one to determine the spin-position difference density (usually simply called 'spin density') $\rho_\alpha(\mathbf{r}) - \rho_\beta(\mathbf{r})$, although sometimes contaminated with magnetic orbit contributions.

Deep inelastic scattering of photons and electrons (Compton scattering), polarised or unpolarised, gives access to a projection $B(\mathbf{u}) = \int \gamma(\mathbf{x}, \mathbf{x}+\mathbf{u})\, d\tau_x$ of the *off-diagonal part* of $\gamma(\mathbf{x}', \mathbf{x}'')$, analogously with or without spin, respectively.[2] In a crystal of sufficient perfection one can finely tune the position of a standing wave field of X-rays by small tilts within the width of each Bragg or Laue reflection. Compton scattering of such standing waves offers even the direct determination of the off-diagonal parts of $\tilde{\gamma}$ because of the position discrimination in the experiment [8.12]–[8.14] (probing phase space).

Finally, *inelastic scattering also at low momentum transfer* provides information about the electron-pair distribution function and hence a piece of the diagonal of $\Gamma^{(2)}$.

2.15. Reconstruction of density matrices

The close relationships between experiments and density matrices have led to attempts to *reconstruct* density matrices from measured data (see also Section 10). These attempts are increasingly successful, particularly during the last few years, and promise to give experimental insight into the 'real internals' of an electronic quantum system.

3. Densities and form factors

At the one-electron level, four observables are the key quantities: the position density $\rho(\mathbf{r})$ (often unsystematically called 'charge density'), the form factor $F(\boldsymbol{\kappa}) = F(\boldsymbol{\chi})$ with $\boldsymbol{\chi} = \hbar\boldsymbol{\kappa}$ of diffraction work, the momentum density $\varpi(\mathbf{p})$ and the reciprocal form factor $B(\mathbf{s})$. For the sake of simplicity, from now on a spin-free formulation is used.

[2] Positron annihilation experiments yield a 1D or 2D projection of the momentum density distorted by the positron wave function. They are not considered here because of their greater theoretical complexity.

The densities ρ and ϖ are the diagonals of the density matrix in the pure position representation, γ, and in the pure momentum representation, $\tilde{\gamma}$, respectively. The form factor F is the 3D Fourier–Dirac transform of ρ and the reciprocal form factor B the one of ϖ:

$$\begin{aligned} F(\chi) &= \int \rho(\mathbf{r}) \exp(\mathrm{i}\,\chi\cdot\mathbf{r}/\hbar)\, \mathrm{d}\tau_r \ , \\ B(\mathbf{s}) &= \int \varpi(\mathbf{p}) \exp(-\mathrm{i}\,\mathbf{p}\cdot\mathbf{s}/\hbar)\, \mathrm{d}\tau_p \ . \end{aligned}$$

Since $\gamma(\mathbf{r}',\mathbf{r}'')$ and $\tilde{\gamma}(\mathbf{p}',\mathbf{p}'')$ are related by a 6D Fourier–Dirac transformation,

$$\begin{aligned} \gamma(\mathbf{r}',\mathbf{r}'') &= (2\pi\hbar)^{-3} \iint \tilde{\gamma}(\mathbf{p}',\mathbf{p}'') \exp\left[\mathrm{i}\,(\mathbf{p}'\cdot\mathbf{r}' - \mathbf{p}''\cdot\mathbf{r}'')/\hbar\right] \mathrm{d}\tau_{p'}\, \mathrm{d}\tau_{p''} \ , \\ \tilde{\gamma}(\mathbf{p}',\mathbf{p}'') &= (2\pi\hbar)^{-3} \iint \gamma(\mathbf{r}',\mathbf{r}'') \exp\left[-\mathrm{i}\,(\mathbf{p}'\cdot\mathbf{r}' - \mathbf{p}''\cdot\mathbf{r}'')/\hbar\right] \mathrm{d}\tau_{r'}\, \mathrm{d}\tau_{r''} \ , \end{aligned}$$

both form factors are projections

$$\begin{aligned} F(\chi) &= \int \tilde{\gamma}(\mathbf{p},\mathbf{p}+\chi)\, \mathrm{d}\tau_p \ , \\ B(\mathbf{s}) &= \int \gamma(\mathbf{r},\mathbf{r}+\mathbf{s})\, \mathrm{d}\tau_r \end{aligned}$$

parallel to the diagonals $\mathbf{p}$ and $\mathbf{r}$, resp., into the subspaces $\{\chi\}$ and $\{\mathbf{s}\}$ orthogonal to the respective diagonal.

When decomposing the density matrix into its NSO contributions,

$$\gamma(\mathbf{r}',\mathbf{r}'') = \sum_j \psi_j(\mathbf{r}')\, n_j\, \psi_j^*(\mathbf{r}'')$$

and

$$\tilde{\gamma}(\mathbf{p}',\mathbf{p}'') = \sum_j \tilde{\psi}_j(\mathbf{p}')\, n_j\, \tilde{\psi}_j^*(\mathbf{p}'') \ ,$$

each orbital contribution to the form factors $B(\mathbf{s})$ and $F(\chi)$ can be interpreted as an orbital autocorrelation function:

$$\begin{aligned} B(\mathbf{s}) &= \sum_j n_j B_j(\mathbf{s}) = \sum_j n_j \int \psi_j(\mathbf{r})\, \psi_j^*(\mathbf{r}+\mathbf{s})\, \mathrm{d}\tau_r \ , \\ F(\chi) &= \sum_j n_j F_j(\chi) = \sum_j n_j \int \tilde{\psi}_j(\mathbf{p})\, \tilde{\psi}_j^*(\mathbf{p}+\chi)\, \mathrm{d}\tau_p \ . \end{aligned}$$

Since the form factors 'look' at the off-diagonal parts of γ and $\tilde{\gamma}$ in a projection, they can 'see' negative regions, provided these are not compensated by positive regions when projected. The form factor F experiences such a compensation regularly because of the dominance of the contribution of the nodeless 1s orbital of every atom, whereas the reciprocal form factor B can show rich nodal

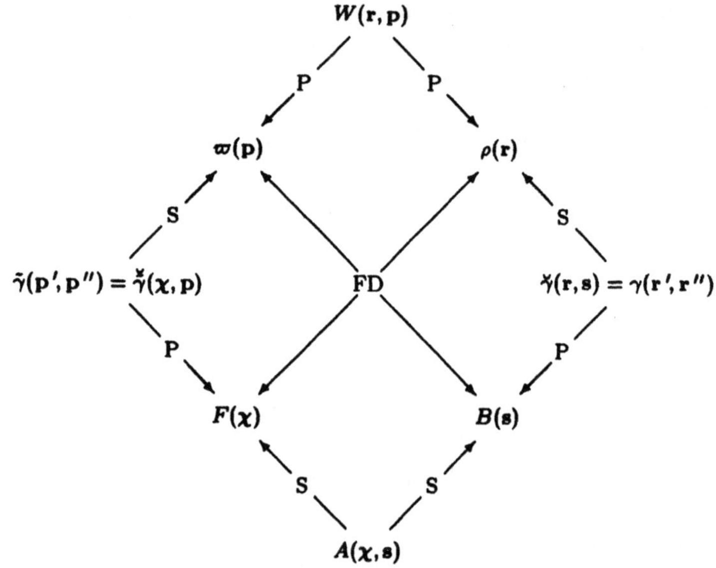

Figure 2. Interrelationships between the position density ρ, the momentum density ϖ, the form factor F, the reciprocal form factor B and the various representations of the one-electron density matrix (cf. Fig. 1). FD stands for a 3D Fourier–Dirac transformation (on the one hand between ρ and F, on the other hand between ϖ and B), S for the selection of a subspace, P for the projection into a subspace. Reconstruction means to overcome the loss of information on selection and/or projection.

structure. As an example: The naïve Madelung-type ionic picture LiH = Li^+ + H^- with two only rigidly Coulomb-interacting 1s orbitals would yield a positive-definite $B(\mathbf{s})$, falsified by the experiment with pronounced negative parts of B at larger s [5.8].

4. Experiments

The four observables ρ, ϖ, F and B are, however, not the primary experimental quantities. The experiments measure intensities proportional to the differential scattering cross-sections.

4.1. *Diffraction*

Only the direction of the solid angle Ω is discriminated:

$$\frac{\mathrm{d}\sigma}{\mathrm{d}\Omega} \sim \begin{cases} |F(\boldsymbol{\kappa})| & \text{dynamical limit} \\ |F(\boldsymbol{\kappa})|^{\alpha(\boldsymbol{\kappa})} & \text{reality, } 1 < \alpha(\boldsymbol{\kappa}) < 2 \\ |F(\boldsymbol{\kappa})|^2 & \text{kinematical limit} \end{cases}$$

with σ = cross-section, $\boldsymbol{\kappa}$ = scattering vector $\mathbf{k}_2 - \mathbf{k}_1$, $\mathbf{k}_1$ and $\mathbf{k}_2$ being the wave vectors of the primary and the scattered beam with normalisation $k_1 = k_2 = 2\pi/\lambda$, $\kappa = 4\pi \sin\theta/\lambda$, θ = Bragg angle. From the overwhelming amount of literature on X-ray diffraction Refs. [4.1]–[4.4] are particularly useful for a first overview with respect to accurate structure factors $F(\boldsymbol{\kappa})$ and position densities $\rho(\mathbf{r})$.

4.2. Compton scattering

The energy $\hbar\omega_2$ of the scattered particle (mostly photon) is discriminated in addition, and within the sudden-impulse approximation

$$\frac{\mathrm{d}^2\sigma}{\mathrm{d}\Omega \mathrm{d}\omega_2} \sim J(\mathbf{q})$$

holds with $J(\mathbf{q})$ = Compton profile = 1D distribution of momentum components = projection

$$J(\mathbf{q}) = \int \varpi(\mathbf{p}) \cdot \delta\left(\frac{\mathbf{p}\cdot\mathbf{q}}{q} - q\right) \mathrm{d}\tau_p$$

of $\varpi(\mathbf{p})$ on a line $\mathbf{q}$ parallel to the scattering vector $\boldsymbol{\kappa}$, $\mathbf{q}$ = initial momentum component of the struck electron in the direction of $\boldsymbol{\kappa}$ with

$$q = \frac{m_e c}{c|\kappa|} \cdot \left(\omega_2 - \omega_1 + \frac{\hbar\omega_1\omega_2}{m_e c^2}(1 - \cos\phi)\right) ,$$

where ϕ is the scattering angle and $c|\kappa| = \sqrt{\omega_1^2 + \omega_2^2 - 2\omega_1\omega_2\cos(\phi)}$. A small, rather personal selection of literature in this field are Refs. [5.1]–[5.20].

4.3. '$\gamma, e\gamma$' or 'X, eX' experiments

Recent experiments that detect and analyse also the ejected electron from a thin solid film in coincidence with the Compton-scattered photon [5.22]–[5.24] yield coincidence intensities

$$I(\Omega_2, \omega_2, \Omega_3) \sim \frac{\mathrm{d}^3\sigma}{\mathrm{d}\Omega_2 \mathrm{d}\omega_2 \mathrm{d}\Omega_3}$$

or even

$$I(\Omega_2, \omega_2, \Omega_3, \omega_3) \sim \frac{\mathrm{d}^4\sigma}{\mathrm{d}\Omega_2 \mathrm{d}\omega_2 \mathrm{d}\Omega_3 \mathrm{d}E_3}$$

with Ω_3 and E_3 = solid angle and energy of the outgoing electron, resp., and thus can measure also other components of the initial electron momentum at the same time, though still with only modest precision.

4.4. (e, 2e)-experiments

The analogous (e, 2e)-experiments [5.28], [5.29] are already performed since more than twenty years on gases and since recently also on thin solid films. They permit high-resolution discrimination in angles and energies and thus to discriminate final states, described as hole states plus outgoing electron, which in some approximation project out the (in the case of gases radial) momentum density of the orbital originally filled in the ground state. The impressive data for solid films map energy-dependent momentum densities $\varpi(\mathbf{p}, \varepsilon)$ and thus bands [5.29] with, however, an only semi-quantitative accuracy because of strong multiple scattering.

4.5. The multiple-scattering problem for electrons as probes

In the description of the $(\gamma, e\gamma)$ and (e, 2e)-experiments one notices that the electrons as projectile or observed particles restrict the investigated systems to gases or thin solid films. This is due to the κ^{-4}-dependence of the Rutherford or Mott cross-section for the scattering of electrons by other charged particles, which reduces the mean-free-path of electrons by strong forward scattering to the order of 10 to 100 Å and limits such experiments seriously in their universal use. In the diffraction of electrons on solids (i. e. the (e, e′)-experiment = LEED or RHEED), the small mean-free-path can be converted into a virtue for studying surfaces; in the inelastic coincidence experiments, however, one of the particles has to come from or to go into the bulk because of momentum conservation, thus making strong multiple scattering unavoidable.

5. Form factors and structure factors

For the evaluation of the form factor $F(\boldsymbol{\kappa}) = |F(\boldsymbol{\kappa})| \cdot \exp(i\varphi(\boldsymbol{\kappa}))$ (or, for crystals, of the discrete 'structure factors' $F(\boldsymbol{\kappa}_{hkl}) \equiv F_{hkl} = |F_{hkl}| \cdot \exp(i\varphi_{hkl})$) from diffraction data, the phase problem has to be solved. As an empirical rule — since for systems with inversion symmetry and with the position origin put to a centre of inversion, $\varphi(\boldsymbol{\kappa})$ can only be 0 or π — the phases resulting for centrosymmetric systems are much more reliable than those for non-centrosymmetric ones, if there are positional parameters in the structure that are not fixed by symmetry. Such parameters have eventually to be determined by a fit to a finite set of form-factor or structure-factor values.

Moduli and phases of form factors and structure factors are often distorted by resonance ('anomalous dispersion') and multiple-diffraction ('Umweganregung' and 'Aufhellung') contributions. The former effect, being essentially of atomic nature, can be taken into account to some extent in the subtraction of an atomic model ('promolecule', see below) with complex atomic form factors $f_j = f_j' + i f_j''$, whereas the latter has to be ruled out experimentally by rotation of the sample around the scattering vector. Publications are to be read carefully with respect to these points.

Our visual and conceptual preference of position space makes it desirable to transform $F(\boldsymbol{\kappa})$ into $\rho(\mathbf{r})$. However, because of the finite set of experimental $F(\boldsymbol{\kappa})$ up to a certain $\kappa_{\max}$ given by the wavelength utilised in the experiment or by thermal vibrations, that transformation can be done only by modelling (usually a multi-centre multipolar expansion of $\rho(\mathbf{r})$, smeared by thermal vibrations), with or without prior subtraction of a model of superimposed free atoms (the already mentioned promolecule). Therefore it is strongly recommended to compare experimental and theoretical data in $\boldsymbol{\kappa}$-space in order to enjoy full accuracy and least bias, although even then the experimental data have already been modified by correcting for the model-dependent thermal vibrations. The modelling transformation to $\rho(\mathbf{r})$ can still be done, if the origin of discrepancies is to be attributed to certain positions in the structure of the system.

Another source of experimental uncertainty in diffraction work on crystals is 'extinction', i. e. whether the scattering cross-section is (almost) proportional to $|F_{hkl}|$ (dynamic diffraction = complete multiple diffraction of a standing wave field in a perfect crystal) or (almost) proportional to $|F_{hkl}|^2$ (kinematic diffraction = independent diffraction by small units, as it is true for gases and liquids and as it is hoped to be the case for sufficiently imperfect crystals).

The proportionality factors between measured intensities and the differential scattering cross-sections, and consequently the conversion factors from intensities to the desired quantities related to the electronic structure of the samples, viz. $F(\boldsymbol{\kappa})$ and $J(\mathbf{q})$, have to be determined by a normalisation procedure. In the case of F, this is $F(0) = Z$, Z being the number of electrons per formula unit or per unit cell. Unfortunately, $F(0)$ is experimentally unaccessible (scattered beam = primary beam), whence the 'scale factor' has to be determined by modelling $F(\boldsymbol{\kappa})$ for small κ. Only the completely different Pendellösung experiment on perfect crystals can yield structure factors free from that uncertainty.

In total, carefully measured and corrected structure factors of favourable samples can have an accuracy of $\approx 1\%$, but quite often different data sets can deviate by as much as 5%.

Most of the remarks about X-ray diffraction apply also to neutron diffraction. Neutrons interact with the nuclei and with the internal magnetisation density of the sample. The latter effect is separated by forming the difference between measurements with opposite external field ('flipping ratio'), leading to magnetic structure factors.

6. Compton profiles

The normalisation of Compton profiles by

$$\int_{-\infty}^{\infty} J(\mathbf{q})\, \mathrm{d}q = Z$$

is much less critical, provided the experimentally finite range of q is so large that the contributions of all valence electrons are included (typically $|q| \leq q_{\max}$, $q_{\max} \geq 5\,p_0$, $p_0 = \hbar/a_0 = m_e c\alpha = 1$ Dumond $= 1$ 'atomic unit of momentum'). In that case a theoretical partial norm

$$\int_{-q_{\max}}^{+q_{\max}} J(\mathbf{q})\,\mathrm{d}q = Z_{\text{partial}} < Z$$

can be used, since the core contributions at $|q| > q_{\max}$ will be hardly affected by the molecular or crystal environment and can therefore be calculated from high-quality atomic wave functions.

Experimental Compton profiles can be (slightly) distorted by two other important effects.

- The first one is partial failure of the 'sudden-impulse approximation', which underlies the proportionality of the scattering cross-section to $J(\mathbf{q})$. It approximates the electronic system by a quasi-free electron gas with the momentum distribution of the real system, but with no binding potential because of its cancellation between initial and final state in the extremely fast scattering process (Δt of the order of attoseconds, Franck–Condon principle with respect to *electronic* motion). The faster the process the better the approximation, which means high energy transfer relative to the electron binding energy and high momentum transfer to be preferred ('deep' inelastic scattering), verified by back scattering of high-energy photons (or, only for molecules and thin solid films, low-angle scattering of high-energy electrons).
- The second experimental distortion is contamination by multiply scattered photons or electrons from the sample with non-zero volume (or non-zero number of scatterers) and has some similarity to the extinction problem of diffraction. Such photons or electrons do not have the proper information from a single well-defined elementary scattering process and form a background that for somewhat complicated reasons can partially coincide in position with the Compton band in the observed spectrum of scattered particles and form a band with a slightly larger width. All experimentalists minimise the multiple-scattering contribution as much as possible by Monte-Carlo simulations or extrapolation of experimental data for differing numbers of scatterers (variable sample thickness or gas pressure) to zero number of scatterers. Nevertheless, some residual contamination usually remains, to a large extent masked by the normalisation procedure.

Both effects influence the data for high-atomic-number materials much more than for 'light' materials, as a general rule.

Some recent experiments with synchrotron radiation and a Cauchois-type analyser show another uncertainty that is overcome in most of the other experiments, viz. from the background to be subtracted. One can hope that this problem will be reduced in the near future.

Since Compton experiments are based on spectral resolution of the scattered particles, the degree of spectral resolution influences the extracted Compton profile $J(\mathbf{q})$. It differs from the infinitely-high-resolution true $J(\mathbf{q})$ by a convolution with the so-called 'residual instrumental function', RIF — residual, because every experimentalist tries to get rid of some of the distortions by the primary resolution in the processing of the data. We discriminate the convolved $J^c(\mathbf{q})$ from the true $J(\mathbf{q})$ by the superscript 'c' and use a pure Gaussian function of standardised width very close to the width of the primary experimental resolution in the q-scale as our RIF with the idea that the information content of an experiment cannot be increased by artificial resolution enhancement and that the RIF should be mathematically simple. Other authors prefer more complicated RIFs for an improvement of the overall shape of the Compton profile. All theoretical data have to convolved with the RIF of the experimental data they are to be compared with. Typical resolutions reported in literature range from 0.6 down to (most recently for very light metals) 0.01 p_0 full width at half maximum (FWHM), depending on which type of analyser has been employed, which, in turn, is a question of the available photon flux and the sample size. For insulators, modest resolution can be quite satisfactory and good statistical precision is much more important, whereas metals make very high resolution desirable for details of the Fermi surface.

With circularly polarised photons, usually generated in synchrotrons or storage rings, spin densities in momentum space are accessible. The spin-compensated background is separated by forming the difference between measurements with opposite magnetisation. This area enjoys increasing interest [5.17], [5.25] since the first experiments about twenty years ago [5.26], [5.27].

7. Reciprocal form factors

The reciprocal form factor $B(\mathbf{s})$ as the position representation of the momentum density $\varpi(\mathbf{p})$ [5.3]–[5.10] is directly accessible from Compton profiles $J(\mathbf{q})$ by a 1D Fourier–Dirac transformation [5.4], [5.5],

$$B(\mathbf{s}) = \int_{-\infty}^{+\infty} J(\mathbf{q}) \exp(-\mathrm{i}qs/\hbar)\,\mathrm{d}q\ , \quad \mathbf{s} \parallel \mathbf{q}\ .$$

In practice, the integration range is finite $(-q_{\max} \cdots + q_{\max})$. The termination error is small for sufficiently large $q_{\max}$ and can be further reduced by adding theoretically calculated atomic-core wings for $|q| > q_{\max}$ and increasing the integration range. Theoretical data should be subject to the same procedure for a full-precision comparison with the experiment.

Since the termination error occurs only in the contribution of the core electrons to the Compton profile, and since the corresponding contribution to the reciprocal form factor is localised to the range $0 \leq s < 1\,\text{Å} \approx 2\,a_0$, it is strongest in this range. However, if a sharp step or bend occur at $\pm q_{\max}$ (where the implicit periodic repetition units of the discrete Fourier transformation touch), their Fourier transform spreads as an artefact also into the pure valence-electron region $s > 1\,\text{Å} \approx 2\,a_0$. In order to minimise that artefact we smooth such features with a convolution function of varying width before the transformation (maximum width at $|q| = q_{max}$, going linearly to zero with $|q| \rightarrow q_{max} - 1\,p_0$). Again, such a procedure has also to be applied to theoretical data to be compared with the experiment.

The finite resolution of the experiment leads to a multiplicative attenuation of $B(\mathbf{s})$ to

$$B^{a}(\mathbf{s}) = B(\mathbf{s}) \cdot \exp(-s^2/S^2)\,,$$

where

$$S = \frac{4\sqrt{\ln 2}\,\hbar}{\Delta q_{FWHM}}\,.$$

The attenuation factor is $\frac{1}{2}$ at a distance

$$s_{1/2} = S\sqrt{\ln 2} = \frac{4\ln 2\,\hbar}{\Delta q_{FWHM}}$$

and $\frac{1}{10}$ at

$$s_{1/10} = S\sqrt{\ln 10} = \frac{4\sqrt{\ln 2 \ln 10}\,\hbar}{\Delta q_{FWHM}}\,,$$

i.e. 2.93 Å $= 5.54\,a_0$ and 5.35 Å $= 10.11\,a_0$, resp., for $\Delta q_{FWHM} = 0.5\,p_0$ as an example. All these relationships assume a purely Gaussian resolution function, to which the experiment can be brought by appropriate data processing. Other authors prefer different 'residual instrumental functions' with a hence different multiplicative transfer function of s from $B(\mathbf{s})$ to $B^{a}(\mathbf{s})$. Because of the merely multiplicative effect of experimental resolution in the $\{\mathbf{s}\}$-domain, adaptation to and changes of the resolution of experimental or theoretical data are most easily performed for or via $B(\mathbf{s})$ in either case.

The distance parameters for a particular momentum resolution provide the quantitative scale for the range of interatomic and interelectronic interactions that can be covered by the experiment under consideration. The parameters have to be compared with interatomic distances in the case of closed-shell systems and of molecules, and with the range of the Fourier transform of the Fermi sphere [5.4],

$$B_{FS}(s) = \frac{3Z_{cond}}{(k_F s)^2}\left[\frac{\sin(k_F s)}{k_F s} - \cos(k_F s)\right]\,,$$

in the case of metals. In both cases one obtains theoretical and experimental insight into the range of electronic interactions whithout the artefact of the infinite range of the canonical orbitals (CMOs and Bloch functions) of single-electron eigenvalue problems à la Hartree–Fock or Kohn–Sham. The contributions of the various canonical orbitals cancel in the observable reciprocal form factor $B(\mathbf{s})$ (as well as in the density matrix $\gamma(\mathbf{r}',\mathbf{r}''))$ to the extent they are unphysical. Localised orbitals (LMOs and Wannier functions) yield exactly the same $B(\mathbf{s})$.

The most serious sources of residual errors in Compton profiles, viz. cross-section, multiple scattering, and — in experiments with Cauchois-type analysers — background, have the fortunate property that their influence is localised in

$B(\mathbf{s})$ to the range $0 \leq s < 2\,\text{Å} \approx 4\,a_0$. Thus they distort the core-plus-valence-electron part $0 \leq s < 1\,\text{Å} \approx 2\,a_0$ and the valence-electron part $1\,\text{Å} \approx 2\,a_0 < s < 2\,\text{Å} \approx 4\,a_0$ including the maximum next-nearest-neighbour interactions (a challenge to reduce the residual errors further), but not the valence-electron range $s > 2\,\text{Å} \approx 4\,a_0$ — apart from a multiplicative normalisation error in the case of excessive multiple scattering. It is that range, where theoretical data can be compared with experimental ones at full experimental precision with the described precautions. The precision can be as high as $10^{-4}Z$ and thus, e. g., a milli-electron for water with $Z = 10$, which enables one to study even hydrogen bonding in detail.

The nodes in $B(\mathbf{s})$ deserve special interest, although generally they are the result of various compensations in the projection of $\gamma(\mathbf{r}', \mathbf{r}'')$: within the statistical error their position is not influenced by the experimental resolution or by normalisation errors (owing to multiple scattering, e. g.). Therefore, deviations of the nodes of theoretical curves are serious deficiencies and cannot be wiped under the experimental carpet.

Other two reasons for observing the nodes in $B(\mathbf{s})$ apply to solids:

- For a Fermi sphere with radius $p_F = \hbar k_F$ and the nodes of its $B(s)$, the relationship $\tan(k_F s) = k_F s$ holds. It is applicable to 'ideal' metals and a rewarding experiment for students: alkali metals yield one conduction electron, aluminium three of them.
- Furthermore, in the independent-particle model, the occupation function (i. e. the occupation numbers of Bloch NSOs) is

$$n(\mathbf{k}) = \sum_{\mathbf{g}} \varpi(\hbar\mathbf{k} + \hbar\mathbf{g})$$

with $\mathbf{g} = 2\pi \cdot (h\mathbf{a}^* + k\mathbf{b}^* + l\mathbf{c}^*)$ as reciprocal-lattice vectors and $\mathbf{k} \in 1^{\text{st}}$ BZ. This relationship clarifies the often ignored difference between the occupation function $n(\mathbf{k})$ and the momentum density $\varpi(\mathbf{p})$ and leads to

$$n(\mathbf{k}) = \sum_{\mathbf{m}} B(\mathbf{m}) \exp(\mathrm{i}\mathbf{k}\cdot\mathbf{m})$$

with $\mathbf{m} = u\mathbf{a} + v\mathbf{b} + w\mathbf{c}$ as another interesting relationship, which allows to calculate experimental occupation functions from $B(\mathbf{s})$ [5.6], [5.7]. Each band contributes such an occupation function to the total one. Filled bands have constant spin-traced $n(\mathbf{k}) = 2$ over the whole first Brillouin zone and therefore merely a share of 2 electrons from $B(0) = Z$. Only a partially filled band possesses $B(\mathbf{m}) \neq 0$ for $\mathbf{m} \neq 0$. The Fourier sum of $B(\mathbf{m})$, i. e. of $B(\mathbf{s})$ with $\mathbf{s}$ at the points $\mathbf{m}$ of the translational lattice, thus yields the Fermi body on top of the constant contribution of all the filled bands. Insulators and semiconductors have only filled bands and therefore $B(\mathbf{m}) = 0$ for all $\mathbf{m} \neq 0$ — the second reason for looking for the nodes of $B(\mathbf{s})$ of solids. If a theoretical independent-particle calculation for a crystalline insulator or semiconductor does not reproduce these nodes, it contains errors.

The virial theorem states rigorously that the (positive) kinetic energy $\langle T\rangle$ of a system of particles interacting via Coulomb forces is equal to the negative total energy of the system with respect to the separated particles, $-E$. In terms of the momentum density $\varpi(\mathbf{p})$,

$$\langle T\rangle = \langle p^2\rangle/2m_e$$

is the trace of the tensor τ of the second moments of $\varpi(\mathbf{p})$ divided by $2m_e$ [6.1]. The tensor τ, in turn, is equal to the negative tensor of the second derivatives of $B(\mathbf{s})$ at $\mathbf{s}=0$ divided by $2m_e$, i.e. in summary

$$E = \left.\frac{\Delta B(\mathbf{s})}{2m_e}\right|_{\mathbf{s}=0}$$

holds, if the relatively small kinetic energy of the nuclei is neglected. Variational optimisation of basis sets thus improves the reciprocal form factor around $\mathbf{s}=0$, but not necessarily for larger s. Experimental $B(\mathbf{s})$ with their high accuracy at large s can for this reason be a very sensitive tool for improving basis sets beyond the necessary energy criterion for theoretical calculations.

8. Momentum densities

In the case of spherical symmetry (atoms, liquids, glasses, powders), there are particularly simple relationships between the momentum density $\varpi(p)$, the Compton profile $J(q)$ and the reciprocal form factor $B(s)$:

$$\begin{aligned}
J(q) &= 2\pi\int_{|q|}^{\infty} p\,\varpi(p)\,\mathrm{d}p = \frac{1}{\pi\hbar}\int_0^{\infty} B(s)\,\cos(qs/\hbar)\,\mathrm{d}s\,,\\
\varpi(p) &= -\frac{1}{2\pi p}\left.\frac{\mathrm{d}J(q)}{\mathrm{d}q}\right|_{q=p} = \frac{1}{2(\pi\hbar)^2}\int_0^{\infty} s^2\,B(s)\,\frac{\sin(ps/\hbar)}{ps}\,\mathrm{d}s\,,\\
\varpi(0) &= -\frac{1}{2\pi}\left.\frac{\mathrm{d}^2J(q)}{\mathrm{d}q^2}\right|_{q=0} = \frac{1}{2\pi^2\hbar^3}\int_0^{\infty} s^2\,B(s)\,\mathrm{d}s\,,\\
B(s) &= 4\pi\hbar\int_0^{\infty} p^2\varpi(p)\,\frac{\sin(ps/\hbar)}{ps}\,\mathrm{d}p = \int_{-\infty}^{+\infty} J(q)\,\cos(qs/\hbar)\,\mathrm{d}q\,.
\end{aligned}$$

The reconstruction of $\varpi(p)$ from its experimental projection $J(q)$ is therefore straightforward, although connected with a substantial deterioration of the signal-to-noise ratio by differentiation or, via $B(s)$, by weighting with s^2. That deterioration is particularly strong for $p=0$. In practice, the increase of noise is limited by a finite discrete q-grid or, equivalently, by a finite integration range over $s^2B(s)$.

In the case of non-spherical symmetry (molecules and crystals), the reconstruction of $\varpi(\mathbf{p})$ is more involved; the various approaches as of 1976 have been summarised by Mijnarends in Chapter 10 of Ref. [5.1]. The most transparent and most versatile method utilises the reciprocal form factor B. $B(\mathbf{s})$ — obtained along various lines $\mathbf{s}$ from directional Compton profiles $J(\mathbf{q})$, $\mathbf{s} \parallel \mathbf{q}$, by 1D Fourier–Dirac transformation — can be converted into the 3D momentum density $\varpi(\mathbf{p})$ by a 3D Fourier–Dirac transformation,

$$\varpi(\mathbf{p}) = (2\pi\hbar)^{-3} \int B(\mathbf{s}) \exp(\mathrm{i}\mathbf{p}\cdot\mathbf{s}/\hbar)\, \mathrm{d}\tau_s \; .$$

Since the data along a set of lines $\mathbf{s}$ do not fill $\{\mathbf{s}\}$-space completely, a sensible interpolation scheme is required. As a vector-space quantity, $B(\mathbf{s})$ is single-centred, and the data points possess the same s-grid in all directions, if (as usual) $q_{\max}$ is the same. The angular interpolation of the data set on spheres with each s-value as its radius in a spherical coordinate system is therefore the natural choice. That interpolation is done by a least-squares fit with spherical harmonics that are symmetry-adapted to the Laue class of the crystal (polyhedral harmonics $X_L(\theta_s, \phi_s)$, $L = (l$, [linear combination of] $m)$, often also called 'lattice harmonics' in spite of the absence of any translation [5.13]). The resulting radial coefficients $b_L(s)$ of

$$B(\mathbf{s}) = \sum_L b_L(s)\, X_L(\theta_s, \phi_s)$$

have then to be 1D-Hankel transformed (kernel $j_l(ps/\hbar)$) into

$$\varpi_L(p) = \frac{1}{2\pi^2}\, \mathrm{i}^l \int_0^\infty b_L(s)\, j_l(ps/\hbar)\, s^2\, \mathrm{d}s$$

for

$$\varpi(\mathbf{p}) = \sum_L \varpi_L(p)\, X_L(\theta_p, \phi_p) \; .$$

The least-squares method enables one to keep track of the experimental errors, which is absolutely necessary because of the complicated nature of all the transformations. The resulting momentum density has, grosso modo, radially the momentum resolution of the Compton experiment and angularly the resolution given by the set of $\mathbf{q}$-directions. Statistical errors pile up in regions not supported by such a direction. Reconstructed momentum densities with reasonable angular resolution are still the exception in experimental Compton work because of the very time-consuming data collection [5.11], [5.21].

The series expansion into polyhedral harmonics is also a general means to calculate the spherical averages $\varpi(p)$, $J(q)$ and $B(s)$ as the $l = 0$ terms from three-dimensional $\varpi(\mathbf{p})$, $J(\mathbf{q})$ and $B(\mathbf{s})$, resp. — either analytically or numerically.

9. Integrals

The matrix elements $\Omega_{i,kl}$ of the operator matrices $\mathbf{\Omega}_i$ are operator integrals over pairs of the basis functions $\phi_k(\mathbf{r})$ or $\tilde{\phi}_k(\mathbf{p})$ of the particular representation used for the density matrix ($\gamma(\mathbf{r}', \mathbf{r}'')$, $\tilde{\gamma}(\mathbf{p}', \mathbf{p}'')$, $W(\mathbf{p}, \mathbf{r})$, or $A(\chi, \mathbf{s})$).

It is unfortunate that formulae for such integrals are scattered over literature. A very good systematic compilation of some of those integrals is due to Kaijser and Smith [7.1], and 'scattering integrals' with Gaussians are presented in Ref. [7.2].

If the $\phi_k(\mathbf{r})$ are Slater-type orbitals (STOs), the $\tilde{\phi}_k(\mathbf{p})$ are higher-order Lorentzians, whereas Gaussian $\phi_k(\mathbf{r})$ lead to $\tilde{\phi}_k(\mathbf{p})$ that are also Gaussian. The densities ρ and ϖ as well as the density matrices γ and $\tilde{\gamma}$ are best calculated simply via basis-function products, whereas (conventional and reciprocal) form factors have the options of Fourier(−Dirac) transforms of basis-function products in the reciprocal representation or of overlap integrals in the same representation (with $\boldsymbol{\chi}$ or $\mathbf{s}$ as additional shift vectors for the second centre). Momentum-space overlap integrals for $F(\boldsymbol{\chi})$ are an exotic species, whereas position-space overlap integrals for $B(\mathbf{s})$ are standard, both for STOs and GTOs. The calculation of $B(\mathbf{s})$ also circumvents the truncation error of the numerical projection of momentum densities to Compton profiles $J(\mathbf{q})$ in those cases where analytical projection integrals are unavailable.

Spherically averaged $J(q)$, however, are better calculated via the momentum density $\varpi(\mathbf{p})$, whenever it can be analytically averaged to $\varpi(p)$.

A yet unsolved challenge (except by very slowly convergent series) are the 'scattering integrals' (Fourier transforms) of two-centre STO products, as needed for the form factors F and Wigner functions W of molecules and solids. Because of the GTO representation of the program CRYSTAL, however, all integrals can there be solved analytically.

10. Experimental density matrices

For systems governed by quantum mechanics the traditional way is to compare the results of a theoretical calculation with the data sets of each type of experiment. Calculated wave functions as the theoretical information carrier are the only common denominator, which, however, may suffer from deficiencies of the theoretical model employed and — most importantly — are, at least hitherto, experimentally inaccessible.

Density matrices have principally the same property, but are much closer to experiments and do not complicate the relationship between observables and information carrier with the construct of a system of information vectors with arbitrary common orientation in single-particle Hilbert space. Particularly the possibility to study the range and the nature of interactions of an electronic system by means of $\gamma(\mathbf{r}',\mathbf{r}'')$ in the pure position representation is a great conceptual and practical advantage. Furthermore, we have found an efficient way to reconstruct density matrices from observables with a finite model and thus to provide a quantum-mechanically correct unified picture for all experimental results [8.4]–[8.10]. The approach of reconstruction is supplemented by the direct experimental determination of $\tilde{\gamma}(\mathbf{p}',\mathbf{p}'')$ with off-diagonal Compton profiles in the case of perfect crystals [8.12]–[8.14]. Once the density matrix is determined, the natural spin orbitals as the system of information vectors can be obtained by solving the eigenvalue equation of the density matrix, if desired.

The first approach to the reconstruction of density matrices dates back to 1969 [8.1], [8.2] and is based on McWeeny's formula [1.4]

$$\mathbf{P}_{\nu+1} = 3\,\mathbf{P}_\nu^2 - 2\,\mathbf{P}_\nu^3\,, \quad \nu \to \infty\,,$$

for the iterative restoration of the idempotency of a density matrix in conjunction with a least-squares fit to observables in which boundary conditions were satisfied by a set of Lagrangian multipliers. Apart from the restriction to idempotency (and thus to independent-particle models) as well as from the restriction to only one coordinate space (mostly $\{\mathbf{r}\}$-space) and hence — as we know in the meantime — to an incorrect treatment of phase space, that approach suffered from extremely slow numerical convergence or even divergence. Therefore there was only slow progress over the years in spite of the power of the pioneering idea. A recent impressive example is the work of Howard et al. on formamide [8.3].

When trying to extend that approach to the simultaneous fit of position- *and* momentum-space data in 1986, we encountered the convergence problem and were forced to find a new way of reconstruction [8.4].

It utilises the eigenvalue equation

$$\mathbf{P} = \mathbf{C}\mathbf{n}\mathbf{C}^\dagger$$

in a finite discrete representation of length M (which is the 'model') with unitary transformation matrices $\mathbf{C}$ from the diagonal density matrix $\mathbf{n}$ in its eigenrepresentation (= occupation-number matrix) to the density matrix $\mathbf{P}$ of the energy eigenstate or of the (e. g. thermally excited) ensemble and determines $\mathbf{C}$ and $\mathbf{n}$ such that the least-squares functional

$$\Sigma \equiv \sum_i w_i\,(\Omega_i - \mathrm{Tr}\,\mathbf{\Omega}_i\mathbf{P})^2$$

is minimised. There, $\mathbf{\Omega}_i$ is the operator matrix corresponding to the experimental data point Ω_i, and w_i is its inverse variance, thus scaling and weighting the terms in the sum properly. The trace relationship $\sum_j n_j = N$ is maintained by shifting electrons between NSOs rather than varying the occupation numbers n_j individually, and the shifts are limited such that neither of the two n_j leaves the interval $[0, 1]$ to ensure ensemble representability. Idempotency is a special case easily introduced by leaving the n_j at the initial values of N times 1 and $M - N$ times 0 for the first stage of the fitting procedure. Any method for constructing unitary $\mathbf{C}$ matrices is permitted; we have started with a decomposition into planar Jacobi rotations (which allows subtle strategies of rotating pairs of NSOs), and Eulerian angles for a single rotational transformation in $M - N$ to M-dimensional space have also been implemented in the meantime [8.10]. The method not only provides $\mathbf{P}$, which can be transformed into any of the continuous representations γ, $\tilde{\gamma}$, W, and A, but simultaneously also $\mathbf{C}$ as the matrix of the NSO vectors.

After successful tests with recovery of the information carrier from theoretically calculated observables in a number of atomic cases [8.4]–[8.10], where we could experience numerically that the density information from only one space hardly supports the correct density matrix beyond the incorrect Kohn–Sham-type idempotent one, we are just attempting to apply the method to a solid (GaAs) with real experimental data (structure factors and Compton profiles) for the first time and to compare the resulting density matrix with the one calculated with the program CRYSTAL [8.11].

11. Acknowledgements

This Chapter has been prepared under contract number CHRX-CT93-0155 of the Human Capital and Mobility Programme of the European Union. The support by that programme as well as the continuous unbureaucratic help of the Fonds der Chemischen Industrie are most gratefully acknowledged.

References

Reduced density matrices

[1.1] K. Husimi, Proc. Phys. Math. Soc. Japan **22**, 264 (1940).
[1.2] P.-O. Löwdin, Phys. Rev. **97**, 1474, 1490, 1509 (1955).
[1.3] R. McWeeny, Proc. Roy. Soc. (London) A **232**, 114 (1955).
[1.4] R. McWeeny, Proc. Roy. Soc. (London) A **235**, 496 (1956).
[1.5] R. McWeeny, Rev. Mod. Phys. **32**, 335 (1960).
[1.6] A. J. Coleman, Rev. Mod. Phys. **35**, 668 (1963).
[1.7] F. Sasaki, Phys. Rev. **138 B**, 1338 (1965).
[1.8] E. R. Davidson, "Reduced Density Matrices in Quantum Chemistry", Academic Press, New York · San Francisco · London 1976.
[1.9] R. Erdahl, V. H. Smith, Jr. (eds.), "Density Matrices and Density Functionals", D. Reidel Publishing Company, Dordrecht · Boston · Lancaster · Tokyo 1987.
[1.10] T. Asthalter, W. Weyrich, Ber. Bunsenges. Phys. Chem. **96**, 1747 (1992).
[1.11] H. Schmider, K. E. Edgecombe, V. H. Smith, Jr., W. Weyrich, J. Chem. Phys. **96**, 8411 (1992).
[1.12] T. Asthalter, W. Weyrich, submitted.

Density-functional theory

[2.1] P. Hohenberg, W. Kohn, Phys. Rev. **136**, B 864 (1964).
[2.2] W. Kohn, L. J. Sham, Phys. Rev. **140**, A 1133 (1965).
[2.3] L. Lam, P. L. Platzman, Phys. Rev. B **9**, 5122 and 5128 (1974).
[2.4] G. E. W. Bauer, Phys. Rev. B **27**, 5912 (1983).
[2.5] R. Erdahl, V. H. Smith, Jr. (eds.), "Density Matrices and Density Functionals", D. Reidel Publishing Company, Dordrecht · Boston · Lancaster · Tokyo 1987.
[2.6] R. G. Parr, W. Yang, "Density-Functional Theory of Atoms and Molecules", International Series of Monographs in Chemistry, Vol. 16, Oxford University Press, New York, and Clarendon Press, Oxford 1989.
[2.7] E. S. Kryachko, E. V. Ludeña, "Energy Density Functional Theory of Many-Electron Systems", Understanding Chemical Reactivity, Vol. 4, Kluwer Academic Publishers, Dordrecht · Boston · London 1990.

[2.8] T.L. Gilbert, Phys. Rev. B **12**, 2111 (1975).
[2.9] J.E. Harriman, Phys. Rev. A **24**, 680 (1981).
[2.10] J.E. Harriman, Phys. Rev. A **34**, 29 (1986).

Density matrices and observables

[3.1] R. Benesch, S.R. Singh, V.H. Smith, Jr., Chem. Phys. Letters **10**, 151 (1971).
[3.2] V.H. Smith, Jr., Phys. Scripta **15**, 147 (1977).
[3.3] W. Weyrich, Habilitationsschrift, Darmstadt 1978.
[3.4] A.J. Thakkar, A.M. Simas, V.H. Smith, Jr., Chem. Phys. **63**, 175 (1981).
[3.5] A.J. Thakkar, A.C. Tanner, V.H. Smith, Jr., in [1.9], p.327.
[3.6] R.A. Bonham, W. Weyrich, J. Brazil. Chem. Soc. **1**, 7 (1990).

Form factors, structure factors and position densities

[4.1] P. Coppens, E.D. Stevens, Adv. Quantum Chem. **10**, 1 (1977).
[4.2] P. Becker (ed.), "Electron and Magnetization Densities in Molecules and Crystals", Plenum Press, New York 1980.
[4.3] P. Coppens, M. Hall (eds.), "Electron Distributions and the Chemical Bond", Plenum Press, New York 1982.
[4.4] P. Coppens, Acta Cryst. A **40**, 184 (1984).

Compton profiles, reciprocal form factors and momentum densities

[5.1] B. Williams (ed.), "Compton Scattering: The Investigation of Electron Momentum Distributions", McGraw Hill, New York 1977.
[5.2] W. Weyrich, Ber. Bunsenges. Phys. Chem. **79**, 1085 (1975).
[5.3] R. Benesch, S.R. Singh, V.H. Smith, Jr., Chem. Phys. Letters **10**, 151 (1971).
[5.4] P. Pattison, B. Williams, Solid State Commun. **20**, 585 (1976).
[5.5] P. Pattison, W. Weyrich, B. Williams, Solid State Commun. **21**, 967 (1977).
[5.6] W. Schülke, phys. stat. sol. (b) **82**, 229 (1977).
[5.7] W. Weyrich, Habilitationsschrift, Darmstadt 1978.
[5.8] P. Pattison, W. Weyrich, J. Phys. Chem. Solids **40**, 213 (1979).
[5.9] W. Weyrich, P. Pattison, B. Williams, Chem. Phys. **41**, 271 (1979).
[5.10] A.J. Thakkar, A.M. Simas, V.H. Smith, Jr., Chem. Phys. **63**, 175 (1981).
[5.11] E. Heuser, Dissertation, Darmstadt 1982.
[5.12] M.J. Cooper, Rep. Progr. Phys. **48**, 415 (1985).
[5.13] E. Heuser-Hofmann, W. Weyrich, Z. Naturforsch. **40a**, 99 (1985).
[5.14] M. Bräuchler, S. Lunell, I. Olovsson, W. Weyrich, Int. J. Quantum Chem. **35**, 895 (1989).
[5.15] T. Asthalter, W. Weyrich, A.H. Harker, A.B. Kunz, R. Orlando, C. Pisani, Solid State Commun. **83**, 725–730 (1992).
[5.16] T. Asthalter, W. Weyrich, Ber. Bunsenges. Phys. Chem. **96**, 1747–1755 (1992).
[5.17] L. Dobrzyński, Z. Naturforsch. **48a**, 266 (1993).
[5.18] T. Asthalter, W. Weyrich, Z. Naturforsch. **48a**, 303 (1993).
[5.19] M. Weisser, W. Weyrich, Z. Naturforsch. **48a**, 315 (1993).
[5.20] R.O. Horenian, W. Weyrich, Z. Naturforsch. **48a**, 325 (1993).
[5.21] M. Weisser, Dissertation, Konstanz 1996.
[5.22] A.J. Rollason, F. Bell, J.R. Schneider, Nucl. Instr. Meth. **A281**, 147 (1989).
[5.23] F. Bell, T. Tschentscher, J.R. Schneider, A.J. Rollason, Z. Naturforsch. **48a**, 273 (1993).
[5.24] T. Tschentscher, J.R. Schneider, F. Bell, Phys. Rev. B **48**, 16965 (1993) and references therein.
[5.25] S.W. Lovesey, Rep. Progr. Phys. (UK) **56**, 257 (1993).
[5.26] N. Sakai, K. Ôno, Phys. Rev. Lett. **37**, 351 (1976).
[5.27] N. Sakai, K. Ôno, J. Phys. Soc. Japan **42**, 770 (1977).
[5.28] M.A. Coplan, J.H. Moore, J.A. Tossell, Z. Naturforsch. **48a**, 358 (1993).

[5.29] E. Weigold, Z. Naturforsch. **48a**, 371 (1993).

Virial theorem

[6.1] W. Weyrich, Ber. Bunsenges. Phys. Chem. **83**, 797 (1979).

Integrals

[7.1] P. Kaijser, V.H. Smith, Jr., Adv. Quantum Chem. **10**, 37 (1977).
[7.2] M. Barua, W. Weyrich, Acta Cryst. A **42**, 257 (1986).

Experimental access to density matrices

[8.1] W.L. Clinton, A. Galli, L. Massa, Phys. Rev. **177**, 1 (1969).
[8.2] iidem, Phys. Rev. **177**, 7 (1969).
[8.3] S.T. Howard, J.P. Huke, P.R. Mallinson, C.S. Frampton, Phys. Rev. B: Condens. Matter **49**, 7124 (1994).
[8.4] H. Schmider, Diplomarbeit, Konstanz 1986.
[8.5] H. Schmider, V.H. Smith, Jr., W. Weyrich, Trans. Amer. Cryst. Assoc. **26**, 125 (1990).
[8.6] H. Meider, Diplomarbeit, Konstanz 1991.
[8.7] H. Schmider, V.H. Smith, Jr., W. Weyrich, J. Chem. Phys. **96**, 8986 (1992).
[8.8] iidem, Z. Naturforsch. **48a**, 211 (1993).
[8.9] iidem, Z. Naturforsch. **48a**, 221 (1993).
[8.10] H. Schmider, Dissertation, Kingston/Ontario 1994.
[8.11] J. Stahn, Diplomarbeit, Konstanz 1995.
[8.12] W. Schülke, Phys. Lett. **83A**, 451 (1981).
[8.13] W. Schülke, U. Bonse, S. Mourikis, Phys. Rev. Lett. **47**, 1209 (1982).
[8.14] W. Schülke, S. Mourikis, Acta Cryst. A **42**, 86 (1986).

Macroscopic Dielectric Polarization: Hartree-Fock Theory

Raffaele Resta

Istituto Nazionale di Fisica della Materia (INFM) and Dipartimento di Fisica Teorica dell'Università di Trieste, Strada Costiera 11, I-34014 Trieste, Italy

May 30, 1996

Summary. A modern theory of macroscopic polarization in crystalline dielectrics has been recently founded. Within this theory, polarization occurs as a geometric quantum phase of the crystalline Bloch orbitals. This modern theory only concerns polarization differences in zero electric field and cope, therefore, with lattice dynamics, piezoelectricity and ferroelectricity. So far, the geometric-phase theory has been formulated and implemented within the density functional theory of Kohn and Sham. In this Chapter I outline the whole theory, focussing on a formulation within the Hartree-Fock framework and discussing a possible implementation in a localized basis set. The final section of this Chapter addresses a somewhat separate issue, namely the computation of a macroscopic dielectric constant from linear-response theory, in a periodic solid.

Key words: Polarization – Dielectrics – Infrared charges – Piezoelectricity – Ferroelectricity – Dielectric permittivity – Hartree-Fock – Electronic states – Geometric phases.

1. Introduction

In a phenomenological description of dielectric media, the concept of macroscopic polarization is the basic one [1]: it is, therefore, quite a surprise to discover that—at the microscopic level—this quantity is even incorrectly *defined* in all textbooks. If one assumes the polarization elements as discrete, à la Clausius-Mossotti, then the polarization mechanism can be safely understood: but such an oversimplified picture does not apply to a real dielectric, where the electronic distribution is continuous and often very delocalized. Typically, textbooks attempt to explain polarization via the dipole moment of a unit cell or something

Correspondence to: R.Resta, e-mail resta@sissa.it

of the kind [2, 3]. If this were correct, one could imagine that the macroscopic polarization is trivially accessible to electronic structure theory: in fact, the microscopic charge distribution is nowadays routinely available from standard computational techniques [4, 5], for crystalline solids at least. But this is not the case at all and macroscopic polarization remained a major challenge for many years [6].

In 1992, there was a breakthrough [7]; in the following year, a modern theory of macroscopic polarization in crystalline dielectrics has been completely established [8, 9]. According to the modern viewpoint, polarization is a physical observable *completely independent* from the periodic charge distribution of the polarized dielectric. In fact—when periodic boundary conditions are used—the charge is obtained from the square modulus of the wavefunctions, whereas polarization is a basic property of their *phase*. A comprehensive account of the modern theory, together with a review of the early calculations, has recently appeared [10]. For an oversimplified outline (in Italian), see [11].

To begin with, let me stress the most fundamental fact: The "absolute" macroscopic polarization of a material has never been measured as a bulk property, independent of sample termination. Instead, the genuine observable bulk properties are typically polarization *derivatives* or even—in the case of ferroelectric crystals—finite differences of polarization between two states of the same solid. The modern theory also concerns polarization differences, $\Delta\mathbf{P}$ and avoids defining absolute polarization at all.

At this point, it is convenient to clearly distinguish between two cases, which prove to be very different from a theoretical standpoint. Case one concerns polarization in an electric field and is, of course, needed to define the dielectric tensor. Case two concerns, instead, polarization in zero field, under which circumstances the piezoelectric tensor [12], the Born effective charge tensor for lattice dynamics [13] and the ferroelectric effect [14] are most usually defined. In *both* cases the polarization difference $\Delta\mathbf{P}$ induced by a given perturbation coincides with the integrated macroscopic current density, which traverses the sample while the perturbation is switched on. The theory addresses precisely this same current, but the state-of-the-art tools available to deal with the two cases are very different. Case two is, by far, the one in much better—I dare say, definitive—shape and the one to which most of the present Chapter is devoted. As for case one, there is still room for important theoretical developments [15, 16]: here, I present (in the last section) only the "traditional" linear-response approach to the polarization induced by a macroscopic electric field.

As illustrated above, the main problem is to define and compute a macroscopic current. In quantum mechanics, the current is basically a property of the phases of the wavefunctions. In the case of vanishing electric field, the modern theory allows us to express the electronic contribution to $\Delta\mathbf{P}$ as a Berry (alias geometric) quantum phase [17]. In [10], the theory is presented at a formally exact level, within the scope of the density functional theory of Kohn and Sham [18, 19]; calculations performed, so far, use the local-density approximation. A generalization, which applies to many-body correlated wavefunctions, has recently appeared [20]. Here, I present the aspects of the theory which are most relevant for its implementation at the Hartree-Fock [5] (HF) level. Needless to say, the density functional formulation of [10] applies almost unchanged to the HF case, upon straightforward replacement of the Kohn-Sham orbitals with the HF ones. However, there is an important conceptual difference: within density functional theory, the many-body wavefunction remains undetermined [18], while within HF, an explicit variational many-body wavefunction is at hand, in the form of a Slater determinant [21]. Furthermore, the non-locality of the HF potential deserves some special attention. In this Chapter, I will present a comprehensive account of the whole theory, in its HF formulation, though relying on [8, 10] to abridge some mathematical passages. The logic of the presentation will be reversed with respect to [9, 10], and will remain close, instead, to the historical development of the theory, which was born with [7, 8]. Units and normalizations will be different from previous papers.

The plan of this Chapter is as follows. In section 2, I define precisely the induced macroscopic polarization in zero field, using as a prototypical example the polarization of a zone-centre transverse-optic phonon in a polar crystal. In section 3, I discuss some features which are peculiar to the HF formulation of the mean-field electronic-structure theory. In section 4, I outline how a polarization difference—alias an integrated macroscopic current—can be evaluated in a boundary-insensitive way in the thermodynamic limit. In section 5, I show how the integrated macroscopic current is transformed into a geometric quantum phase [17], using a simplified one-band one-dimensional formulation and I discuss some unusual features of the physical observable so defined. In section 6, I provide three alternative and formally equivalent, geometric-phase expressions for the polarization difference, $\Delta \mathbf{P}$, in a crystalline solid, whose ground state is known at the HF level. In section 7, I discuss the numerical algorithm used in actual implementations, which have been performed within a density functional theory framework (and not within HF) so far. In section 8, I give the explicit expressions to be evaluated when the crystalline eigenstates are expanded over a localized basis set. In the final section of this Chapter, section 9, I outline a quite different theoretical approach—established well before the modern theory discussed in the other sections—which allows evaluation of the macroscopic polarization linearly induced by any perturbation, even in presence of a macroscopic field: the calculation of a macroscopic dielectric constant is used as the illustrative example.

2. Polarization differences in zero field

To focus on these ideas, I refer to a specific example: the polarization difference, $\Delta \mathbf{P}$, induced by a relative sublattice displacement in a polar crystal. If the displacement is rigid, the perturbation is, in fact, a zone-centre optic phonon. If, furthermore, one chooses the boundary condition of vanishing macroscopic field, then the crystalline potential remains periodic at any values of the displacement: this applies to a *transverse* optic phonon in the long-wavelength (zone-centre) limit. To linear order in the displacement, this $\Delta \mathbf{P}$ defines the Born (or transverse) effective charge tensor [13, 22]. General considerations [23] show that the Born tensor equivalently defines the force which a macroscopic field exerts, to linear order, upon a given (undisplaced) atom. Therefore, the Born tensors allow evaluation of the infrared activity of zone-centre phonon modes. The Born tensors are the simplest application of the present theory.

The molecular analogues of the Born tensors go under the name of displacement polar tensors, alias dipole-moment derivatives [24]. Not surprisingly, even in molecules, these yield the infrared activity and they are sometimes expressed as the bare nuclear charge times a "shielding tensor" [25]. However, this obvious analogy gives me occasion to stress, instead, the paramount difference between a molecule and a solid, as far as macroscopic electrostatics is concerned. When the charge distribution is *finite*, there is a one-to-one correspondence between charge distribution and electrostatic potential, since the solution of Poisson's equation is unique, under the hypothesis that the potential vanishes at infinity. If one considers, instead, a periodic charge distribution, extended to all space, then the charge density alone does *not* uniquely determine the solution of Poisson's equation. Additional boundary conditions must be explicitly assumed. If we choose (as usual) the periodic solution of Poisson's equation, then we are assuming vanishingly small macroscopic field: other choices are in principle possible. Incidentally, it is worth noticing that even when the field is taken as zero, the crystal potential remains undetermined by an arbitrary constant (see *e.g.* p. 18 in [5]). Of course, the concept of an infinite solid is an idealization and one could find disturbing and unphysical the need to deal with the macroscopic field as with an additional boundary condition. But, in fact, this is not the case, since, in many experimental circumstances, it is precisely the field which is under experimental control. In particular, the field is zero whenever the solid is perturbed while being kept inside a shorted capacitor and polarization differences are, in fact, measured via the currents flowing across the shorting wire [11]. Under such circumstances, one is, indeed, studying a bulk property of an open system, where charge is injected from one face and drained from the opposite one. I stress once more that the most fundamental definitions of the Born tensors [13], of the piezoelectric tensor [12], and of ferroelectric polarization [14] are given at zero field.

In view of the above considerations, our main object of investigation is the polarization difference, $\Delta\mathbf{P}$, between two states of the same solid, assuming that the perturbation conserves crystal periodicity. Within the Born-Oppenheimer approximation, we may separate the nuclear and electronic contributions as:

$$\Delta\mathbf{P} = \Delta\mathbf{P}_{\text{nucl}} + \Delta\mathbf{P}_{\text{el}} \tag{1}$$

The former term refers to point charges and is trivial: therefore, we focus only on the latter term, in the following. It is, however, worth noticing that each of the two terms in equation 1 depends on the choice of the origin in the crystal cell, while only their sum is a *macroscopic* (*i.e.* origin-independent) bulk observable. The two states of the solid are connected by a continuous transformation of the electronic Hamiltonian, which I parametrize with a dimensionless scalar variable λ, chosen to have the values of 0 and 1 at the initial and final states, respectively. In the above example, λ parametrizes the phonon amplitude in convenient units. In a more general case, λ can be thought of as defined in a multidimensional parameter space.

3. Aspects of the Hartree-Fock method

We work within the HF scheme; atomic Hartree units are adopted throughout. The effective periodic potential in the Fock operator depends on λ, both in its bare (electron-ion) and self-consistent (Coulomb and exchange) terms:

$$F^{(\lambda)} = \frac{1}{2}p^2 + V^{(\lambda)}, \tag{2}$$

where $V^{(\lambda)}$ is *non-local* due to exchange and possibly to pseudopotentials as well [4]. The HF orbitals have the Bloch form at any λ:

$$\psi_s^{(\lambda)}(\mathbf{r};\boldsymbol{\kappa}) = e^{i\boldsymbol{\kappa}\cdot\mathbf{r}} u_s^{(\lambda)}(\mathbf{r};\boldsymbol{\kappa}) \tag{3}$$

where the u's are cell-periodic functions of $\mathbf{r}$. We assume the ψ's and the u's are normalized to one over the crystal cell. It is, furthermore, convenient to choose the phases in such a way that $\psi_s^{(\lambda)}(\mathbf{r};\boldsymbol{\kappa})$ is a periodic function of $\boldsymbol{\kappa}$ in the reciprocal lattice: besides this, the phases of the Bloch functions at different $\boldsymbol{\kappa}$ points are unrelated and completely arbitrary. This fact is usually referred to as gauge freedom.

The Fock operator, equation 2, is usually diagonalized one $\boldsymbol{\kappa}$ vector at a time: at a fixed $\boldsymbol{\kappa}$, the spectrum is discrete and only the lowest $n/2$ states are relevant to the crystalline ground-state. We make in fact the explicit hypothesis that our solid remains an insulator—in the sense that its HF gap does not close—for any λ value: supposing there are n electrons per unit cell, we have then $n/2$ doubly occupied bands at any $\boldsymbol{\kappa}$. As usual in quantum mechanics, the spectrum of an operator is determined by the boundary conditions. Diagonalization of equation 2 at a given $\boldsymbol{\kappa}$ simply means that we impose quasi-periodic conditions at the elementary-cell boundary. This secular problem can be mapped into an equivalent one, involving the u's instead of the ψ's: in fact, the u's are eigenfunctions of the effective Fock operator:

$$\tilde{F}^{(\lambda)}(\boldsymbol{\kappa}) = \frac{1}{2}(\mathbf{p}+\boldsymbol{\kappa})^2 + \tilde{V}^{(\lambda)}(\boldsymbol{\kappa}) \tag{4}$$

with the same eigenvalues $\epsilon_s^{(\lambda)}(\boldsymbol{\kappa})$ as the ψ's. The periodic potential is *not* the same because of non-locality: the explicit relationship is:

$$\tilde{V}^{(\lambda)}(\boldsymbol{\kappa})(\mathbf{r},\mathbf{r}') = e^{i\boldsymbol{\kappa}\cdot(\mathbf{r}'-\mathbf{r})} V^{(\lambda)}(\mathbf{r},\mathbf{r}') \tag{5}$$

I wish to point out, however, the dual role of the boundary conditions under such transformation. At a given $\boldsymbol{\kappa}$, the ψ's are eigenstates of a $\boldsymbol{\kappa}$-independent operator, equation 2, with $\boldsymbol{\kappa}$-dependent (quasi-periodic) boundary conditions; the u's are instead eigenstates of a $\boldsymbol{\kappa}$-dependent operator, equation 4, with $\boldsymbol{\kappa}$-independent (periodic) boundary conditions on the elementary cell.

Since we will be interested in the current, we need to discuss the velocity operator: it has, of course, a different form depending on whether it acts on the ψ's or the u's. Furthermore, because of the non-local potential, the HF velocity even depends on λ:

$$\mathbf{v}^{(\lambda)} = i[F^{(\lambda)},\mathbf{r}] = \mathbf{p} + i[V^{(\lambda)},\mathbf{r}] \tag{6}$$

$$\tilde{\mathbf{v}}^{(\lambda)}(\boldsymbol{\kappa}) = i[\tilde{F}^{(\lambda)},\mathbf{r}] = \mathbf{p} + \boldsymbol{\kappa} + i[\tilde{V}^{(\lambda)}(\boldsymbol{\kappa}),\mathbf{r}] \tag{7}$$

Notice that, in the above equations, $\mathbf{p} = -i\nabla_{\mathbf{r}}$ is the *canonical* momentum operator in the Schrödinger representation. One further identity is easily verified and will be most useful in the following:

$$\tilde{\mathbf{v}}^{(\lambda)}(\boldsymbol{\kappa}) = \nabla_{\boldsymbol{\kappa}} \tilde{F}^{(\lambda)}(\boldsymbol{\kappa}) \tag{8}$$

In the rest of this Chapter, the u's play the most fundamental role. The (spin-integrated) reduced one-body density matrix is then written as:

$$\gamma^{(\lambda)}(\mathbf{r},\mathbf{r}') = \frac{2\Omega}{(2\pi)^3} \sum_{v=1}^{n/2} \int_{\mathrm{BZ}} d\boldsymbol{\kappa}\, e^{i\boldsymbol{\kappa}\cdot(\mathbf{r}-\mathbf{r}')} u_v^{(\lambda)}(\mathbf{r};\boldsymbol{\kappa}) u_v^{(\lambda)*}(\mathbf{r}';\boldsymbol{\kappa}) \tag{9}$$

where Ω is the cell volume and BZ is the Brillouin zone, whose volume is $(2\pi)^3/\Omega$. The one-body observables of the crystalline electronic ground state, at a given λ, are simply expressed in terms of $\gamma^{(\lambda)}$. Macroscopic polarization is a notable exception, in that it is one-body but it is *not* a function of $\gamma^{(\lambda)}$ in the usual sense. Essentially, one would like to express $\Delta\mathbf{P}_{\mathrm{el}}$ in terms of $\gamma^{(1)} - \gamma^{(0)}$, but this is impossible since the relevant phase information is destroyed in the wavefunction products of equation 9. In fact—when periodic boundary conditions are adopted—$\Delta\mathbf{P}_{\mathrm{el}}$ is *not* the expectation value of any quantum-mechanical operator!

In practical implementations of the HF method to a solid, all BZ integrations are approximated as a finite sum over a discrete $\boldsymbol{\kappa}$-point set. This heuristic approach has some significance as a matter of principle.

Suppose we use an uniform mesh with $N = N_1 N_2 N_3$ points along the primitive $\mathbf{b}_i$ vectors in reciprocal space:

$$\boldsymbol{\kappa}_{j_1,j_2,j_3} = \frac{j_1}{N_1}\mathbf{b}_1 + \frac{j_2}{N_2}\mathbf{b}_2 + \frac{j_3}{N_3}\mathbf{b}_3, \quad j_i = 0,\ldots,N_i - 1 \tag{10}$$

then the discretization is:

$$\frac{\Omega}{(2\pi)^3} \int_{\mathrm{BZ}} d\boldsymbol{\kappa} \rightarrow \frac{1}{N} \sum_{j_1=0}^{N_1-1} \sum_{j_2=0}^{N_2-1} \sum_{j_3=0}^{N_3-1} \tag{11}$$

and the density matrix of equation 9 becomes identical to the one of a large *finite* system, made of N cells and nN electrons, with overall periodic boundary conditions and whose nondegenerate ground eigenfunction is taken as a single Slater determinant [21]. We may also write this wavefunction Ψ as the antisymmetrized product of N small determinants of size n, *i.e.*:

$$\Psi^{(\lambda)} = \frac{1}{\sqrt{n!}} \mathsf{A} \prod_{j_1,j_2,j_3} |\psi_1^{(\lambda)}(\boldsymbol{\kappa}_{j_1,j_2,j_3})\overline{\psi}_1^{(\lambda)}(\boldsymbol{\kappa}_{j_1,j_2,j_3}) \cdots \psi_{n/2}^{(\lambda)}(\boldsymbol{\kappa}_{j_1,j_2,j_3})\overline{\psi}_{n/2}^{(\lambda)}(\boldsymbol{\kappa}_{j_1,j_2,j_3})| \tag{12}$$

where A is the antisymmetrizer and the electron coordinates are omitted, from now on, to simplify the notation. The determinants of the ψ functions are simply related—through the plane-wave factors of equation 3—to determinants of the u's. Since I will need these Slater determinants in the following, I introduce a special notation for them:

$$|\Phi^{(\lambda)}(\boldsymbol{\kappa})\rangle = \frac{1}{\sqrt{n!}} |u_1^{(\lambda)}(\boldsymbol{\kappa})\overline{u}_1^{(\lambda)}(\boldsymbol{\kappa}) \ \ldots \ u_{n/2}^{(\lambda)}(\boldsymbol{\kappa})\overline{u}_{n/2}^{(\lambda)}(\boldsymbol{\kappa})| \tag{13}$$

where the u's are the $n/2$ lowest periodic eigenfunctions of equation 4.

4. Polarization difference as a macroscopic current

As stressed in the Introduction, we evaluate the electronic polarization difference, $\Delta\mathbf{P}_{\rm el}$, as an integrated macroscopic current. Explicitly, if the perturbation is adiabatically switched on in a time Δt, *i.e.* $\lambda=\lambda(t)$, with $\lambda(0)=0$ and $\lambda(\Delta t)=1$, then:

$$\Delta\mathbf{P}_{\rm el} = \int_0^1 d\lambda\, \mathbf{P}'_{\rm el}(\lambda) = \int_0^{\Delta t} dt\, \mathbf{j}(t) \tag{14}$$

with $\mathbf{j}(t)=\mathbf{P}'_{\rm el}(\lambda(t))\lambda'(t)$. Therefore, without loss of generality, we may think of λ as an adiabatic time in appropriate units, such that $\mathbf{j}=\mathbf{P}'_{\rm el}$. When we attempt to evaluate such current, we discover that the mean value of the current on the adiabatic istantaneous crystalline ground state vanishes at any λ. In fact from equation 8 we have:

$$\langle u_v^{(\lambda)}(\boldsymbol{\kappa})|\, \tilde{\mathbf{v}}^{(\lambda)}(\boldsymbol{\kappa})\, |u_v^{(\lambda)}(\boldsymbol{\kappa})\rangle = \langle u_v^{(\lambda)}(\boldsymbol{\kappa})|\, \nabla_{\boldsymbol{\kappa}}\tilde{F}^{(\lambda)}(\boldsymbol{\kappa})\, |u_v^{(\lambda)}(\boldsymbol{\kappa})\rangle = \nabla_{\boldsymbol{\kappa}}\epsilon_v^{(\lambda)}(\boldsymbol{\kappa}) \tag{15}$$

where the second equality is due to the Hellmann-Feynman theorem, applied to the Fock operator. Because of the time-reversal symmetry ($\boldsymbol{\kappa} \rightarrow -\boldsymbol{\kappa}$), the Brillouin-zone integral of equation 15 vanishes. It should be realized that this is the fundamental reason why $\gamma^{(\lambda)}$ does not provide the macroscopic polarization, while it provides, instead, the more common adiabatic one-body observables.

We have, therefore, to consider the next leading term in the adiabatic expansion, which requires us to use time-dependent perturbation theory [26] or similar means [7, 10]. The result takes the form of a so-called Kubo formula:

$$\begin{aligned}\mathbf{P}'_{\rm el}(\lambda) \;&=\; -\frac{4}{(2\pi)^3} \\ &\times {\rm Im}\sum_{v,c}\int_{\rm BZ} d\boldsymbol{\kappa}\, \frac{\langle u_v^{(\lambda)}(\boldsymbol{\kappa})|\, \tilde{\mathbf{v}}^{(\lambda)}(\boldsymbol{\kappa})\, |u_m^{(\lambda)}(\boldsymbol{\kappa})\rangle\langle u_m^{(\lambda)}(\boldsymbol{\kappa})|\partial u_v^{(\lambda)}(\boldsymbol{\kappa})/\partial\lambda\rangle}{\epsilon_m^{(\lambda)}(\boldsymbol{\kappa}) - \epsilon_v^{(\lambda)}(\boldsymbol{\kappa})}\end{aligned} \tag{16}$$

where the v index runs from 1 to $n/2$ and the c index, from $n/2+1$ to ∞. Again, due to time-reversal symmetry, the BZ integral is purely imaginary. What is remarkable in equation 16 is the fact that the integrated value of this nonadiabatic current is geometric in nature, in that it does not depend on the actual time evolution, but only on the the path traversed in λ-space (either one- or multi-dimensional). If we now substitute equation 8 into equation 16, the sum over the virtual HF orbitals can be eliminated; in fact:

$$|\nabla_{\boldsymbol{\kappa}} u_s^{(\lambda)}(\boldsymbol{\kappa})\rangle = \sum_{s'\neq s} |u_{s'}^{(\lambda)}(\boldsymbol{\kappa})\rangle \frac{\langle u_{s'}^{(\lambda)}(\boldsymbol{\kappa})|\, \nabla_{\boldsymbol{\kappa}}\tilde{F}^{(\lambda)}(\boldsymbol{\kappa})\, |u_s^{(\lambda)}(\boldsymbol{\kappa})\rangle}{\epsilon_s^{(\lambda)}(\boldsymbol{\kappa}) - \epsilon_{s'}^{(\lambda)}(\boldsymbol{\kappa})} \tag{17}$$

After a few manipulations, one transforms equation 16 into the much simpler expression:

$$\mathbf{P}'_{\rm el}(\lambda) = -\frac{4}{(2\pi)^3}{\rm Im}\sum_{v=1}^{n/2}\int_{\rm BZ} d\boldsymbol{\kappa}\, \langle\nabla_{\boldsymbol{\kappa}} u_v^{(\lambda)}(\boldsymbol{\kappa})|\frac{\partial}{\partial\lambda}u_v^{(\lambda)}(\boldsymbol{\kappa})\rangle \tag{18}$$

We thus cast equation 14 as a four-dimensional integral in the $(\boldsymbol{\kappa},\lambda)$ space:

$$\Delta \mathbf{P}_{\rm el} = -\frac{4}{(2\pi)^3}\,{\rm Im} \int_0^1 d\lambda \int_{\rm BZ} d\boldsymbol{\kappa}\, \sum_{v=1}^{n/2} \langle \nabla_{\boldsymbol{\kappa}} u_v^{(\lambda)}(\boldsymbol{\kappa}) | \frac{\partial}{\partial \lambda} u_v^{(\lambda)}(\boldsymbol{\kappa}) \rangle \tag{19}$$

5. Outline of the Berry phase formulation

The following step requires a further nontrivial transformation [8]. For the sake of simplicity, I will illustrate it on the ultrasimplified one-dimensional case, with only one doubly occupied band: equation 19 then takes the form:

$$\Delta P_{\rm el} = -\frac{2}{\pi}\,{\rm Im} \int_0^1 d\lambda \int_{-\frac{\pi}{a}}^{-\frac{\pi}{a}} d\kappa\, \langle \frac{\partial}{\partial \kappa} u^{(\lambda)}(\kappa) | \frac{\partial}{\partial \lambda} u^{(\lambda)}(\kappa) \rangle \tag{20}$$

where a is the one-dimensional lattice periodicity. Equation 20 represents a two-dimensional integral in the (κ, λ) plane, over the domain shown in Fig. 1.

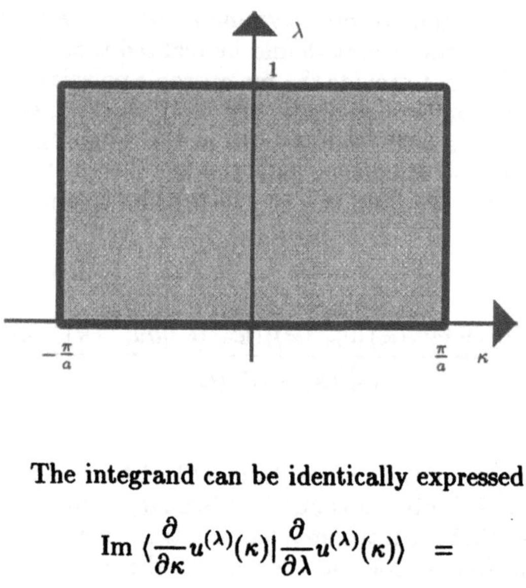

Fig. 1. Integration domain in the (κ, λ) plane used in equation 20. Using Stokes's theorem, the integral is transformed into a circuit integral over the boundary, equation 26.

The integrand can be identically expressed as the curl of a vector field, *i.e.*:

$$\begin{aligned} {\rm Im}\, \langle \frac{\partial}{\partial \kappa} u^{(\lambda)}(\kappa) | \frac{\partial}{\partial \lambda} u^{(\lambda)}(\kappa) \rangle &= \\ &= -\frac{\imath}{2} \Big[\frac{\partial}{\partial \kappa} \langle u^{(\lambda)}(\kappa) | \frac{\partial}{\partial \lambda} u^{(\lambda)}(\kappa) \rangle \\ &\quad - \frac{\partial}{\partial \lambda} \langle u^{(\lambda)}(\kappa) | \frac{\partial}{\partial \kappa} u^{(\lambda)}(\kappa) \rangle \Big] \end{aligned} \tag{21}$$

This immediately suggests the exploitation of Stokes's theorem, upon defining the linear differential form:

$$d\varphi = -\imath \Big[\langle u^{(\lambda)}(\kappa) | \frac{\partial}{\partial \kappa} u^{(\lambda)}(\kappa) \rangle\, d\kappa + \langle u^{(\lambda)}(\kappa) | \frac{\partial}{\partial \lambda} u^{(\lambda)}(\kappa) \rangle\, d\lambda \Big] \tag{22}$$

Such a differential form is known as a Berry connection [17] and gives the phase variation of $u^{(\lambda)}(\kappa)$ for an infinitesimal variation of κ and λ. This property is easily verified. For instance, taking a κ variation at constant λ, the phase change is:

$$e^{\iota \Delta\varphi} = \frac{\langle u^{(\lambda)}(\kappa)|u^{(\lambda)}(\kappa+\Delta\kappa)\rangle}{|\langle u^{(\lambda)}(\kappa)|u^{(\lambda)}(\kappa+\Delta\kappa)\rangle|} \tag{23}$$

$$\Delta\varphi = \text{Im}\ln\langle u^{(\lambda)}|u^{(\lambda)}(\kappa+\Delta\kappa)\rangle \tag{24}$$

Supposing φ is a differentiable function of κ, and using the conservation of the norm, one gets:

$$d\varphi = -\iota\langle u^{(\lambda)}|\frac{\partial}{\partial\kappa}u^{(\lambda)}(\kappa)\rangle\, d\kappa \tag{25}$$

to be compared with equation 22.

We thus arrive at recasting equation 20 as the circuit integral of the connection, equation 21, over the boundary of the two-dimensional domain of Fig. 1:

$$\Delta P_{\text{el}} = -\frac{1}{\pi}\oint d\varphi \tag{26}$$

Circuit integrals of the kind in equation 26 are known as Berry phases [17]. Finally, it can be proved that the contributions of the two vertical sides in Fig. 1 cancel and therefore, the polarization difference can be written as:

$$\Delta P_{\text{el}} = \frac{\iota}{\pi}\int_{-\frac{\pi}{a}}^{\frac{\pi}{a}} d\kappa \left[\langle u^{(0)}(\kappa)|\frac{\partial}{\partial\kappa}u^{(0)}(\kappa)\rangle - \langle u^{(1)}(\kappa)|\frac{\partial}{\partial\kappa}u^{(1)}(\kappa)\rangle\right] \tag{27}$$

This is the outstanding major result of [8] and is the basis of the present approach. We notice that it is a two-point formula, in that it involves only the wavefunctions at the initial and final values of λ and not at the intermediate ones, at variance with all the formulas considered above.

I have previously observed that—because of gauge freedom—the phase difference between wavefunctions at different κ values is completely arbitrary, except when the κ vectors differ by a reciprocal lattice vector. Therefore, the *integrand* in equation 27 has a large arbitrariness and is nonphysical, while its integrated value is instead a physical observable. This is a common feature when dealing with quantum geometric phases (alias Berry phases) [17], whose archetype is the well-known Aharonov-Bohm phase [27], discovered in 1959. The occurrence of nontrivial geometric phases in the band structure of solids was first discovered in 1989 by Zak [28]. From a mathematical standpoint, the situation is the same as in classical electromagnetism, where the vector potential has a large gauge freedom and is nonphysical, while its circuit integral along a given contour is, instead, gauge-invariant and observable, being equal to the flux of the magnetic field across a surface enclosed by the contour.

6. Polarization as a geometric quantum phase

The Berry phase transformation has been illustrated, in the previous section, for the ultrasimplified one-dimensional case. Applying this finding to the three-dimensional, multi-band case, one starts from the four-dimensional integral of equation 19 and—using Stokes's theorem—transforms it into a three-dimensional integral. The analogue of equation 27 is then:

$$\Delta\mathbf{P}_{\text{el}} = -\frac{2\iota}{(2\pi)^3}\int_{\text{BZ}} d\boldsymbol{\kappa} \sum_{v=1}^{n/2} [\,\langle u_v^{(1)}(\boldsymbol{\kappa})|\nabla_{\boldsymbol{\kappa}} u_v^{(1)}(\boldsymbol{\kappa})\rangle - \langle u_v^{(0)}(\boldsymbol{\kappa})|\nabla_{\boldsymbol{\kappa}} u_v^{(0)}(\boldsymbol{\kappa})\rangle\,] \tag{28}$$

This expression can be transformed into a more elegant one, where a Berry connection between many-body states explicitly appear. I will, in fact, prove that equation 28 is identical to:

$$\Delta \mathbf{P}_{\text{el}} = -\frac{\iota}{(2\pi)^3} \int_{\text{BZ}} d\boldsymbol{\kappa} \ [\ \langle \Phi^{(1)}(\boldsymbol{\kappa}) | \nabla_{\boldsymbol{\kappa}} \Phi^{(1)}(\boldsymbol{\kappa}) \rangle - \langle \Phi^{(0)}(\boldsymbol{\kappa}) | \nabla_{\boldsymbol{\kappa}} \Phi^{(0)}(\boldsymbol{\kappa}) \rangle\ ] \tag{29}$$

where $|\Phi^{(\lambda)}(\boldsymbol{\kappa})\rangle$ are the single-determinant wavefuctions defined in equation 13, built up of n lattice-periodical spinorbitals. Incidentally, equation 29 provides the link between the HF theory of polarization presented here and its generalization to the correlated case, which is due to Ortiz and Martin [20].

To proceed with the proof, a good starting point is to define the phase difference between two many-body determinants, thus generalizing equation 24. For a $\boldsymbol{\kappa}$ variation at constant λ, the phase change is:

$$\Delta\varphi = \text{Im ln } \langle \Phi^{(\lambda)}(\boldsymbol{\kappa}) | \Phi^{(\lambda)}(\boldsymbol{\kappa} + \Delta\boldsymbol{\kappa}) \rangle \tag{30}$$

Supposing $|\Phi^{(\lambda)}(\boldsymbol{\kappa})\rangle$ is a differentiable function of $\boldsymbol{\kappa}$, the differential of equation 30 is:

$$d\varphi = -\iota \langle \Phi^{(\lambda)}(\boldsymbol{\kappa}) | \nabla_{\boldsymbol{\kappa}} \Phi^{(\lambda)}(\boldsymbol{\kappa}) \rangle \cdot d\boldsymbol{\kappa} \tag{31}$$

which defines the relevant (many-electron) Berry connection. I now wish to express this Berry connection in terms of the u orbitals, starting from equation 30. In that equation, we have the overlap between two Slater determinants, each built out of an orthonormal set, but where the two sets overlap. According to a well-known theorem, the overlap between the two determinants equals the determinant of the overlap matrix between the two sets of orbitals. We have, here, a minor complication, since the $|\Phi\rangle$'s are determinants of n spinorbitals, while I wish to use the $n/2 \times n/2$ overlap matrix $S(\boldsymbol{\kappa}, \boldsymbol{\kappa} + \Delta\boldsymbol{\kappa})$ amongst *space* orbitals, whose elements are:

$$S^{(\lambda)}_{v,v'}(\boldsymbol{\kappa}, \boldsymbol{\kappa} + \Delta\boldsymbol{\kappa}) = \langle u^{(\lambda)}_v(\boldsymbol{\kappa}) | u^{(\lambda)}_{v'}(\boldsymbol{\kappa} + \Delta\boldsymbol{\kappa}) \rangle = \int_{cell} d\mathbf{r}\ u^{(\lambda)*}_v(\mathbf{r}; \boldsymbol{\kappa}) u^{(\lambda)}_{v'}(\boldsymbol{\kappa} + \Delta\boldsymbol{\kappa}, \mathbf{r}) \tag{32}$$

The identity is, therefore, written with a second power:

$$\langle \Phi^{(\lambda)}(\boldsymbol{\kappa}) | \Phi^{(\lambda)}(\boldsymbol{\kappa} + \Delta\boldsymbol{\kappa}) \rangle = \ \det^2 S^{(\lambda)}(\boldsymbol{\kappa}, \boldsymbol{\kappa} + \Delta\boldsymbol{\kappa}) \tag{33}$$

I now exploit another well-known identity [29], valid for any matrix A:

$$\det \exp A = \exp \text{tr } A \tag{34}$$

which, when applied to $A = \ln S$, transforms equation 30 into:

$$\Delta\varphi = 2 \text{ Im ln det } S^{(\lambda)}(\boldsymbol{\kappa}, \boldsymbol{\kappa} + \Delta\boldsymbol{\kappa}) = 2 \text{ Im tr ln } S^{(\lambda)}(\boldsymbol{\kappa}, \boldsymbol{\kappa} + \Delta\boldsymbol{\kappa}) \tag{35}$$

Taking the gradient with respect to $\Delta\boldsymbol{\kappa}$ and noticing that $S^{(\lambda)}(\boldsymbol{\kappa}, \boldsymbol{\kappa})$ coincides with the identity, we find that equation 29 is equivalent to:

$$\Delta \mathbf{P}_{\text{el}} = -\frac{2\iota}{(2\pi)^3} \int_{\text{BZ}} d\boldsymbol{\kappa} \ \text{tr} \ \{ \nabla_{\boldsymbol{\kappa}'} S^{(1)}(\boldsymbol{\kappa}, \boldsymbol{\kappa}') - \nabla_{\boldsymbol{\kappa}'} S^{(0)}(\boldsymbol{\kappa}, \boldsymbol{\kappa}') \} \Big|_{\boldsymbol{\kappa}'=\boldsymbol{\kappa}} . \tag{36}$$

Using the definition of the overlap matrices S, we can immediately verify that equation 36 is also identical to our starting expression, equation 28, thus completing our proof. I anticipate that the overlap matrices, S, are the key ingredients of numerical implementations. Let me end this section by stressing once more that the integrands in equations 28, 29 and 36 are gauge-dependent and unphysical, while the integrated values are gauge-invariant and physically observable.

7. The numerical algorithm

In practical calculations, the BZ integral must be replaced with a discrete sum, as in equation 11. Typically, eigenfunctions, at a given $\boldsymbol{\kappa}$, are obtained by numerical diagonalization over a finite basis: the phase is then chosen essentially at random by the diagonalization routine. It is, therefore, of paramount importance to choose an algorithm where such phase arbitrariness is harmless.

We observe that the BZ integrations can be performed equivalently over a unit cell in the reciprocal lattice; furthermore, in order to use the mesh of equation 10, we perform the linear change of variables:

$$\boldsymbol{\kappa} = \zeta_1 \mathbf{b}_1 + \zeta_2 \mathbf{b}_2 + \zeta_3 \mathbf{b}_3 \tag{37}$$

The component of $\Delta \mathbf{P}_{\mathrm{el}}$ along $\mathbf{b}_3$, say, is then, from equation 29:

$$\mathbf{b}_3 \cdot \Delta \mathbf{P}_{\mathrm{el}} = -\frac{\iota}{\Omega} \int d\zeta_1 d\zeta_2 d\zeta_3 \, [\, \langle \Phi^{(1)}(\boldsymbol{\zeta}) | \frac{\partial}{\partial \zeta_3} \Phi^{(1)}(\boldsymbol{\zeta}) \rangle - \langle \Phi^{(0)}(\boldsymbol{\zeta}) | \frac{\partial}{\partial \zeta_3} \Phi^{(0)}(\boldsymbol{\zeta}) \rangle \,] \tag{38}$$

and the integral is performed over the unit cube in the $\boldsymbol{\zeta}$ variable. We perform such an integral as an integral in the (ζ_1, ζ_2) plane and an integral along ζ_3, in succession. It is easy to realize that the inner integral is a line integral of the Berry connection, equation 31, in the $\boldsymbol{\zeta}$ variable, over a suitable segment of unit length:

$$-\iota \int_0^1 d\zeta_3 \, \langle \Phi^{(\lambda)}(\boldsymbol{\zeta}) | \frac{\partial}{\partial \zeta_3} \Phi^{(\lambda)}(\boldsymbol{\zeta}) \rangle = \int d\varphi \tag{39}$$

We approximate the differential as a finite phase difference, among states $|\Phi\rangle$ at the neighbouring points $\boldsymbol{\zeta}_{j_1,j_2,j_3}$ and $\boldsymbol{\zeta}_{j_1,j_2,j_3+1}$: we use equation 30 and we introduce the notation for the phase difference at neighbouring points of the grid (at fixed λ):

$$\Delta \varphi^{(\lambda,3)}_{j_1,j_2,j_3} = \operatorname{Im} \ln \langle \Phi^{(\lambda)}(\boldsymbol{\zeta}_{j_1,j_2,j_3}) | \Phi^{(\lambda)}(\boldsymbol{\zeta}_{j_1,j_2,j_3+1}) \rangle \tag{40}$$

and the obvious analogues for the remaining two components with $i \neq 3$. With these notations, the discretization of the three-dimensional integrals in equation 38 is simply:

$$\mathbf{b}_i \cdot \Delta \mathbf{P}_{\mathrm{el}} = -\frac{\iota N_i}{N \Omega} \sum_{j_1,j_2,j_3} (\, \Delta \varphi^{(1,i)}_{j_1,j_2,j_3} - \Delta \varphi^{(0,i)}_{j_1,j_2,j_3} \,) \tag{41}$$

generalized to an arbitrary component. This is the expression which is actually coded in the calculations performed so far (in the framework of density functional theory), where the phase difference among two determinants at different $\boldsymbol{\zeta}$ points (alias $\boldsymbol{\kappa}$ points) is evaluated as the phase of the determinant of the overlap matrix amongst the single-particle orbitals, as in the first equality in equation 35.

Next I am going to show why equation 41 is not affected by the strong gauge arbitrariness discussed above. Let us consider again the component $i = 3$ and focus on the sum over j_3 at fixed values of (j_1, j_2). Further, I use the simplified notation $|\Phi^{(\lambda)}(\boldsymbol{\zeta}_{j_1,j_2,j_3})\rangle = |\Phi^{(\lambda)}_{j_3}\rangle$: a typical sum over j_3 is, therefore:

$$\sum_{j_3=0}^{N_3-1} \Delta\varphi^{(\lambda,3)}_{j_1,j_2,j_3} = \sum_{j_3=0}^{N_3-1} \text{Im ln } \langle \Phi^{(\lambda)}_{j_3} | \Phi^{(\lambda)}_{j_3+1} \rangle = \text{Im ln } \prod_{j_3=0}^{N_3-1} \langle \Phi^{(\lambda)}_{j_3} | \Phi^{(\lambda)}_{j_3+1} \rangle \quad (42)$$

The very crucial step is that only N_3 independent diagonalizations, *not* N_3+1, must be used: otherwise, the random phase introduced by the diagonalization routine destroys the result! The key point is that Bloch states, whose index differ by exactly N_3, correspond to $\boldsymbol{\kappa}$ vectors differing by exactly the reciprocal vector $\mathbf{b}_3$. Since the Bloch orbitals ψ are taken as *periodical* in reciprocal space, we have the important relationship for u:

$$u^{(\lambda)}_n(\mathbf{r}; \boldsymbol{\kappa}+\mathbf{b}_3) = e^{-i\mathbf{b}_3\cdot\mathbf{r}} u^{(\lambda)}_n(\mathbf{r}; \boldsymbol{\kappa}) \quad (43)$$

This implies the identity:

$$\langle \Phi^{(\lambda)}_{N_3-1} | \Phi^{(\lambda)}_{N_3} \rangle = \langle \Phi^{(\lambda)}_{N_3-1} | e^{-i\mathbf{b}_3\cdot\mathbf{r}} | \Phi^{(\lambda)}_0 \rangle \quad (44)$$

where the shorthand notation means that the monoelectronic operator acts upon all the single-particle orbitals in the many-electron determinant $|\Phi\rangle$. The relationship in equation 44 *must* be numerically exploited in order to guarantee meaningful results. equation 42 is then equivalent to:

$$\begin{aligned} &\text{Im ln } \langle \Phi^{(\lambda)}_0 | \Phi^{(\lambda)}_1 \rangle \langle \Phi^{(\lambda)}_1 | \Phi^{(\lambda)}_2 \rangle \cdots \langle \Phi^{(\lambda)}_{N_3-1} | e^{-i\mathbf{b}_3\cdot\mathbf{r}} | \Phi^{(\lambda)}_0 \rangle = \\ &= \text{Im ln tr } \{ e^{-i\mathbf{b}_3\cdot\mathbf{r}} | \Phi^{(\lambda)}_0 \rangle \langle \Phi^{(\lambda)}_0 | \Phi^{(\lambda)}_1 \rangle \langle \Phi^{(\lambda)}_1 | \cdots | \Phi^{(\lambda)}_{N_3-1} \rangle \langle \Phi^{(\lambda)}_{N_3-1} | \} \quad (45) \end{aligned}$$

where the trace involves products of operators, such as $|\Phi^{(\lambda)}_{j_3}\rangle\langle\Phi^{(\lambda)}_{j_3}|$. Each of them is nothing other than the projector over the occupied spinorbitals, at a given $\boldsymbol{\kappa}$ and is, therefore, unique (gauge-invariant) and unambiguous.

8. Localized basis sets

The calculations performed so far have implemented the previous equations directly, where the overlap matrices, equation 32, have been obtained from the self-consistent local-density crystalline orbitals in a reciprocal-space mesh. Furthermore, the variational basis sets which have been used, so far, are either simple plane waves in a pseudopotential framework [4] or linear augmented plane waves (LAPW) [30, 31]. Here, I give the explicit formulation in the case of a localized basis set. It is also understood that the crystalline orbitals in the present case are the eigenstates of the self-consistent Fock operator [5]: it is not necessary to use the same reciprocal mesh for the Berry calculation as is used in achieving self-consistency.

According to the discussion in the previous section, the main quantities to be evaluated are overlap matrices of the kind in equation 32. Suppose we are expanding the crystalline HF orbitals in the localized basis set, $\varphi_\mu(\mathbf{r})$ centred at sites $\mathbf{s}_\mu$, where $\mu = 1, \ldots, M$ is a basis label in the primitive cell. If $\mathbf{T}_m$ is a generic lattice translation, the occupied Bloch HF orbitals have the form:

$$\psi^{(\lambda)}_v(\mathbf{r}; \boldsymbol{\kappa}) = \sum_{\mu=1}^{M} \alpha^{(\lambda)}_{v,\mu}(\boldsymbol{\kappa}) \sum_m e^{i\boldsymbol{\kappa}\cdot\mathbf{T}_m} \varphi_\mu(\mathbf{r} - \mathbf{s}_\mu - \mathbf{T}_m) \quad (46)$$

where the α-coefficients are provided by the diagonalization routine at a given $\boldsymbol{\kappa}$. A typical overlap matrix element is, therefore:

$$\langle u_v^{(\lambda)}(\boldsymbol{\kappa})|u_{v'}^{(\lambda)}(\boldsymbol{\kappa}+\Delta\boldsymbol{\kappa})\rangle = \sum_{\mu,\mu'=1}^{M} \alpha_{v,\mu}^{(\lambda)*}(\boldsymbol{\kappa})\alpha_{v',\mu'}^{(\lambda)}(\boldsymbol{\kappa}+\Delta\boldsymbol{\kappa})$$

$$\times \sum_m \mathrm{e}^{\imath(\boldsymbol{\kappa}+\Delta\boldsymbol{\kappa})\cdot\mathbf{T}_m} \langle\varphi_\mu(\mathbf{r}-\mathbf{s}_\mu)|\, \mathrm{e}^{-\imath\Delta\boldsymbol{\kappa}\cdot\mathbf{r}}\, |\varphi_{\mu'}(\mathbf{r}-\mathbf{s}_{\mu'}-\mathbf{T}_m)\rangle \tag{47}$$

Besides the α-coefficients, the basic ingredients are, therefore, the matrix elements of suitable plane waves over the localized basis.

9. Macroscopic fields and dielectric constants

The problem of an extended solid in a finite macroscopic field is a very difficult one—even as a matter of principle [32]—and is presently an active area of research [15, 16]. The difficulty can be traced back to the fact that, in the presence of a macroscopic field, the electrons are in a *nonperiodic* self-consistent potential and therefore, the orbitals no longer have the Bloch form. Even worse, the Hamiltonian is no longer bounded from below and the spectrum undergoes a *nonanalytic* qualitative change. Notice that such a statement applies not only to a solid, but even to an isolated hydrogen atom!

At the moment of writing, the well-established theoretical methods and the first-principle calculations rely heavily on linear-response theory and therefore, they concern only crystal properties which are *linear* in the field magnitude: this is the case for calculations of the macroscopic dielectric constants [33, 34]. Basically, it is possible to perform self-consistent calculations for a crystal in a macroscopic field, which are correct to first-order in the field magnitude, without trading away the major simplification of dealing with pure Bloch states. These methods have been developed earlier than the geometric phase approach discussed above but have a close link with it. I stress, however, that linear-response calculations do *not* presently involve the evaluation of a geometric phase.

So far, linear-response methods in solids have only been used in the framework of density functional theory: I refer to [35] for a comprehensive account of the most powerful approach, amongst the several which have been proposed and implemented in recent years. In these notes, I give an abridged presentation, focussing on the features of a possible HF implementation and discussing only the special case of the macroscopic response to a macroscopic field: the linear-response methods have been proven capable of dealing with more general cases as well (phonon spectra [35], piezoelectricity [36] amongst others).

The key idea is essentially the same as in the so-called coupled HF schemes, quite common in quantum chemistry for dealing with finite systems (atoms and molecules) [37, 24]. Let us start, therefore, with a large and finite system, having discrete HF orbitals ψ_i, which vanish outside the system: the thermodynamic limit will be taken only at the end. We write the electronic charge density as:

$$\rho_{\mathrm{el}}(\mathbf{r}) = -\sum f_i|\psi_i(\mathbf{r})|^2 \tag{48}$$

where f_i is an occupancy factor (either 2 or 0), selecting the doubly-occupied HF orbitals.

The crystalline observable of interest is the macroscopic dielectric tensor, defined as*:

$$\varepsilon_\infty = 1 + 4\pi \frac{\partial \mathbf{P}_{\text{el}}}{\partial \mathbf{E}} \tag{49}$$

where $\mathbf{P}$ is the macroscopic electronic polarization linearly induced by the (screened) field $\mathbf{E}$; in a cubic material, such a tensor obviously reduces to a constant. Using equation 48, we wish to evaluate:

$$\frac{\partial \mathbf{P}_{\text{el}}}{\partial \mathbf{E}} = -\frac{1}{\mathcal{V}} \sum_i f_i \langle \psi_i | \mathbf{r} | \frac{\partial \psi_i}{\partial \mathbf{E}} \rangle + \text{c.c.} \tag{50}$$

where $\mathcal{V}$ is the volume of the finite sample. It proves useful to transform the dipole matrix elements, using the identity:

$$\langle \psi_i | \mathbf{r} | \psi_j \rangle = -\iota \frac{\langle \psi_i | \mathbf{v} | \psi_j \rangle}{\epsilon_i - \epsilon_j}, \quad i \neq j \tag{51}$$

where ϵ_i are the HF eigenvalues and the HF velocity operator $\mathbf{v}$ is the same as in equation 6. Straightforward manipulations transform equation 50 into:

$$\frac{\partial \mathbf{P}_{\text{el}}}{\partial \mathbf{E}} = \frac{\iota}{\mathcal{V}} \sum_{i,j} \frac{f_j - f_i}{\epsilon_j - \epsilon_i} \langle \psi_i | \mathbf{v} | \psi_j \rangle \langle \psi_j | \frac{\partial \psi_i}{\partial \mathbf{E}} \rangle \tag{52}$$

where I have exploited the fact that $|\partial \psi_i / \partial \mathbf{E}\rangle$ may be chosen as orthogonal to $|\psi_i\rangle$. Notice that equation 52 is precisely the finite-system analogue of the Kubo formula previously displayed in equation 16 for an infinite periodic system (and for a lattice-periodical perturbation).

In order to proceed further, we need a self-consistent scheme providing the derivatives of the HF orbitals and of the HF potential with respect to the perturbation. The first step consists of writing the first-order corrections to the orbitals as:

$$\Delta \psi_i = \sum_{j \neq i} \psi_j \frac{\langle \psi_j | \Delta V | \psi_i \rangle}{\epsilon_i - \epsilon_j} \tag{53}$$

As a second step, we look for an independent relationship providing instead ΔV in terms of the orbital variations $\Delta \psi_i$: iterating over these two steps, the self-consistency goal is reached. As usual, I split the self-consistent HF potential into:

$$V = V_{\text{ext}} + V_{\text{H}} + V_{\text{x}} \tag{54}$$

where V_{ext} is the bare potential due to the ions and to the (unscreened) perturbation, while V_{H} and V_{x} are the usual Hartree and exchange self-consistent terms, respectively. Within HF, the exchange potential is non-local. The two self-consistent terms are quadratic in the HF orbitals and therefore, their linear variation has a very simple form. For instance, the first-order Hartree term is:

* I discuss here ε_∞, which accounts for the purely electronic component of dielectric screening. In solid state textbooks, ε_∞ is usually referred to as the "static high-frequency" dielectric constant: this means measured at frequencies much higher than the lattice vibrations and much lower than electronic excitations.

$$\Delta V_{\mathrm{H}}(\mathbf{r}) = \sum_i f_i \int d\mathbf{r}' \frac{1}{|\mathbf{r}-\mathbf{r}'|} [\, \psi_i^*(\mathbf{r}')\Delta\psi_i(\mathbf{r}') + \text{c.c.}\,] \tag{55}$$

When a macroscopic field $\mathbf{E}$ acts on the solid, it proves useful to write the self-consistent HF perturbation potential, ΔV, in a slightly different way from equation 54, explicitly separating the macroscopic field from the microscopic components. The potential of the *screened* macroscopic field—due to both the bare and the self-consistent terms in equation 54—is written as $-e\mathbf{E}\cdot\mathbf{r}$, while the remaining *microscopic* term is a lattice-periodical (though non-local) operator in the thermodynamic limit. Therefore, we write the first-order variation of equation 54 as:

$$\Delta V = -e\mathbf{E}\cdot\mathbf{r} + \Delta V^{(\mathrm{micro})} \tag{56}$$

At this point, we may take the thermodynamic limit of all the above expressions. In fact, all of the equations which are needed in order to implement the method have a simple and well-defined expression for the infinite periodic crystal in the thermodynamic limit, where the index i is identified with the band index and the Bloch vector altogether and the orbitals are expanded over a given basis set. The self-consistent loop is performed cycling over equations 56 and 53, while the value of the screened field $\mathbf{E}$ is kept constant during the iteration. Whenever equation 56 is inserted in equation 53, the matrix elements of $\mathbf{r}$ are transformed to the velocity form, using equation 51 again, while the elements of $\Delta V^{(\mathrm{micro})}$ are directly evaluated from the Bloch orbitals, as *e.g* in equation 55.

As anticipated above, linear-response methods have not been implemented so far in a periodic solid at the HF level, while a rather large experience has been gathered on implementations within a density functional framework instead, both at the local-density-approximation level [33, 35] and beyond [34]. In most calculations, a plane-wave expansion of the orbitals is used; in a few recent papers, other basis sets are used as well [38, 39].

In atoms and molecules, linear-response methods have a well-established record of implementations at the HF level and beyond [24]. No major complications are expected for a crystalline solid, by analogy with similar density functional implementations, when the formulation presented in this work is used. For the particular case of computing a macroscopic dielectric constant, linear response is essentially the only viable tool (for an alternative, though less effective, approach, see [40]).

Acknowledgments

This work was written during a stay at the Institute for Theoretical Physics, University of California at Santa Barbara, where I have profited of many discussions with several colleagues about the topics of this work. The research was supported in part by National Science Foundation under Grant No. PHY89-04035.

References

1. L.D. Landau and E.M. Lifshitz, *Electrodynamics of Continuous Media* (Pergamon Press, Oxford, 1984).
2. C. Kittel, *Introduction to Solid State Physics*, 6th. edition (Wiley, New York, 1986).

3. N.W. Ashcroft and N.D. Mermin, *Solid State Physics* (Saunders, Philadelphia, 1976).
4. W.E. Pickett, Computer Phys. Reports **9**, 115 (1989).
5. C. Pisani, R. Dovesi, and C. Roetti, *Hartree-Fock Ab-Initio Treatment of Crystalline Systems* (Springer, Berlin, 1988).
6. R.M. Martin, Phys. Rev. B **9**, 1998 (1974).
7. R. Resta, Ferroelectrics **136**, 51 (1992).
8. R.D. King-Smith and D. Vanderbilt, Phys. Rev. B **47**, 1651 (1993).
9. R. Resta, Europhys. Lett. **22**, 133 (1993).
10. R. Resta, Rev. Mod. Phys. **66**, 899 (1994).
11. R. Resta, Il Nuovo Saggiatore 9(5/6), 79 (1993).
12. R.M. Martin, Phys. Rev. B **5**, 1607 (1972).
13. M. Born and K. Huang, *Dynamical Theory of Crystal Lattices* (Oxford University Press, Oxford, 1954).
14. M.E. Lines and A.M. Glass, *Principles and Applications of Ferroelectrics and Related Materials* (Clarendon Press, Oxford,
15. R.W. Nunes and D. Vanderbilt, Phys. Rev. Lett. **73**, 712 (1994).
16. A. Dal Corso and F. Mauri, Phys. Rev. B **50**, 5756 (1994).
17. *Geometric Phases in Physics*, edited by A. Shapere and F. Wilczek (World Scientific, Singapore, 1989).
18. *Theory of the Inhomogeneous Electron Gas*, edited by S.Lundqvist and N.H. March (Plenum, New York, 1983).
19. R.O. Jones and O. Gunnarsson, Rev. Mod. Phys. **61**, 689 (1989).
20. G. Ortíz and R.M. Martin, Phys. Rev. B **49**, 14202 (1994).
21. R. McWeeny, *Methods of Molecular Quantum Mechanics*, Second Edition (Academic, London, 1992).
22. R. Pick, M.H. Cohen, and R.M. Martin, Phys. Rev. B **1**, 910 (1970).
23. K. Huang, Proc. Roy. Soc. **A203**, 178 (1950).
24. R.D. Amos, in: *Ab-Initio Methods in Quantum Chemistry - I*, edited by K.P. Lawley (Wiley, New York, 1987), p. 99.
25. P. Lazzeretti and R. Zanasi, J. Chem. Phys. **83**, 1218 (1985).
26. D.J. Thouless, Phys. Rev. B **27**, 6803 (1983).
27. Y. Aharonov and D. Bohm, Phys. Rev. **115**, 485 (1959); reprinted in: *Geometric Phases in Physics*, edited by A. Shapere and F. Wilczek (World Scientific, Singapore, 1989), p. 104.
28. J. Zak, Phys. Rev. Lett. **62**, 2747 (1989).
29. L.I. Schiff, *Quantum mechanics* (McGraw-Hill, Singapore, 1968).
30. O.K. Andersen, O. Jepsen and M. Sob, in: *Electronic Band Structure and Its Applications*, edited by M. Yussouf (Springer, Berlin, 1987), p. 1
31. H.J.F. Jansen and A.J. Freeman, Phys. Rev. B. **30**, 561 (1984).
32. R. Nenciu, Rev. Mod. Phys. **63**, 91 (1991).
33. S. Baroni and R. Resta, Phys. Rev. B **33**, 7017 (1986).
34. A. Dal Corso, S. Baroni, and R. Resta, Phys. Rev. B **49**, 5323 (1994).
35. P. Giannozzi, S. de Gironcoli, P. Pavone, and S. Baroni, Phys. Rev. B **43**, 7231 (1991).
36. S. de Gironcoli, S. Baroni, and R. Resta, Phys. Rev. Lett. **62**, 2853 (1989).
37. A. Dalgarno, Adv. Phys. **11**, 281 (1962).
38. Rici Yu and H. Krakauer, Phys. Rev. B **49**, 4467 (1994); Cheng-Zhang Wang, Rici Yu, and H. Krakauer, Phys. Rev. Lett. **72**, 368 (1994).
39. F. Aryasetiawan and O. Gunnarsson, Phys. Rev. B **49**, 16214 (1994).
40. K. Kunc and R. Resta, Phys. Rev. Lett. **51**, 686 (1983); R. Resta and K. Kunc, Phys. Rev. B **34**, 7146 (1986).

The Hubbard Models and Superconductivity

Mario Rasetti

Dipartimento di Fisica e Unità INFM Politecnico di Torino, C.so Duca degli Abruzzi, I-10129 Torino, Italy.

May 30, 1996

Summary. We discuss concisely how the conventional Hubbard model and its generalizations (including phonons) might be good candidates for an explanation of high T_c superconductivity, reviewing its properties both on crystals defined by conventional point groups and on non-Euclidean lattices. We also study the metal-insulator transition of the Hubbard model by treating the hopping term between adjacent 1-D chains in the frame of a fermionic linearization scheme and keeping the full model on the chains. A general equation for the critical point is worked out in terms of correlation functions of the one-dimensional model, assuming that the transition is second-order, at zero temperature.

Key words: High T_c superconductivity – Hubbard model – Metal-insulator transition – Non-Euclidean lattices

1. The Hubbard model

In the so-called *tight-binding* scheme, the hamiltonian for electrons of the band generated by the atomic shell of ϕ_σ (where $\phi_\sigma(\mathbf{r}-\mathbf{R}_j)$ denotes the orbital wave function for a (conduction) electron of spin σ ($\sigma = \uparrow, \downarrow$ on the atom in lattice site $\mathbf{j}$ (at $\mathbf{R}_j$)) writes:

$$H = \sum_{i,j} \sum_{\sigma} t_{i,j}^{(\sigma)} a_{i,\sigma}^{\dagger} a_{j,\sigma} + \frac{1}{2} \sum_{i,j,k,\ell} \sum_{\sigma,\sigma'} V_{i,j;k,\ell}^{(\sigma,\sigma')} a_{i,\sigma}^{\dagger} a_{j,\sigma'}^{\dagger} a_{\ell,\sigma'} a_{k,\sigma} \tag{1}$$

where:

$$t_{ij}^{(\sigma)} = < i,\sigma | h(\mathbf{r}) | j,\sigma > \equiv \int d\mathbf{r} \phi_\sigma^*(\mathbf{r}-\mathbf{R}_i) h(\mathbf{r}) \phi_\sigma(\mathbf{r}-\mathbf{R}_j) \tag{2}$$

is the *hopping amplitude* $\left[h(\mathbf{r}) \doteq -\frac{\hbar^2}{2m}\Delta_{\mathbf{r}} + \mathcal{V}(\mathbf{r};\{\mathbf{R}_\ell\})\right.$ denotes the single-particle electronic hamiltonian, $\mathcal{V}$ being the potential "seen" by a single electron in the lattice$\left.\right]$. The first term in equation 1 is then nothing but the customary band hamiltonian, whereas:

$$\begin{aligned} V^{(\sigma,\sigma')}_{i,j;k,\ell} &= \langle i,\sigma\,;\,j,\sigma'|V_C(\mathbf{r},\mathbf{r}')|k,\sigma\,;\,\ell\sigma'\rangle \\ &\equiv e^2\int d\mathbf{r}\,d\mathbf{r}'\,\phi^*_\sigma(\mathbf{r}-\mathbf{R}_i)\,\phi^*_{\sigma'}(\mathbf{r}'-\mathbf{R}_j) \\ &\quad\times\frac{1}{|\mathbf{r}-\mathbf{r}'|}\phi_\sigma(\mathbf{r}-\mathbf{R}_k)\,\phi_{\sigma'}(\mathbf{r}-\mathbf{R}_\ell) \end{aligned} \tag{3}$$

describes instead coulomb interaction between electrons.

The conventional wisdom [1] is to make a first, drastic approximation of equation 1 which consists of assuming, on the one hand, that the values of $t^{(\sigma)}_{ij}$ differ from zero only if sites $\mathbf{R}_i$ and $\mathbf{R}_j$ are nearest neighbours in the lattice Λ, independent of σ (*i.e.* the probability amplidute is the same for electrons with spin *up* or *down*), and all equal (translation invariance); and on the other hand, in keeping only those interaction terms corresponding to $i = j = k = \ell$, assuming that these are equal while all others are negligible. The resulting hamiltonian is usually referred to as the Hubbard hamiltonian [2, 3]:

$$H = -t\sum_{<i,j>}\sum_\sigma\left(a^\dagger_{i,\sigma}a_{j,\sigma}a^\dagger_{j,\sigma}a_{i,\sigma}\right) + U\sum_j n_{j,\uparrow}n_{j,\downarrow} \tag{4}$$

where $-t$ (the minus sign is conventional) represents the common value of all non-vanishing $t^{(\sigma)}_{ij}$, whereas, analogously, U is the value of the $\mathcal{V}$'s assumed to be non-zero. In view of the Pauli principle and of the assumed strong localization of electrons, among the latter one finds, of course, only terms with $\sigma' = -\sigma$.

In a seminal paper, in which the model, which has ever since borne his name, was proposed in its full extension, Hubbard [4] performs a detailed analysis of the most general hamiltonian suitable to describe a system of itinerant interacting electrons in a d-dimensional lattice Λ of N sites. In doing so, he puts into play five parameters describing as we shall see, respectively:

i) the *density-density* interactions: $U \equiv\langle \mathbf{i},\sigma;\mathbf{i},-\sigma|V_C(\mathbf{r})|\mathbf{i},\sigma;\mathbf{i},-\sigma\rangle$ and $V \equiv\langle \mathbf{i},\sigma;\mathbf{j},\sigma'|V_C(\mathbf{r},\mathbf{r}')|\mathbf{i},\sigma;\mathbf{j},\sigma'\rangle$;
ii) the *density-bond* interactions: $X \equiv\langle \mathbf{i},\sigma;\mathbf{i},-\sigma|h(\mathbf{r})|\mathbf{i},\sigma;\mathbf{j},\sigma\rangle$);
iii) the *bond-bond* interactions $\tilde{J} \equiv\langle \mathbf{i},\sigma;\mathbf{j},\sigma'|h(\mathbf{r})|\mathbf{i},\sigma';\mathbf{j},\sigma\rangle$ and $\tilde{J}' \equiv\langle \mathbf{i},\sigma;\mathbf{i},-\sigma|h(\mathbf{r})|\mathbf{j},-\sigma;\mathbf{j}\sigma\rangle$;

where $\mathbf{i}$ and $\mathbf{j}$ are assumed to be nearest neighbours (*n.n.*) in Λ.

The resulting hamiltonian has the form:

$$\begin{aligned} H' &= -t\sum_{<\mathbf{i},\mathbf{j}>}\sum_\sigma\left(a^\dagger_{\mathbf{i},\sigma}a_{\mathbf{j},\sigma} + a^\dagger_{\mathbf{j},\sigma}a_{\mathbf{i},\sigma}\right) + U\sum_{\mathbf{i}} n_{\mathbf{i},\Uparrow}n_{\mathbf{i},\Downarrow} \\ &\quad +V\sum_{<\mathbf{i},\mathbf{j}>}\sum_{\sigma,\sigma'} n_{\mathbf{i},\sigma}n_{\mathbf{j},\sigma'} + X\sum_{<\mathbf{i},\mathbf{j}>}\sum_\sigma\left(a^\dagger_{\mathbf{i},\sigma}a_{\mathbf{j},\sigma} + a^\dagger_{\mathbf{j},\sigma}a_{\mathbf{i},\sigma}\right)(n_{\mathbf{i},-\sigma} + n_{\mathbf{j},-\sigma}) \end{aligned}$$

$$+\tilde{J}\sum_{<\mathbf{i},\mathbf{j}>}\sum_{\sigma,\sigma'} a^\dagger_{\mathbf{i},\sigma}a^\dagger_{\mathbf{j},\sigma'}a_{\mathbf{i},\sigma'}a_{\mathbf{j},\sigma}$$

$$+\tilde{J}'\sum_{<\mathbf{i},\mathbf{j}>}\left(a^\dagger_{\mathbf{i},\Uparrow}a^\dagger_{\mathbf{i},\Downarrow}a_{\mathbf{j},\Downarrow}a_{\mathbf{j},\Uparrow}+a^\dagger_{\mathbf{j},\Uparrow}a^\dagger_{\mathbf{j},\Downarrow}a_{\mathbf{i},\Downarrow}a_{\mathbf{i},\Uparrow}\right) \tag{5}$$

where the index σ labels the electron spin ($\sigma \in \{\uparrow,\downarrow\}$). The d-dimensional lattice Λ of $N = N_1N_2\ldots N_d$ sites, is assumed to be cubic. The operators $a^\dagger_{\mathbf{i},\sigma}$ and $a_{\mathbf{i},\sigma}$ are fermionic creation and annihilation operators $\left(\{a_{\mathbf{i},\sigma},a_{\mathbf{j},\sigma'}\}\right.$ $= 0$, $\{a^\dagger_{\mathbf{i},\sigma},a_{\mathbf{j},\sigma'}\} = \delta_{\mathbf{i},\mathbf{j}}\,\delta_{\sigma,\sigma'}$, $\left.n_{\mathbf{i},\sigma} \doteq a^\dagger_{\mathbf{i},\sigma}a_{\mathbf{i},\sigma}\right)$ over Λ; $<\mathbf{i},\mathbf{j}>$ denotes ($n.n.$) sites in Λ.

The extended hamiltonian in equation 5 provides, on the one hand, a more realistic description of the hopping processes, by attributing different probability amplitudes to hops carrying an electron from a doubly-occupied site to an empty site, from a doubly- to a singly-occupied site, from one singly-occupied state to another and from a singly-occupied state to an empty site; on the other hand, Coulomb interaction is accounted for not only locally, but in its long-range features (we shall, in the following, consider only a range equal to the nearest neighbour distance).

All of this implies that two parameters are no longer sufficient and at least four more are needed: besides t and U, we define an amplitude to describe long-range Coulomb interactions, $V \equiv< i,\sigma\,;\,j,\sigma'|V_C(\mathbf{r},\mathbf{r}')|i,\sigma\,;\,j,\sigma' >$; a parameter accounting for different hopping probabilities for different site occupation:

$$\begin{aligned} X &\equiv\ < i,\sigma\,;\,i,-\sigma|h(\mathbf{r})|i,\sigma\,;\,j,\sigma > \\ &\doteq \int d\mathbf{r}\phi^*_\sigma(\mathbf{r}-\mathbf{R}_i)h(\mathbf{r})\phi_\sigma(\mathbf{r}-\mathbf{R}_j)|\phi_{-\sigma}(\mathbf{r}-\mathbf{R}_j)|^2 \end{aligned}$$

and two for second-order hopping processes: *exchange* (an electron hops from site i to site j while ***simultaneously*** another electron jumps from site j to site i):

$$\begin{aligned} \tilde{J} &\equiv\ < i\sigma\,;\,j,\sigma'|h(\mathbf{r})|i,\sigma'\,;\,j,\sigma > \\ &\doteq \int d\mathbf{r}\phi^*_\sigma(\mathbf{r}-\mathbf{R}_i)\phi^*_{\sigma'}(\mathbf{r}-\mathbf{R}_j)h(\mathbf{r})\phi_{\sigma'}(\mathbf{r}-\mathbf{R}_i)\phi^*_\sigma(\mathbf{r}-\mathbf{R}_j) \end{aligned}$$

and *pair hopping* (two electrons hop "together" from a doubly-occupied to an empty neighbouring site)

$$\begin{aligned} \tilde{J}' &\equiv\ < i\sigma\,;\,i,-\sigma|h(\mathbf{r})|j,-\sigma\,;\,j,\sigma > \\ &\doteq \int d\mathbf{r}\phi^*_\sigma(\mathbf{r}-\mathbf{R}_i)\phi^*_{-\sigma}(\mathbf{r}-\mathbf{R}_i)h(\mathbf{r})\phi_\sigma(\mathbf{r}-\mathbf{R}_j)\phi_{-\sigma}(\mathbf{r}-\mathbf{R}_j) \end{aligned}$$

One assumes, for simplicity, that those, among the above coefficients, which involve two sites i and j are zero, unless i and j are nearest neighbours in Λ.

The possibility of a realistic description of the system spectrum relies, of course, on an accurate estimate of the parameters t, U, V, $\tilde{J}$ and $\tilde{J}'$.

By a thorough analysis of the data, Hubbard estimated that U ranges between 20 and 30 eV, V between 2 and 3 eV, X is of the order of 0.5 eV and $\tilde{J}$, $\tilde{J}'$ are of the order of 0.025 eV. It is for this reason that he confirms the claim that it should be sufficient, at least in the study of transition metals, to keep only U, suggesting a hamiltonian of the form given by equation 4 (*i.e.* the first line in equation 5). In view of its use in the frame of a grand-canonical ensemble, we shall add to such a hamiltonian, the term $-\mu N_e$, where μ is the chemical potential and $N_e = \sum_{\mathbf{i}} \sum_{\sigma} n_{\mathbf{i},\sigma}$, the total number of electrons:

$$H_{Hub} = -\mu \sum_{\mathbf{i}} \sum_{\sigma} n_{\mathbf{i},\sigma} - t \sum_{<\mathbf{i},\mathbf{j}>} \sum_{\sigma} \left(a^{\dagger}_{\mathbf{i},\sigma} a_{\mathbf{j},\sigma} + a^{\dagger}_{\mathbf{j},\sigma} a_{\mathbf{i},\sigma} \right) + U \sum_{\mathbf{i}} n_{\mathbf{i},\Uparrow} n_{\mathbf{i},\Downarrow} \quad (6)$$

The "standard" Hubbard model, given by equation 6, provides the simplest description of a system of itinerant interacting electrons on a lattice, by assuming that the itinerant electrons interact only via an on-site Coulomb repulsion term U. With the discovery of the novel high-T_c superconductors, such a model was universally considered to be a good candidate to describe two of their most important features: antiferromagnetism and the possibility of undergoing a correlation-induced Mott-Hubbard [4, 5] *metal-insulator transition* (*MIT*). The latter was expected solely on the basis of the competition between the two phases with dominant repulsion (insulator) and hopping (metal) terms. On the other hand, an exact solution of the model is known only for the ground state in one dimension [6], in which case the system has been shown to be insulating – at half-filling (we define *filling* as the average number of electrons, of any spin, per lattice site, $n_0 \equiv < N_e >/N$, so that *half-filling* means $n_0 = 1$) – for any $U > 0$.

In dimensions greater than one, the *MIT* transition has to be studied by approximate treatments, non-perturbative in the interaction. Different approaches have been proposed [4, 7, 8], supporting the appearance of a continuous metal-insulator transition at $T = 0$ as a function of U. Also in the limit of infinite dimension, numerical calculations [9, 10] show that there is a MIT for finite and non-vanishing U. Nevertheless, none of these methods is capable, at the same time, of matching the exact result in the limit where the number of dimensions becomes one. Moreover, there is no description available of the phase space for generic filling and for non-vanishing temperature. In the final part of this paper, we shall show how the latter goals can be achieved by considering a d-dimensional Hubbard hamiltonian and using one of the above schemes (referred to as *CFLS*: *cluster fermionic linearization* [11]) only to approximate the hopping between neighbouring 1-D chains, the hamiltonian of the latter being complete and treated exactly.

The first goal we deal with is, naturally, that of finding the spectrum of equation 6. We do that by resorting to one of the most important tools of quantum mechanics, that of exploiting the symmetry properties of the system to more easily derive its features.

At half-filling, the model defined by equation 6 has an $su(\tilde{2}) \oplus su(2)$ global symmetry algebra [12, 13] (more precisely, the symmetry group is $SO(4) \sim SU(\tilde{2}) \otimes SU(2)/\mathbb{Z}_2$, because locally the Casimir operator of $su(\tilde{2})$ is proportional to the Cartan element of $su(2)$, and analogously, the Casimir of $su(2)$ is proportional to the Cartan of $su(\tilde{2})$, which implies that, in every state, spin and pseudo-spin are either both integral or both half-integral), generated by:

$$J_+ = \sum a^{\dagger}_{\mathbf{i},\Uparrow} a_{\mathbf{i},\Downarrow} \quad \text{and} \quad J_- = \sum a^{\dagger}_{\mathbf{i},\Downarrow} a_{\mathbf{i},\Uparrow} = (J_+)^{\dagger}$$

$$J_z = \frac{1}{2}\sum_{\mathbf{i}} (n_{\mathbf{i},\Uparrow} - n_{\mathbf{i},\Downarrow}) \quad \text{for} \quad \textbf{magnetic}\, su(\tilde{2}) \tag{7}$$

(number-preserving) and:

$$K_+ = \sum_{\mathbf{j}} e^{i\mathbf{G}\cdot\mathbf{j}} a^\dagger_{\mathbf{j},\Uparrow} a^\dagger_{\mathbf{j},\Downarrow} \quad \text{and} \quad K_- = \sum_{\mathbf{j}} e^{-i\mathbf{G}\cdot\mathbf{j}} a_{\mathbf{j},\Downarrow} a_{\mathbf{j},\Uparrow} = (K_+)^\dagger$$

$$K_z = \frac{1}{2}\sum_{\mathbf{j}} (n_{\mathbf{i},\Uparrow} + n_{\mathbf{i},\Downarrow} - 1) \quad \text{for} \quad \textbf{superconductive}\, su(2) \tag{8}$$

(spin-conserving), where $\mathbf{G}$ is a vector whose d components are all equal to π. It is immediate to check that the two $su(2)$ algebras above are mutually orthogonal, *i.e.* $[J_\alpha, K_\beta] = 0$, $\forall \alpha, \beta = \pm, z$. Moreover, the algebra $\mathbb{Z}_2$, which is generated by:

$$\mathcal{T} = \frac{1}{\sqrt{N}} \sum_{\mathbf{l}} e^{i\,\mathbf{G}\cdot\mathbf{l}} \left(a_{\mathbf{l},\Uparrow} + a^\dagger_{\mathbf{l},\Uparrow} \right) e^{i\pi n_{\mathbf{l},\Downarrow}} \tag{9}$$

simply interchanges the two algebras:

$$\mathcal{T} a_{\mathbf{j},\Uparrow} \mathcal{T}^{-1} = a^\dagger_{\mathbf{j},\Uparrow} \quad \text{and} \quad \mathcal{T} a_{\mathbf{j},\Downarrow} \mathcal{T}^{-1} = e^{i\,\mathbf{G}\cdot\mathbf{j}}\, a^\dagger_{\mathbf{j},\Downarrow}$$

The symmetry described amounts to saying that for $\mu = \frac{1}{2}U$ (the condition for half-filling) both $[K_\alpha, H_{Hubb}] = 0$ and $[J_\beta, H_{Hubb}] = 0$.

Resorting to such symmetry Yang [14] constructed eigenstates of the Hubbard hamiltonian characterized by off-diagonal long-range order (ODLRO), labelled in terms of two quantum numbers, one of which is related just to the existence of superconductivity. Such eigenstates, however, were shown to have energy higher than that of the ground state. On the other hand, Korepin *et al.* [15] were able to prove, by using this symmetry, that the lowest weight states of the algebra, annihilated by the lowering operators of $su(\tilde{2})$ and $su(2)$, are of the type introduced in the solution to the one dimensional model by Lieb and Wu [6] resorting to a Bethe *Ansatz*. They proposed, therefore, an "$SO(4)$-extended" Bethe Ansatz for the Hubbard model, whereby a complete set of states is generated by application of the raising operators to such lowest weight states. It is worth mentioning also that the very general model, described by equation 5, has itself a global symmetry, holding in the special case when $t = X = \tilde{J} = -\frac{1}{2}\tilde{J}' = -4V$, realized by the superalgebra $u(2|2)$ and, thanks to the latter, is then exactly solvable in one dimension [16]. The local superalgebras whereby such global symmetry is generated had been earlier introduced by Montorsi, Rasetti and Solomon [11], just in connection with the possible interpretation of the superconductive phase transition in the novel high-T_c copper-oxide compounds as a spontaneous breaking of supersymmetry.

Returning to the customary Hubbard model, it so happens that switching on a phonon field, although breaking the $su(2)$ symmetry, may restore it as a quantum group $\mathcal{U}_q(su(2)) \sim [su(2)]_q$ symmetry, where the deformation parameter q is related to the strength λ of the electron-phonon coupling. Moreover, the filling at which symmetry is restored depends on λ and the symmetric eigenstates of the hamiltonian (which exhibit ODLRO) have energy lower than both the corresponding (η-paired) states of Yang and the usual BCS states.

In order to prove this intriguing feature, we note first that the grand-canonical hamiltonian H_{Hub} has the form $H_{Hub} = H_{el}^{(loc)} + H_{el}^{(hop)}$. The first factor represents the local contributions to the electrons energy:

$$H_{el}^{(loc)} = \sum_{\mathbf{j}\in\Lambda} (-\mu(n_{\mathbf{j},\Uparrow} + n_{\mathbf{j},\Downarrow}) + U\, n_{\mathbf{j},\Uparrow} n_{\mathbf{j},\Downarrow}) \tag{10}$$

whereas the second is the non-local kinetic part of the energy:

$$H_{el}^{(hop)} = -t \sum_{<\mathbf{j},\mathbf{k}>} \sum_{\sigma\in\{\Uparrow,\Downarrow\}} \left(a_{\mathbf{j},\sigma}^{\dagger} a_{\mathbf{k},\sigma} + a_{\mathbf{k},\sigma}^{\dagger} a_{\mathbf{j},\sigma} \right) \tag{11}$$

We shall focus our attention on the amplitude t. This parameter (actually $t_{\mathbf{i},\mathbf{j}}$, assumed to be bond-independent in view of the system's translational invariance) is the hopping integral of the band electrons, given by: [2]

$$t = \int d\mathbf{r}\, \phi^*(\mathbf{r} - R_{\mathbf{j}}) \left[-\frac{\hbar^2}{2m} \Delta + \mathcal{V}(\mathbf{r}, \{R_{\mathbf{l}}\}) \right] \phi(\mathbf{r} - R_{\mathbf{k}}) \tag{12}$$

where $\phi(\mathbf{r} - R_{\mathbf{j}})$ is the atomic wave function for an electron of mass m with respect to the ion at $R_{\mathbf{j}}$ in Λ and $\mathcal{V}(\mathbf{r}, \{R_{\mathbf{l}}\})$, the single electron potential.

By inclusion of phonons, treated here simply as an ensemble of independent Einstein oscillators [17] with frequency ω, describing the ionic vibrations, the Hubbard hamiltonian becomes:

$$H = H_{el}^{(loc)} + H_{ph} + H_{el-ph}^{(hop)} \tag{13}$$

where:

$$H_{ph} = \sum_{\mathbf{j}} \left(\frac{p_{\mathbf{j}}^2}{2M} + \frac{1}{2} M \omega^2 x_{\mathbf{j}}^2 \right) \tag{14}$$

with M denoting the ionic mass and $x_{\mathbf{j}}$ and $p_{\mathbf{j}}$, the local ion displacement and momentum operators, respectively $\left([x_{\mathbf{j}}, p_{\mathbf{k}}] = i\hbar\delta_{\mathbf{j},\mathbf{k}}\right)$, commuting with the fermi operators. Once more:

$$H_{el-ph}^{(hop)} = \sum_{<\mathbf{j},\mathbf{k}>} \sum_{\sigma} \left(t_{\mathbf{j},\mathbf{k}}\, a_{\mathbf{k},\sigma}^{\dagger} a_{\mathbf{j},\sigma} + \text{h.c.} \right) \tag{15}$$

but with $t_{\mathbf{j},\mathbf{k}}$, given by the obvious generalization [18] of equation 12, taking into account the fact that the orbitals should now be centered at the displaced ion sites:

$$t_{\mathbf{j},\mathbf{k}} = \int d\mathbf{r} \phi^*(\mathbf{r} - R_{\mathbf{j}} - x_{\mathbf{j}}) \left[-\frac{\hbar^2}{2m} \Delta + \mathcal{V}(\mathbf{r}, \{R_{\mathbf{l}} - x_{\mathbf{l}}\}) \right] \phi(\mathbf{r} - R_{\mathbf{k}} - x_{\mathbf{k}} \tag{16}$$

and can no longer be assumed bond-independent.

The novel feature here is that the hopping amplitude, $t_{\mathbf{j},\mathbf{k}}$, defined by equation 16 is itself an operator, depending on the phonon degrees of freedom and commuting with fermionic operators.

The hopping hamiltonian equation 15 appears thus to be itself the sum of two terms: $H_{el-ph}^{(hop)} = H_{el-ph}^{(loc)} + H_{el-ph}^{(non-loc)}$. The former, due to the direct contribution of $\mathcal{V}(\mathbf{r},\{R_\mathbf{l}\})$ to equation 16, consists of purely local terms (of which, for the sake of simplicity, but with absolutely no loss of generality, we keep only the first [19, 20]):

$$H^{(loc)}_{el-ph} \approx -\lambda \sum_{\mathbf{j}} (n_{\mathbf{j},\Uparrow} + n_{\mathbf{j},\Downarrow})\, x_\mathbf{j}$$

which describes the coupling between electrons and ion dipoles induced by phonons, with strength:

$$\lambda_\mathbf{l} = -\int d\mathbf{r}\phi^*(\mathbf{r}-R_\mathbf{j}-x_\mathbf{j})\,(\nabla_{R_\mathbf{l}}\mathcal{V}(\mathbf{r},\{R_\mathbf{l}\}))\,\phi(\mathbf{r}-R_\mathbf{l}-x_\mathbf{l}) \approx \lambda$$

Note that $\lambda \neq 0$ dynamically breaks the $su(2)$, <u>not</u> the $su(\tilde{2})$ symmetry.

$H^{(non-loc)}_{el-ph}$ contains instead the contribution of the ion displacements to the hopping amplitude $t_{\mathbf{j},\mathbf{k}}$ induced in equation 16 by the Laplacian:

$$\begin{aligned} \tilde{t}_{\mathbf{j},\mathbf{k}} &= -\frac{\hbar^2}{2m}\int d\mathbf{r}\nabla\phi^*(\mathbf{r}-R_\mathbf{j}-x_\mathbf{j})\cdot\nabla\phi(\mathbf{r}-R_\mathbf{k}-x_\mathbf{k}) && (17)\\ &= t\,\exp\{\zeta\,(x_\mathbf{j}-x_\mathbf{k})\}\,\exp\{i\kappa\,(p_\mathbf{j}-p_\mathbf{k})\} && (18)\end{aligned}$$

In equation 18, the explicit exponential asymptotic form of $\phi(\mathbf{r}-R_\mathbf{k})$ from the ion core, was used at the *r.h.s.* The constants ζ and κ are real, depending on the single particle potential $\mathcal{V}(\mathbf{r},\{R_\mathbf{l}\})$ and on the interatomic distance $\delta = |R_\mathbf{j}-R_\mathbf{k}|$. The former simply gauges the exponential decay of the wave function $\phi(\mathbf{r})$ in the overlap region, from the ion cores (where the oscillatory behaviour of the wave-function, together with the exponentially small value of the *n.n.* orbital, give instead a negligible contribution to the integral). The factor, $(p_\mathbf{j}-p_\mathbf{k})$, obviously, accounts for the relative variation of the electron momentum with respect to that of the vibrating ions. One has thus:

$$H^{(non-loc)}_{el-ph} = t\sum_{<\mathbf{j},\mathbf{k}>}\sum_\sigma\Big(\exp\{\zeta\,(x_\mathbf{j}-x_\mathbf{k})\}\,\exp\{i\kappa\,(p_\mathbf{j}-p_\mathbf{k})\}\,a^\dagger_{\mathbf{k},\sigma}a_{\mathbf{j},\sigma} + \text{h.c.}\Big) \qquad (19)$$

Before discussing fully the quantum symmetry, to which this interaction leads, we consider a somewhat reduced problem – which is, however, very inspiring with respect to the more general question – dealing only with the hamiltonian first:

$$\tilde{H} = H_{ph} + H^{(loc)}_{el} + H^{(hop)}_{el} + H^{(loc)}_{el-ph} \qquad (20)$$

which one obtains by neglecting non-local electron-phonon terms. Only later on shall we explicitly take into account both of the contributions to $H^{(hop)}_{el-ph}$. It is interesting to point out that the hamiltonian in equation 20 coincides with the one considered *e.g.* in [21], to account for the role possibly played by phonons in giving rise to pairs (*bipolarons*) in superconductors.

The spectrum of equation 20 is typically investigated [21] by resorting to the so-called Lang-Firsov [22] transformation:

$$\mathcal{R} = \exp\left[i\xi\sum_\mathbf{j} p_\mathbf{j}(n_{\mathbf{j},\Uparrow} + n_{\mathbf{j},\Uparrow})\right]$$

which formally "rotates away", for a suitable choice of the parameter ξ, the local electron-phonon interaction term. $\mathcal{R}$ leaves the pure phononic part H_{ph} unchanged, while it leads to a new electronic effective hamiltonian H'_{el} of the form:

$$\begin{aligned} H'_{el} &= -\mu' \sum_{\mathbf{j}} (n_{\mathbf{j},\Uparrow} + n_{\mathbf{j},\Downarrow}) + U' \sum_{\mathbf{j}} n_{\mathbf{j},\Uparrow} n_{\mathbf{j},\Downarrow} \\ & \quad -t \sum_{<\mathbf{j},\mathbf{k}>,\sigma} \left\{ \exp\left[-i\xi(p_{\mathbf{j}} - p_{\mathbf{k}})\right] a^{\dagger}_{\mathbf{j},\sigma} a_{\mathbf{k},\sigma} + \text{h.c.} \right\} \end{aligned} \tag{21}$$

where $\mu' = \mu + \dfrac{\lambda^2}{2M\omega^2}$, $U' = U - \dfrac{\lambda^2}{M\omega^2}$ and ξ now has the value $\xi = \dfrac{\lambda}{\hbar M\omega^2}$ (note that $\xi \to 0$ when $\lambda \to 0$). The hopping operator in H'_{el} is customarily dealt with by solid state physicists by treating the phonon-dependent term within a mean-field scheme [21], namely by replacing $\exp[-i\xi(p_{\mathbf{j}} - p_{\mathbf{k}})]$ with e^{-g^2}, for some real g to be determined self-consistently. One may note that, even though such an approximation manifestly restores the full $SO(4)$ symmetry, in that it recovers a renormalized Hubbard model, it hides completely the dynamical effect of phonons.

On the other hand, upon defining the new nilpotent operators:

$$f_{\mathbf{j},\sigma} \doteq a_{\mathbf{j},\sigma} \exp\left(-i\xi \sum_{\mathbf{k}} \sum_{\sigma'} n_{\mathbf{j},\sigma'} p_{\mathbf{k}} \right)$$

the hamiltonian in equation 21, after setting $\nu_{\mathbf{j},\sigma} = f^{\dagger}_{\mathbf{j},\sigma} f_{\mathbf{j},\sigma}$ (the number operator for $f_{\mathbf{j},\sigma}$) reads:

$$\begin{aligned} H'_{el} &= -\mu' \sum_{\mathbf{j}} (\nu_{\mathbf{j},\Uparrow} + \nu_{\mathbf{j},\Downarrow}) + U' \sum_{\mathbf{j}} \nu_{\mathbf{j},\Uparrow} \nu_{\mathbf{j},\Downarrow} \\ & \quad -t \sum_{<\mathbf{j},\mathbf{k}>} \sum_{\sigma} \left(f^{\dagger}_{\mathbf{j},\sigma} f_{\mathbf{k},\sigma} + \text{h.c.} \right) \end{aligned}$$

This latter form for H'_{el} is suggestive, in that it has once more the same formal structure as the Hubbard hamiltonian, but with the operators $f_{\mathbf{j},\sigma}$ replacing the electron operators $a_{\mathbf{j},\sigma}$. The set $f_{\mathbf{j},\sigma}$ on the other hand, has the properties of fermions only in part. Indeed, the relations connecting them are the following:

$$\begin{aligned} f^{\dagger}_{\mathbf{j},\sigma} f_{\mathbf{k},\sigma'} + q_{\mathbf{j},\mathbf{k}}\, f_{\mathbf{k},\sigma'} f^{\dagger}_{\mathbf{j},\sigma} &= \delta_{\mathbf{j},\mathbf{k}} \delta_{\sigma,\sigma'} \\ f_{\mathbf{j},\sigma} f_{\mathbf{k},\sigma'} + q_{\mathbf{j},\mathbf{k}}\, f_{\mathbf{k},\sigma'} f_{\mathbf{j},\sigma} &= 0 \end{aligned} \tag{22}$$

where $q_{\mathbf{j},\mathbf{k}} = \exp[i\xi(p_{\mathbf{j}} - p_{\mathbf{k}})]$ is an operator (which, however, commutes with all values of $f_{\mathbf{i},\sigma}$ and can, therefore, be dealt with as a c-number). Relations of the type in equation 22 generate a fermionic version of the Gel'fand-Fairlie [23] quantum algebra and reduce to the usual anticommutation relations for fermions only in the limit $\xi \to 0$. It is intriguing that they reproduce the deformed commutation relations of anyonic oscillators [24] on a lattice, which coincide with those of fermions only for $\mathbf{j} = \mathbf{k}$ ($q_{\mathbf{j},\mathbf{j}} \equiv 1$).

On the basis of the above considerations, we conjectured that a global symmetry of the model, described by the complete H in equation 13, might be restored at the *quantum* level. We looked, therefore, for a global symmetry algebra of the form $su(\tilde{2}) \oplus [su(2)]_q$. This could indeed be done, by adopting a scheme essentially identical to that used by Yang and Zhang in [12], namely constructing first the local $[su(2)_{\mathbf{j}}]_q$, $\mathbf{j} \in \Lambda$, and requiring that each commutes with the $\mathbf{j}$-th term of the local hamiltonian:

$$H^{(loc)} = H_{ph} + H_{el}^{(loc)} + H_{el-ph}^{(loc)}$$

then extending these successively to a global $[su(2)]_q$ generating – if possible – a novel symmetry of the whole hamiltonian H. The complete proof of existence of such a global *quantum* symmetry, intended in the sense of deformation of the universal enveloping algebra of the symmetry algebra $so(4)$, dealt with as a full-fledged Hopf algebra, is given in references [25, 26]. We review it here, with only a few additional comments on the algebraic structure.

Dealing first with the local symmetry generators, we set:

$$\begin{aligned} K_{\mathbf{j}}^{(+)} &= e^{-i\mathbf{G}\cdot\mathbf{j}} e^{-i\phi p_{\mathbf{j}}} a_{\mathbf{j},\Uparrow}^{\dagger} a_{\mathbf{j},\Downarrow}^{\dagger}, \; K_{\mathbf{j}}^{(-)} = K_{\mathbf{j}}^{(+)\dagger} \text{ and} \\ K_{\mathbf{j}}^{(z)} &= \frac{1}{2}(n_{\mathbf{j},\Uparrow} + n_{\mathbf{j},\Downarrow} - 1) \end{aligned} \tag{23}$$

The three operators $K_{\mathbf{j}}^{(\gamma)}$, $\gamma = \pm, z$, generate a quantum $[su(2)]_q$ algebra as well [27]:

$$[K_{\mathbf{j}}^{(z)}, K_{\mathbf{j}}^{(\pm)}] = \pm K_{\mathbf{j}}^{(\pm)} \text{ and } [K_{\mathbf{j}}^{(+)}, K_{\mathbf{j}}^{(-)}] = \frac{\sinh\left(2\alpha K_{\mathbf{j}}^{(z)}\right)}{\sinh(\alpha)} \equiv \left[\left[2\alpha K_{\mathbf{j}}^{(z)}\right]\right]_q \tag{24}$$

with $q = e^{-\alpha}$, $\alpha \in \mathbb{C}$ arbitrary, as a $su(2)$ Lie algebra ($\alpha = 0$; in which case, the *r.h.s.* of equation 24 equals $2K_{\mathbf{j}}^{(z)}$).

We can immediately check that $H_{\mathbf{j}}^{(loc)}$ commutes with all $K_{\mathbf{j}}^{(\gamma)}$ $\gamma = \pm, z$, provided the following two relations are satisfied:

$$\mu = \frac{1}{2}\left(U - \frac{2\lambda^2}{M\omega^2}\right) \text{ and } \phi = \frac{2\lambda}{\hbar M\omega^2} \equiv 2\xi$$

(observe that, for $\lambda = 0$, the latter leads to the realization of the $su(2)$ of Yang and Zhang, whereas the former reduces to the condition of half-filling). Otherwise, the local hamiltonian commutes only with $K_{\mathbf{j}}^{(z)}$ and the local symmetry reduces to $u(1)$.

Extension of the above local algebras to the whole lattice, gives rise to a physically non-trivial symmetry of H only in the quantum algebra case.

If the system, described by hamiltonian H, is assumed to be endowed with a symmetry equipped with true Hopf algebraic structure, then this implies that the correct global extension of the local (whether quantum or not) algebras should be performed in terms of coproduct. As in the general Lie-Hopf case, for $[su(2)]_q$, the Cartan element $K^{(z)}$ is also primitive and its coproduct is simply given by: $\Delta(K^{(z)}) = \mathbb{1} \otimes K^{(z)} + K^{(z)} \otimes \mathbb{1}$, whereas for the off-diagonal elements:

$$\Delta(K^{(+)}) = e^{\alpha K^{(z)}} \otimes K^{(+)} + K^{(+)} \otimes e^{-\alpha^* K^{(z)}} \quad \text{and} \quad \Delta(K^{(-)}) = \left[\Delta(K^{(+)})\right]^{\dagger}$$

contrary to the non-deformed case, where they are primitive as well.

The extension to the whole Λ of the local $[su(2)]_q$ is then given by the iterated coproduct $\hat{K}^{(\gamma)} = \Delta^{(\nu)}(K^{(\gamma)})$, $\gamma = \pm, z$ (where $\Delta^{(\ell)} \equiv \Delta\left(\Delta^{(\ell-1)}\right)$, $\ell \geq 2$; $\Delta^{(1)} \equiv \Delta$. Explicitly:

$$\hat{K}^{(z)} = \sum_{\mathbf{j}} \mathbb{1} \otimes \cdots \mathbb{1} \otimes K^{(z)} \otimes \mathbb{1} \cdots \otimes \mathbb{1} = \sum_{\mathbf{j}} K_{\mathbf{j}}^{(z)} \tag{25}$$

$$\begin{aligned} \hat{K}^{(+)} &= \sum_{\mathbf{j}} e^{i\mathbf{G}\cdot\mathbf{j}} e^{\alpha K^{(z)}} \otimes \cdots \otimes e^{\alpha K^{(z)}} \otimes K^{(+)} \otimes e^{-\alpha^* K^{(z)}} \otimes \cdots \otimes e^{-\alpha^* K^{(z)}} \\ &= \sum_{\mathbf{j}} e^{i\mathbf{G}\cdot\mathbf{j}} \prod_{\mathbf{k}<\mathbf{j}} e^{\alpha K_{\mathbf{k}}^{(z)}} K_{\mathbf{j}}^{(+)} \prod_{\mathbf{k}>\mathbf{j}} e^{-\alpha^* K_{\mathbf{k}}^{(z)}} \quad ; \quad \hat{K}^{(-)} = \left[\hat{K}^{(+)}\right]^\dagger \end{aligned} \tag{26}$$

The tensor products in equation 25 [see equation 26] consist of N factors, the only one of which not equal to the identity $\mathbb{1}$ [or to the exponential of $K^{(z)}$] is at position $\mathbf{j}$. Moreover, the notation $\mathbf{k} > \mathbf{j}$ or $\mathbf{k} < \mathbf{j}$ refers to the same arbitrary order of the lattice sites adopted in the product.

The algebra $[su(2)]_q$, defined in equations 25 and 26, commutes with $H^{(loc)}$. In order for H to have the whole $su(\tilde{2}) \oplus [su(2)]_q$ global symmetry, it should be $\left[H_{el-ph}^{(non-loc)}, [su(2)]_q\right] = 0$. Due to equations 25 and 23, $\hat{K}^{(z)}$ always commutes with H. The q-symmetry then holds if $\left[H_{el-ph}^{(non-loc)}, \hat{K}^{(\pm)}\right] = 0$. Explicit calculation finally shows that this happens if the parameter α has a non-vanishing real part $\mathrm{Re}\,\alpha = \dfrac{2\zeta\kappa}{\hbar}$ and $\kappa = -\xi$. Note that $\zeta = 0$ implies $\alpha = 0$, thus the quantum symmetry can be ignored only if the effect of ion displacements on the hopping amplitude is neglected. It should be remarked, on the other hand, that setting $\zeta = 0$ would generate a hamiltonian which, by a simple unitary transformation (essentially the inverse of Lang-Firsov's), could be mapped into one describing uncorrelated phonons and electrons. It should, moreover, be emphasized that the conditions for q-symmetry are not very stringent in that they impose only two constraints among four physical parameters: λ, U, ω and n_0 ($\neq 1$, through μ), via the two atomic parameters ζ and κ. The parameter ζ alone has the role of fixing the quantum group deformation parameter q.

We consider first the family of η-paired states $|\psi_n>$, the analog of Yang's paired states [14], of the form:

$$|\psi_n> = \frac{1}{\sqrt{\mathcal{N}_\eta}} \left(\eta^\dagger\right)^n |\mathrm{vac}>$$

with $\mathcal{N}_\eta{}^{-1} = \dfrac{n!\,N!}{(N-n)!}$, but with the operator η, depending on the set of variational parameters w and $\{\vartheta_{\mathbf{j}}\}$, given by:

$$\eta^\dagger = \sum_{\mathbf{j}} e^{i\vartheta_{\mathbf{j}}} a_{\mathbf{j},\Uparrow}^\dagger a_{\mathbf{j},\Downarrow}^\dagger e^{iwp_{\mathbf{j}}} \text{ where } w, \vartheta_{\mathbf{j}} \in \mathbb{R}$$

Direct calculation shows that the state, among $|\psi_n>$, which minimizes the energy expectation $\mathcal{E}_n =< \psi_n|H|\psi_n>$ with respect to w, has $w = 2\xi\ (\equiv \phi)$ and the same energy expectation value as the q-symmetric eigenstates:

$$|\phi_n> = \frac{1}{\sqrt{\mathcal{N}}} \left(\hat{K}^{(+)}\right)^n |\text{vac}> \text{ of } H, \text{ with } \mathcal{N}^{-1} = \frac{[n]_q! [N]_q!}{[N-n]_q!} \tag{27}$$

Both Yang's η-paired state [14] and the BCS ground state (corresponding to $w = 0$ and $\vartheta_{\mathbf{j}} = 0\,, \forall \mathbf{j} \in \Lambda$) have, therefore, energy $\mathcal{E}_n + \frac{2\lambda^2}{M\omega^2}$ larger than $\mathcal{E}_n$.

Moreover, if the symmetry does not hold (for example if the filling has a different value than that required by the condition for μ), yet $\left[H, \hat{K}^{(\pm)}\right] = \delta\mathcal{E}\, \hat{K}^{(\pm)}$, with $\delta\mathcal{E} = U - 2\mu - 4\frac{\lambda^2}{M\omega^2}$, and starting from any eigenstate $|\mathcal{E}>$ of H one can construct a sequence of eigenstates in the (unnormalized) form:

$$[\hat{K}^{(\pm)}]^n |\mathcal{E}>,\ n \leq N$$

with energy eigenvalues:

$$\mathcal{E}'_n = \mathcal{E} + n\delta\mathcal{E}(< \mathcal{E}$$

provided that $\mu + 2\frac{\lambda^2}{M\omega^2} > \frac{1}{2}U$). It is very interesting that this condition is the same that guarantees U_{eff} negative in the approach of Ranninger *et al.* of reference [17].

Finally, as in the case analyzed by Yang, the state $|\phi_n>$ exhibits ODLRO [28], in that, for large $(\mathcal{O}(N))|\mathbf{r} - \mathbf{s}|$, the expectation value $< \phi_n | a^\dagger_{\mathbf{s},\Uparrow} a^\dagger_{\mathbf{s},\Downarrow} a_{\mathbf{r},\Downarrow} a_{\mathbf{r},\Uparrow} | \phi_n >$ equals:

$$e^{i\mathbf{G}\cdot(\mathbf{r}-\mathbf{s})} e^{\alpha\left(N+1-|\mathbf{r}+\mathbf{s}|\right)} \frac{\left[\!\left[{N \atop n-1} \right]\!\right]_q}{\left[\!\left[{N \atop n} \right]\!\right]_q} \neq 0$$

The states $|\phi_n>$ manifestly have *pairing*. The degenerate eigenstates $|\phi_n>$ of the hamiltonian in equation 13, when the q-symmetry condition is satisfied by the parameters, have energy lower than η-paired and *BCS* states (which are not eigenstates of H). Switching off the phonon field thus raises the paired states energy indeed.

2. Itinerant Interacting Electrons on Non-Euclidean Lattices

In one dimension, the algebra, generated by relations of the type in equations 7 and 8, is a sub-algebra of the more extended structure realized by the operators:

$$\begin{aligned} \mathcal{E}_j^{(n)} &\equiv a^\dagger_{j,\Uparrow} a_{j+n,\Downarrow} \\ \mathcal{F}_j^{(n)} &\equiv a^\dagger_{j,\Downarrow} a_{j+n,\Uparrow} \\ \mathcal{H}_j^{(n)} &\equiv a^\dagger_{j,\Uparrow} a_{j+n,\Uparrow} - a^\dagger_{j,\Downarrow} a_{j+n,\Downarrow} \end{aligned} \tag{28}$$

together with those obtained from these by the canonical transformation in equation 9 – that we shall denote here by a *tilde*. We can immediately check that:

$$J_+ \equiv \mathcal{E}_0 \equiv \sum_j \mathcal{E}_j^{(0)}, \quad J_- \equiv \mathcal{F}_0 \equiv \sum_j \mathcal{F}_j^{(0)} \quad \text{and} \quad J_z \equiv 2\mathcal{H}_0 \equiv \sum_j \mathcal{H}_j^{(0)}$$

and:

$$K_+ \equiv \tilde{\mathcal{E}}_0 \equiv \sum_j \tilde{\mathcal{E}}_j^{(0)}, \quad K_- \equiv \tilde{\mathcal{F}}_0 \equiv \sum_j \tilde{\mathcal{F}}_j^{(0)} \quad \text{and} \quad K_z \equiv 2\tilde{\mathcal{H}}_0 \equiv \sum_j \tilde{\mathcal{H}}_j^{(0)}$$

It was shown by Uglov and Korepin [29] that the set of six operators $\{\mathcal{E}_\lambda, \mathcal{F}_\lambda, \mathcal{H}_\lambda\}_{\lambda=0,1}$, with:

$$\begin{aligned}
\mathcal{E}_1 &\equiv \sum_j \left(\mathcal{E}_j^{(1)} - \mathcal{E}_j^{(-1)}\right) - \frac{U}{2t}\sum_{i<j}\left(\mathcal{E}_i^{(0)}\mathcal{H}_j^{(0)} - \mathcal{E}_j^{(0)}\mathcal{H}_i^{(0)}\right) \\
\mathcal{F}_1 &\equiv \sum_j \left(\mathcal{F}_j^{(1)} - \mathcal{F}_j^{(-1)}\right) + \frac{U}{2t}\sum_{i<j}\left(\mathcal{F}_i^{(0)}\mathcal{H}_j^{(0)} - \mathcal{F}_j^{(0)}\mathcal{H}_i^{(0)}\right) \\
\mathcal{H}_1 &\equiv \frac{1}{2}\left[\sum_j \left(\mathcal{H}_j^{(1)} - \mathcal{H}_j^{(1)}\right) + \frac{U}{t}\sum_{i<j}\left(\mathcal{E}_i^{(0)}\mathcal{F}_j^{(0)} - \mathcal{E}_j^{(0)}\mathcal{F}_i^{(0)}\right)\right]
\end{aligned} \tag{29}$$

provide a faithful representation of the $s\ell(2)$-Yangian [30], with deformation parameter $\varepsilon = \left(\frac{U}{t}\right)^2$:

- $\{\mathcal{E}_0, \mathcal{F}_0, \mathcal{H}_0\}$ generate $s\ell(2)$:

$$[\mathcal{E}_0, \mathcal{F}_0] = 2\mathcal{H}_0 \;, \; [\mathcal{H}_0, \mathcal{E}_0] = \mathcal{E}_0 \;, \; [\mathcal{H}_0, \mathcal{F}_0] = -\mathcal{F}_0 \;; \tag{30}$$

- $\{\mathcal{E}_1, \mathcal{F}_1, \mathcal{H}_1\}$ form a vector representation of equation 30:

$$\begin{array}{lll}
[\mathcal{E}_0, \mathcal{F}_1] = 2\mathcal{H}_1, & [\mathcal{F}_0, \mathcal{E}_1] = -2\mathcal{H}_1, & [\mathcal{H}_0, \mathcal{E}_1] = \mathcal{E}_1, \\
[\mathcal{E}_0, \mathcal{H}_1] = -\mathcal{E}_1, & [\mathcal{F}_0, \mathcal{H}_1] = \mathcal{F}_1, & [\mathcal{H}_0, \mathcal{F}_1] = -\mathcal{F}_1, \\
[\mathcal{E}_0, \mathcal{E}_1] = 0, & [\mathcal{F}_0, \mathcal{F}_1] = 0, & [\mathcal{H}_0, \mathcal{H}_1] = 0;
\end{array}$$

- the deformed Serre relations hold:

$$\begin{aligned}
[\mathcal{H}_1, [\mathcal{E}_1, \mathcal{F}_1]] &= \varepsilon\left(\{\mathcal{H}_0|\mathcal{E}_0|\mathcal{F}_1\} - \{\mathcal{H}_0|\mathcal{F}_0|\mathcal{E}_1\}\right) \\
[\mathcal{E}_1, [\mathcal{H}_1, \mathcal{E}_1]] &= \varepsilon\left(\{\mathcal{E}_0|\mathcal{H}_0|\mathcal{E}_1\} - \{\mathcal{E}_0|\mathcal{E}_0|\mathcal{H}_1\}\right) \\
[\mathcal{F}_1, [\mathcal{H}_1, \mathcal{F}_1]] &= \varepsilon\left(\{\mathcal{F}_0|\mathcal{H}_0|\mathcal{F}_1\} - \{\mathcal{F}_0|\mathcal{F}_0|\mathcal{H}_1\}\right) \\
2[\mathcal{H}_1, [\mathcal{H}_1, \mathcal{E}_1]] + [\mathcal{E}_1, [\mathcal{E}_1, \mathcal{F}_1]] &= \varepsilon\Big(2\{\mathcal{H}_0|\mathcal{H}_0|\mathcal{E}_1\} - 2\{\mathcal{H}_0|\mathcal{E}_0|\mathcal{H}_1\} \\
&\quad + \{\mathcal{E}_0|\mathcal{E}_0|\mathcal{F}_1\} - \{\mathcal{E}_0|\mathcal{F}_0|\mathcal{E}_1\}\Big) \\
2[\mathcal{H}_1, [\mathcal{H}_1, \mathcal{F}_1]] + [\mathcal{F}_1, [\mathcal{F}_1, \mathcal{E}_1]] &= \varepsilon\Big(2\{\mathcal{H}_0|\mathcal{H}_0|\mathcal{F}_1\} - 2\{\mathcal{H}_0|\mathcal{F}_0|\mathcal{H}_1\} \\
&\quad + \{\mathcal{F}_0|\mathcal{F}_0|\mathcal{E}_1\} - \{\mathcal{F}_0|\mathcal{E}_0|\mathcal{F}_1\}\Big)
\end{aligned}$$

where we have introduced the curly bracket symbol to denote the fully symmetrized product: $\{\mathcal{O}_1 \| \mathcal{O}_2 \| \mathcal{O}_3\} \doteq \frac{1}{3!} \sum_{\pi \in \mathcal{P}} \mathcal{O}_{\pi(1)} \mathcal{O}_{\pi(2)} \mathcal{O}_{\pi(3)}$, $\mathcal{P}$ indicating the set of all inequivalent permutations of the three operators $\mathcal{O}_\alpha$, $\alpha = 1,2,3$. The generators with tilde satisfy exactly the same relations and commute with those without tilde.

It is an interesting feature of the Hubbard model at half – filling in 1-D:

$$\begin{aligned} H_{Hub}^{(\mathrm{h.f.})} &= -t \sum_j \sum_{\sigma=\Uparrow,\Downarrow} \left(a_{j,\sigma}^\dagger a_{j+1,\sigma} + a_{j+1,\sigma}^\dagger a_{j,\sigma} \right) \\ &\quad + U \sum_j \left(n_{j,\Uparrow} - \tfrac{1}{2} \right) \left(n_{j,\Downarrow} - \tfrac{1}{2} \right) \end{aligned} \tag{31}$$

that $H_{Hub}^{(\mathrm{h.f.})}$ commutes with all the six×two operators generating the Yangian symmetry $\{\mathcal{E}_\lambda, \mathcal{F}_\lambda, \mathcal{H}_\lambda, \tilde{\mathcal{E}}_\lambda, \tilde{\mathcal{F}}_\lambda, \mathcal{H}_\lambda \,|\, \lambda = 0, 1\}$. Thus, at least in one dimension, a *quantum* global symmetry characterizes the Hubbard model, even in the absence of phonons.

Such quantum symmetry survives and plays a particularly interesting role when the Hubbard model is defined over *finite* lattices whose generating group of translations is *not abelian*. Before discussing the Yangian quantum symmetry of these systems, we briefly review some basic notions concerning such lattices (it is a slight abuse of language to refer to these structures as lattices, because they are finite: for example, the *fullerene*, that belongs to this family, and of which it is the forefather, is in fact a molecule, not a lattice. However, provided no confusion arises, we shall adopt that term in all cases).

It is the recent discovery of superconductivity in the *fullerenes* [31, 32], in particular in doped C_{60}, together with the growing interest in the physical properties of this new allotropic form of carbon that has revived the interest in studying both the dynamical properties – electronic and vibrational – of systems defined over *non-Euclidean* lattices [33] and the possible existence of novel crystallographic structures with non-abelian symmetry.

Among the non-Euclidean lattices, the truncated icosahedron, characteristic of C_{60} is the first to play a relevant role. The main feature of such a lattice, common, incidentally, to all lattices of the hierarchy to be described in the following, is that it is both a finite, finitely presented group (more precisely, the Sylvester graph of a group) and a manifold (namely, a discrete collection of points embedded in a differentiable variety). In the case of C_{60}, the group will be denoted here as $\mathcal{G}_{60}$ and the manifold as Σ_{60} (a truncated icosahedron embedded in the sphere $S^{(2)}$). The former property means that the points of Σ_{60} are in one-to-one correspondence with the elements of $\mathcal{G}_{60}$ and can be obtained from each other, by discrete transformations of $S^{(2)}$, generated by "*words*" in $\mathcal{G}_{60}$(forming a discrete subgroup of the continuous group of automorphisms of $S^{(2)}$ itself, isomorphic with the group of automorphisms of Σ_{60}, *i.e.* a discrete subgroup of $SU(2)$).

The fullerene structures can be generalized [33], to both open and closed lattices retaining the same high regularity (namely, they are Sylvester graphs of finite, finitely presented groups), yet having different topology. It is plausible that such lattices exist in nature, possibly in forms where the global symmetry is partially broken, or embedded in periodic or pseudo-periodic tubular structures, obtained by extending the generating group by a non-trivial homology. This has been recently confirmed by numerical simulations.

The icosahedral lattice can be thought of as an element of a hierarchy of non-Euclidean lattices identifiable with the Sylvester graphs of finitely presented, finite groups [34]. The notion of non-Euclidean lattices belongs to a generalized view of crystallography, defined over curved space, quite different from that familiar to physicists and crystallographers.

The concept of tesselation of non-Euclidean (non-flat) space has played an important role in mathematics ever since the pioneering work of Felix Klein [34, 35]. A tesselation is a covering of the ambient space, without gaps and overlaps, by *tiles*. In the Euclidean case, these are polytopes (polygons in 2-D) congruent with each other. The notion of congruence naturally relates the concept of tesselation to the existence of a group $\mathbf{T}$ of rigid motions, mapping the space onto itself. It is through such a group-theoretical view that tesselation can be defined over non-Euclidean space. The applications which have been relevant, so far, for the physics of non-Euclidean tesselations were related to the groups of motions referred to as *triangular elliptic groups* [36]. We shall show how *hyperbolic triangular groups* can also play a role in dealing with the new structures of the fullerene family.

The *triangle groups* have reflections along the sides of triangular plaquettes as generators. Such plaquettes are the elementary tiles whereby any tesselation of the non-Euclidean plane can be constructed.

First, one associates a triangle, Δ, with angles π/l, π/m and π/n with any triple of integers ≥ 2, say $\{l, m and n\}$. Clearly, if $\delta = (1/l + 1/m + 1/n) - 1$ is zero, the triangle is Euclidean, whereas it is spherical for $\delta > 0$ and hyperbolic if $\delta < 0$. We denote the operations of reflection along the side of Δ opposite to the angle π/l, π/m and π/n by L, M and N, respectively. Note that while the triangle sides are straight lines in the Euclidean case, they are *geodesics* if $\delta \neq 0$. The triangle group $\mathbf{T}^{\#}(l,m,n)$ is the group generated by L, M and N.

The subgroup $\mathbf{T}(l,m,n)$ of $\mathbf{T}^{\#}(l,m,n)$ consists of the orientation preserving motions in $\mathbf{T}^{\#}$. The symbol, Δ, denotes the *basic triangle* of $\mathbf{T}^{\#}$ and it is the canonical fundamental region of both $\mathbf{T}^{\#}$ and $\mathbf{T}$. *Canonical* refers, here, to the property that every tile of the corresponding tesselation is the image of a particular fixed tile under the action of exactly one element of $\mathcal{G}$. We shall discuss, in the following, two lattices, related respectively with $\mathbf{T}(2,3,5)$ ($\mathcal{G}_{60}$) and $\mathbf{T}(2,3,7)$ ($\mathcal{G}_{168}$).

The triple $\{2,3,5\}$ gives *deficit* $\delta = \dfrac{1}{30} > 0$ and $\mathbf{T}^{\#}(2,3,5)$ is, therefore, an instance of spherical triangle group. It was shown by Klein that its subgroup $\mathbf{T}(2,3,5)$ consists of all orientation preserving rotations of the sphere $S^{(2)}$ which carry the icosahedron inscribed in $S^{(2)}$ into itself. It follows that $\mathbf{T}(2,3,5)$ has order 60, is isomorphic with the alternating group over 5 symbols $\mathcal{A}_5$ and is generated by rotations u and v, defined by $u \doteq LM$, $v \doteq LN$, $uv \doteq MN$, which satisfy $u^5 = v^2 = (uv)^3 = id$.

Analogously, for $\{2,3,7\}$, the deficit is $\delta = -\dfrac{1}{42} < 0$ and the corresponding triangle group $\mathbf{T}^{\#}(2,3,7)$ is hyperbolic. $\mathbf{T}^{\#}(2,3,7)$ is defined by the *local* relations $L^2 = M^2 = N^2 = id.$ and $(LM)^2 = (MN)^3 = (NL)^7 = id.$

The subgroup $\mathbf{T}(2,3,7)$, is generated by $u \doteq NL$, $v \doteq LM$, with relations $u^7 = v^2 = (uv)^3 = id$.

The above structure of finite, finitely presented group characteristics of the triangular groups, induces a connection with the theory of Riemann surfaces. The property one has to keep in mind, for this purpose, is that if $\mathcal{G}$ is a discontinuous group of non-Euclidean motions in two dimensions, with a canonical fundamental region D, and $\mathcal{H}$ is a subgroup of index j in $\mathcal{G}$, then the fundamental region of $\mathcal{H}$ consists of the disjoint union of j congruent replicas of D. $\mathbf{T}(2,3,7)$ has a normal subgroup of lowest index, whose index in $\mathbf{T}$ itself is 168, isomorphic with the fundamental group, Φ_3, of a Riemann surface of topological genus $g = 3$. The fundamental region D_3 of Φ_3 is, therefore, tesselated by 168 replicas of the triangular fundamental region of $\mathbf{T}(2,3,7)$. D_3 is, in the Poincaré disk [34] realization, the regular 14-gon in the shown in Fig. 1a. The corresponding Riemann surface of genus 3, which is closed and orientable, is obtained by identifying the sides of D_3 in pairs, according to the rule $(2\nu+1) \mapsto (2\nu+6)\,[\,mod\ 14]$, $\nu = 1,\ldots,14$ (Fig. 1b shows the step-by-step construction of the surface). The quotient $\mathbf{T}/\Phi_3$ is isomorphic with the simple group $PSL(2,\mathbb{Z}_7)$ (where $\mathbb{Z}_7 \equiv \mathbb{Z}/7\mathbb{Z}$ is the Galois field of order 7, *i.e.* the field of integers modulo 7).

The above scheme leads quite naturally to the following general construction, which allows us to deal in a unified way with both the elliptic and the hyperbolic cases.

The *modular group* $\mathcal{M}$ is isomorphic with $PSL(2,\mathbb{Z}) \sim SL(2,\mathbb{Z})/\pm\mathbb{I}$, where $SL(2,\mathbb{Z})$, the crystallographic group on the psuedosphere, is a discrete subgroup of $SU(1,1)$. $\mathcal{M}$ is the group of 2×2 matrices with integer entries and unit determinant:

$$\mathcal{M} = \left\{ M = \begin{pmatrix} a & b \\ c & d \end{pmatrix} \middle| a,b,c,d \in \mathbb{Z}; ad - bc = 1 \right\}$$

($\mathbb{Z}$ denotes the ring of integers), acting as a group of transformations on the Lobachevskiĭ plane, of which the Poincaré disk is a conformal image:

$$\mathcal{L} = \left\{ z = x + iy \middle| z \in \mathbb{C}, y > 0, ds^2 = \frac{dx^2 + dy^2}{y^2} \right\}$$

via fractional linear (Möbius) transformations (which we denote by the same symbol):

$$M = \begin{pmatrix} a & b \\ c & d \end{pmatrix} \Longleftrightarrow M \circ z = \frac{az+b}{cz+d}$$

$\mathcal{M}$ is obviously discrete and therefore, discontinuous in $\mathcal{L}$, which is mapped onto itself by its elements. Therefore, $\mathcal{M}$ can also be thought of as the group with *presentation* [37, 38]:

$$\mathcal{M} \sim < U, V | V^2, (UV)^3 > \tag{32}$$

where the generators:

$$V = \begin{pmatrix} 0 & 1 \\ -1 & 0 \end{pmatrix} \quad \text{and} \quad U = \begin{pmatrix} 1 & 1 \\ 0 & 1 \end{pmatrix}$$

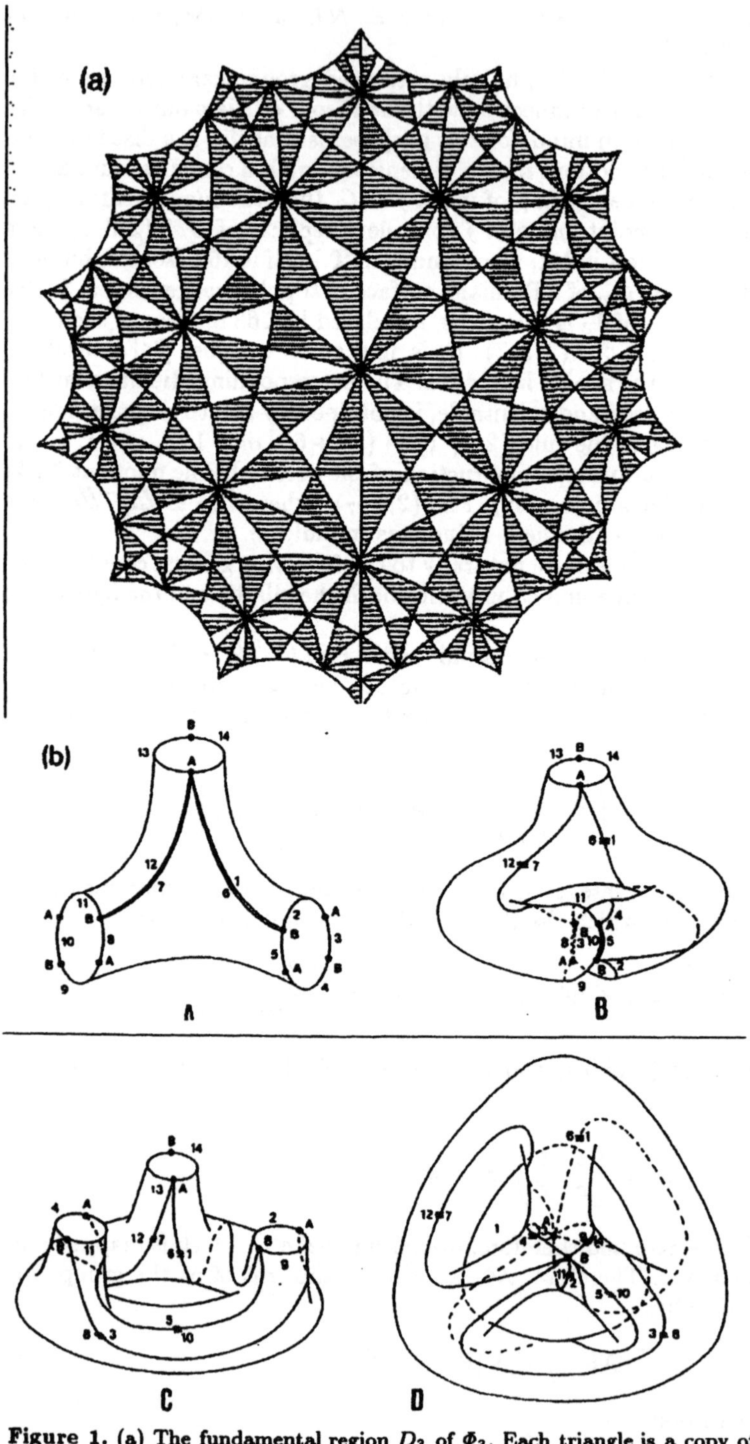

Figure 1. (a) The fundamental region D_3 of Φ_3. Each triangle is a copy of the fundamental region of $\mathbf{T}(2,3,7)$; (b) Step-by-step folding of D_3 in Γ_3. The sides of D_3 are labelled successively counterclockwise from 1 to 14.

correspond, respectively, in the Möbius realization, to:

$$V : z \to -\frac{1}{z} \quad \text{and} \quad U : z \to z+1$$

$PSL(2,\mathbb{Z})$ is, therefore, isomorphic to the triangle group $\mathbf{T}(2,3,\infty)$ also. $\mathcal{M}$ has a principal congruence of invariant subgroups $\{\mathcal{M}_p \sim PSL(2,\mathbb{Z}_p) \,|\, p = \text{odd prime}\}$ defined by:

$$\mathcal{M}_p = \left\{ M_0 \in \mathcal{M} \middle| M_0 = \begin{pmatrix} \pm 1 & 0 \\ 0 & \pm 1 \end{pmatrix} (\text{mod}\, p) \right\}$$

One can define the factor between $\mathcal{M}$ and $\mathcal{M}_p$:

$$\mathcal{G}_\omega \doteq \mathcal{M}/\mathcal{M}_p \quad \text{which is} \left\{ \begin{array}{l} \text{a finite group of order } \omega, \\ \text{a Sylvester graph, } \Sigma_p \text{, embedded in a manifold } \Gamma_g \end{array} \right.$$

where:

$$\omega = \frac{1}{2} p\,(p^2 - 1)$$

is both the number of elements of $\mathcal{G}_\omega$ and the number of points of Σ_p, and:

$$g = \frac{1}{4!}(p+2)(p-3)(p-5)$$

is the topological *genus* of the differentiable manifold Γ_g. In other words, the Sylvester graph Σ_g is nothing but the orbit under $\mathcal{G}_\omega$ of an arbitrary point in the canonical fundamental region of the triangle group, $\mathbf{T}$, corresponding to the tesselation of Γ_g, induced by $\mathbf{T}^{\#}(2,3,p)$.

Recalling now Euler's theorem, which states that for any polyhedron with $\mathcal{V}$ vertices, $\mathcal{F}$ faces, $\mathcal{E}$ edges, embedded in a surface of topological genus g, the relation:

$$\mathcal{V} + \mathcal{F} - \mathcal{E} = 2(1-g)$$

holds; together with the second of relations in equation 32 – whereby Σ_g has to be made in part of hexagons – one can easily obtain, for given g and ω, the presentation of $\mathcal{G}_\omega$ from purely combinatorial considerations.

The case corresponding to $p = 5$, in this scheme, is the *icosahedral group* $\mathcal{G}_{60}$, whose graph, Σ_5, provides a regular non-abelian lattice (a truncated icosahedron) on the sphere $S^{(2)}$ ($g = 0$). For $\omega = 60$ and $g = 0$, due to the definition of $\mathcal{M}_5$, another type of plaquette is necessary, besides the hexagon: the pentagon; the total number of pentagons is 12, whereas there are 20 hexagons arranged in such a way that each pentagon is surrounded by 5 hexagons and each hexagon is adjacent to 3 pentagons and 3 hexagons, alternating. Based on equation 32, there follows that $\mathcal{G}_{60}$ has the presentation:

$$\mathcal{G}_{60} \sim < U, V \,|\, V^2, (UV)^3, U^5 > \tag{33}$$

Fig. 2 shows the lattice of C_{60}, Σ_5, embedded in $\mathbb{R}^3$, with the action of the generators explicitly indicated (it is obviously the relation $U^5 = \mathbb{I}$ which generates the pentagons). Such a structure was to be expected noticing that $\delta > 0$ implies

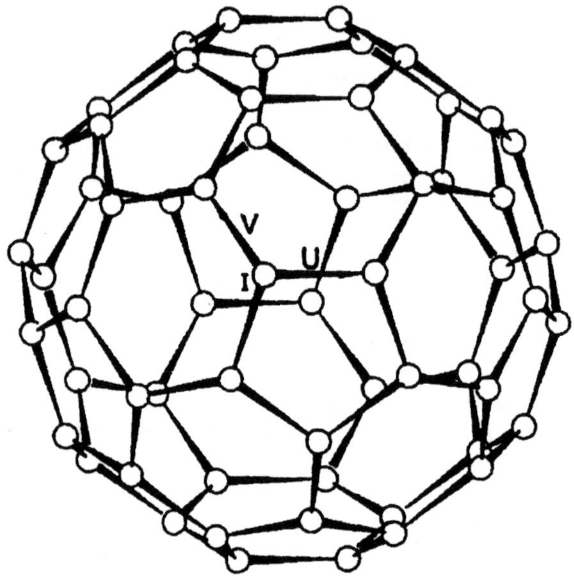

Figure 2. The graph Σ_5, characteristic of the fullerene C_{60}.

positive curvature, as may also be argued from observing that Γ_0 consists in the disjoint union of 30 adjoining pairs of spherical triangles of equal deficit, $\frac{1}{30}$, or of 12 spherical *caps* (each consisting of a pentagonal plaquette plus the 5 bonds emerging from each vertex), any one of which spans a solid angle $\frac{\pi}{3}$ and has positive curvature.

A similar construction for $\mathcal{G}_{168}$ with the application, once more, of the Euler's theorem, straightforwardly shows that the lattice Σ_7 has 24 eptagonal and 56 hexagonal plaquettes; each eptagon being surrounded by 7 hexagons. The presentation of $\mathcal{G}_{168}$ is, therefore:

$$\mathcal{G}_{168} \sim < U, V \mid V^2, (UV)^3, U^7, (VU^4)^4 > \qquad (34)$$

It is worth noticing that, as the group manifold has genus higher than 0, yet the group is finite, a new *global* relation appears, $(VU^4)^4 = \mathbb{I}$, which guarantees the closure of the homology group of Γ_3 (which is isomorphic with the surface shown in Fig. 1b). If one constructs the lattice – as in the case of $\mathcal{G}_{60}$ – from an elementary constituent, in this case an eptagonal plaquette with 7 extra bonds, emerging from each vertex, it exhibits negative curvature.

The lattice Σ_7 is obtained from the disk of Fig. 1a by selecting an arbitrary point and finding its orbit under the whole $\mathcal{G}_{168}$ [the orbit consists on one point in each of the 168 copies of the fundamental region of $\mathbf{T}(2,3,7)$], and then folding the resulting structure, as in Fig. 1b. Σ_7 cannot be isometrically and conformally embedded in $\mathbb{R}^3$, therefore, a real molecule with 168 carbon atoms with the full symmetry $\mathcal{G}_{168}$ is likely to be found in nature, only if the proximity-induced interactions generated when the surface Γ_3 is allowed to have self-intersections, preserve the topology of the set of points of Σ_7, yet leading to a physical configuration with minimal free energy. Nonetheless, the

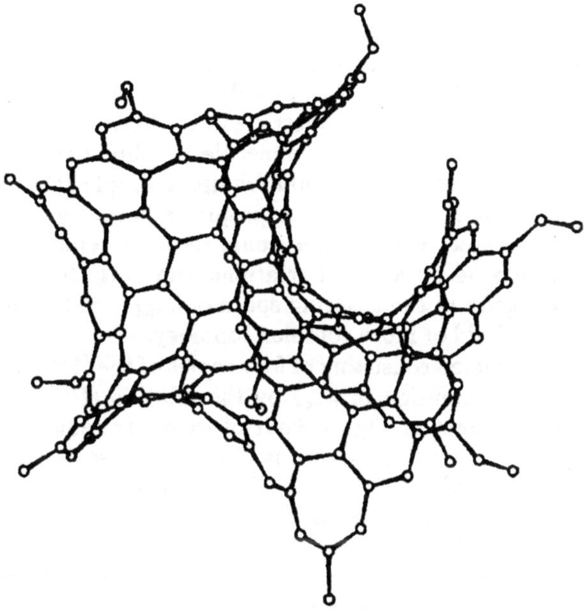

Figure 3. An orbit of $\mathcal{G}_{red}$ isometrically and conformally embedded in $\mathbb{R}^3$.

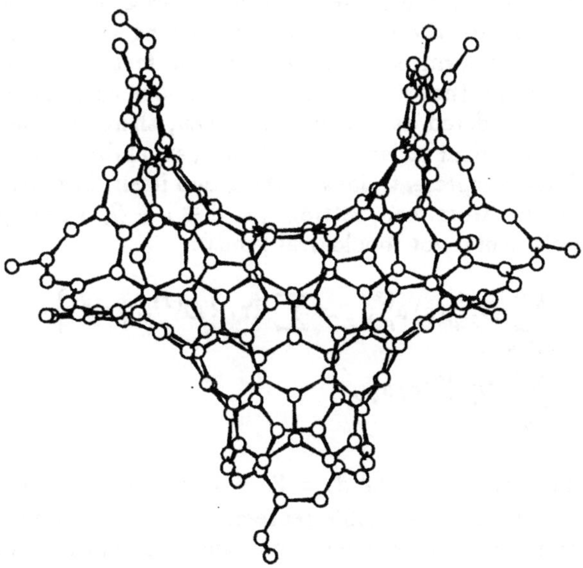

Figure 4. The graph of Fig. 3 after implementation of the identifications (A) in the scheme of Fig. 1b.

isometric and conformal embedding in $\mathbb{R}^3$ of the fundamental region D_3 of $\mathbf{T}(2,3,7)$ (*i.e.* the lattice generated by $\mathcal{G}_{red} =< U, V \mid V^2, (UV)^3, U^7 >$, with no identifications of the sides of D_3) has the shape shown in Fig. 3. Therefore, if one performs only the first step (A) of the folding of D_3 into Γ_3 (see Figs. 1a and 1b), generating the structure shown in Fig. 4, one can easily check that this shape can be obtained with a minimum amount of isometric and conformal distorsion. Thus, it appears plausible that, instead of single "molecules", tubular structures of the *schwartzite* type may exist. The latter would require a simple cobordism operation, inducing the extension of $\mathcal{G}_{red}$ by the appropriate homology group. The homological structure generated by the latter would result in replacing the last of the relations in equation 34 by a set of relations imposing the closure of an extended fundamental region in the covering space of $\mathcal{G}_{168}$ – consisting of several copies of D_3 – in a manifold of more complex topology.

For example, a fundamental region consisting of four copies of the (tesselated) polygon D_3 of Fig. 1a, and the extension of $\mathcal{G}_{red}$ by the relation $(U^3VU^2V)^3$ generate the basic structural element of Fig. 5. Four of these elements fill the primitive cell of a complex lattice structure shown in Fig. 6 – whose skeleton is a "superdiamond" three-dimensional lattice. The basic unit of such a structure can be thought of as being obtained from Σ_7 itself by identifying four cycles in $\mathcal{G}_{168}$ generated by the relations $(U^2V)^4$ and cutting Γ_3 along these cycles. The construction of the tubular infinite lattice is finally realized by glueing different units along the cycles themselves, which thus become generators of the homology of the new structure. Analysis of the lattice dynamics shows that this structure is more stable than the fullerene molecule in that the specific excess formation energy per atom, necessary to create it, is substantially lower than for C_{60}.

We return, once again to the Hubbard model, defining it over "lattices" Λ of the hierarchy Σ_p. In the case $p = 5$, solution of this model would be directly relevant to the electronic physics of the fullerene, while the case of $p = 7$ would provide the necessary insight into the problem of establishing the stability of complex tubular lattices of the form described above. The fact that Λ is the Sylvester graph Σ_p of $\mathcal{G}_\omega$ implies that its sites are in one-to-one correspondence with the elements of $\mathcal{G}_\omega$, considered as a group. One can, therefore, perform harmonic analysis on any dynamical variable defined on Λ, expanding it in terms of a complete set of irreducible representations of the group itself. In particular, the fermionic annihilation operators read, setting $\mathbf{j} \in \Lambda \equiv g \in \mathcal{G}_\omega$ (note: g will henceforth denote group elements, not topological genus):

$$\begin{aligned} a_{\mathbf{j},\sigma} \equiv a_{g,\sigma} &= \sum_{J;\lambda,\mu} D^{(J)}_{\lambda,\mu}(g)\, a^{(J)}_{\mu,\lambda;\sigma} = \sum_J \mathrm{Tr}_J \left\{ \mathbf{D}^{(J)}(g)\, \mathbf{a}^{(J)}_\sigma \right\} \\ a^{(J)}_{\lambda,\mu;\sigma} &= \sum_{g\in\mathcal{G}_\omega} D^{(J)*}_{\mu,\lambda}(g)\, a_{g,\sigma} \end{aligned} \tag{35}$$

where $\mathbf{D}^{(J)}(g)$ denotes the matrix (of rank $d_J = 2J + 1$ and elements $D^{(J)}_{\lambda,\mu}(g)$) representing g in the J-th unitary irreducible representation. One should note that the most convenient representation here is not that customarily adopted in crystallography, but the representation, obviously unitarily related to the standard one, naturally generated by the group presentation (for instance, in the case of $\mathcal{G}_{60}$, such representation is given in reference [39]). In equation 35, the sum over J ranges over all bosonic irreducible representations of $\mathcal{G}_\omega$, whereas indices λ and μ range from 1 to d_J. The set $\{D^{(J)}_{\lambda,\mu}(g)\}$ is normalized, in such a way that completeness and orthogonality relations read:

$$\sum_{g\in\mathcal{G}_\omega} D^{(J)*}_{\lambda,\mu}(g)D^{(J')}_{\lambda',\mu'}(g) = \delta_{J,J'}\delta_{\lambda,\lambda'}\delta_{\mu,\mu'} \quad \text{and} \quad \sum_{J;\lambda,\mu} D^{(J)*}_{\lambda,\mu}(g)D^{(J)}_{\lambda,\mu}(g') = \delta_{g,g'}$$

Due to equation 36, the transformed operators $a^{(J)}_{\lambda,\mu;\sigma}$ (matrix elements of $\mathbf{a}^{(J)}_\sigma$) are, of course, fermionic themselves:

$$\begin{aligned} \left\{a^{(J)}_{\lambda,\mu;\sigma}, a^{(J')\dagger}_{\lambda',\mu';\sigma'}\right\} &= \delta_{J,J'}\delta_{\lambda,\lambda'}\delta_{\mu,\mu'}\delta_{\sigma,\sigma'}, & \left\{a^{(J)}_{\lambda,\mu;\sigma}, a^{(J')}_{\lambda',\mu';\sigma'}\right\} &= 0 \\ n^{(J)}_{\lambda,\mu;\sigma} \equiv a^{(J)\dagger}_{\lambda,\mu;\sigma}a^{(J)}_{\lambda,\mu;\sigma} &\Longrightarrow & \left[n^{(J)}_{\lambda,\mu;\sigma}\right]^2 &= n^{(J)}_{\lambda,\mu;\sigma} \end{aligned} \tag{36}$$

$\mathcal{G}_\omega$ has a central extension $\mathcal{G}_{2\omega}$, whereby one could deal with the spin content of the fermionic operators as well.

Harmonic analysis, of course, does not, in general, lead to an exactly solvable model in dual space. Indeed the hamiltonian in the dual space variables $a^{(J)}_{\lambda,\mu;\sigma}$, has the same complexity in the interaction term as the hopping term has in ambient space. However, it is worth reporting the form of H in the group representation space, in that it has a suggestive, as well as intriguing, structure:

$$\begin{aligned} H &= \sum_J \mathrm{Tr}_J \left\{\left(-\mu\mathbb{I}^{(J)} - t\Delta^{(J)}\right)\left(\mathbf{X}^{(J)}_\Uparrow + \mathbf{X}^{(J)}_\Downarrow\right)\right\} \\ &\quad + U\sum_J \mathrm{Tr}_J \left\{\mathcal{J}[\{j_\alpha\}]\mathbf{a}^{(j_1)\dagger}_\Uparrow \mathbf{a}^{(j_2)}_\Uparrow \mathbf{a}^{(j_3)\dagger}_\Downarrow \mathbf{a}^{(j_4)}_\Downarrow\right\} \end{aligned} \tag{37}$$

where the rank-8 tensor $\mathcal{J}$ is defined, in terms of the 3-j symbols of $\mathcal{G}_\omega$ (these are known, at present, only for $\mathcal{G}_{60}$[40]), by:

$$\begin{aligned} \mathcal{J}_{\{\lambda_\beta,\mu_\gamma\}}[\{j_\alpha\}] &= \\ &= \begin{pmatrix} j_1 & j_2 & J \\ \lambda_1 & \lambda_2 & \lambda \end{pmatrix}^* \begin{pmatrix} j_1 & j_2 & J \\ \mu_1 & \mu_2 & \mu \end{pmatrix} \begin{pmatrix} j_3 & j_4 & J \\ \lambda_3 & \lambda_4 & \lambda \end{pmatrix}^* \begin{pmatrix} j_3 & j_4 & J \\ \mu_3 & \mu_4 & \mu \end{pmatrix} \end{aligned}$$

(here α, β and γ range from 1 to 4; λ_α, μ_α denote the matrix indices of $\mathbf{a}^{(j_\alpha)}_\sigma$ (or its conjugate) and sum over repeated indices is implied), while:

$$X^{(J)}_{\lambda,\mu;\sigma} \equiv \left(\mathbf{X}^{(J)}_\sigma\right)_{\lambda,\mu} = \sum_\eta a^{(J)\dagger}_{\eta,\lambda;\sigma}a^{(J)}_{\eta,\mu;\sigma}$$

and:

$$2\Delta^{(J)} = \mathbf{D}^{(J)}(V) + \mathbf{D}^{(J)}(V^{-1}) + \mathbf{D}^{(J)}(S) + h.c.$$

The form of the hamiltonian in equation 37, is the equivalent for the Sylvester graph lattice $\mathcal{G}_\omega$ of the customary Fourier transformed form of the hamiltonian (see equation 6), holding for Euclidean lattices. A promising feature of the non-abelian lattice structures, considered above, is that the construction starting with equation 28, valid there only for infinite 1-D lattices, can be extended to them. The two basic ingredients are the following. One defines first:

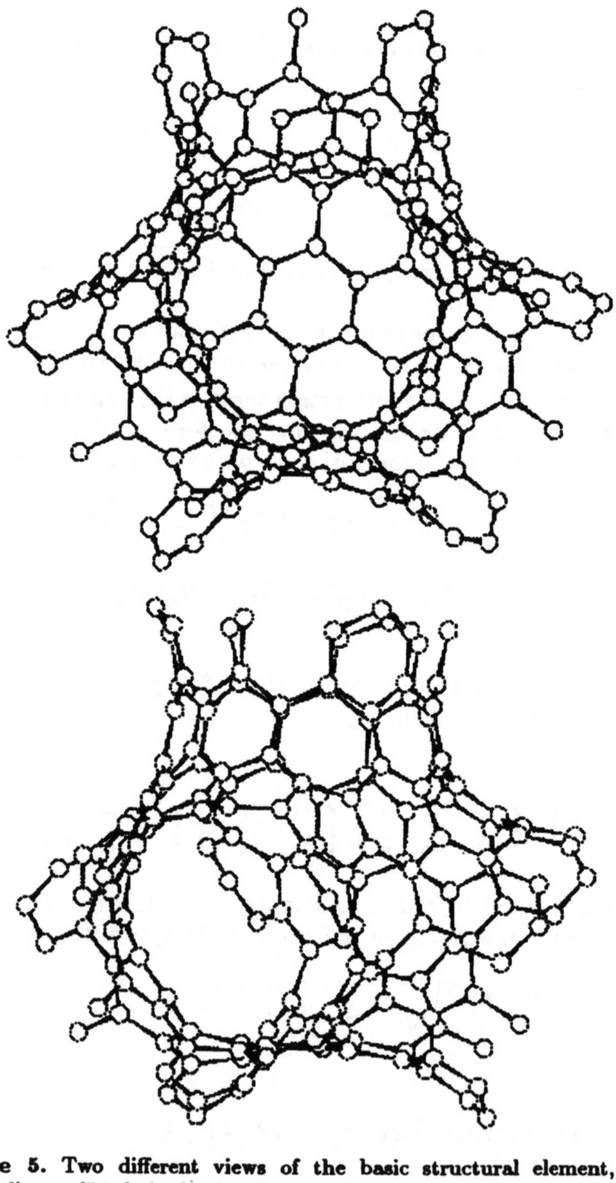

Figure 5. Two different views of the basic structural element, of order 168, of the "superdiamond" tubular lattice shown in Fig. 6.

$$
\begin{aligned}
\mathcal{E}_{\mathbf{j}}^{(\mathbf{n})} &\equiv \mathcal{E}_{g}^{(g_0)} \Rightarrow a^{\dagger}_{g,\Uparrow} a_{g_0 \circ g,\Downarrow} \\
\mathcal{F}_{\mathbf{j}}^{(\mathbf{n})} &\equiv \mathcal{F}_{g}^{(g_0)} \Rightarrow a^{\dagger}_{g,\Downarrow} a_{g_0 \circ g,\Uparrow} \\
\mathcal{H}_{\mathbf{j}}^{(\mathbf{n})} &\equiv \mathcal{H}_{g}^{(g_0)} \Rightarrow a^{\dagger}_{g,\Uparrow} a_{g_0 \circ g,\Uparrow} - a^{\dagger}_{g,\Downarrow} a_{g_0 \circ g,\Downarrow}
\end{aligned} \tag{38}
$$

where $g, g_0 \in \mathcal{G}_\omega$. The operators analogous to equation 29 can then be easily written in the space dual to $\mathcal{G}_\omega$ of harmonic transforms (see equation 35), in that one has:

$$
\begin{aligned}
Y_{g_0,\sigma} \equiv \sum_{g \in \mathcal{G}_\omega} a^{\dagger}_{g,\sigma} a_{g_0 \circ g,-\sigma} &= \sum_{J;\lambda,\mu} D^{(J)}_{\lambda,\mu}(g_0) \sum_{\nu} a^{(J)\dagger}_{\nu,\mu;\sigma} a^{(J)}_{\nu,\lambda;-\sigma} \\
&= \sum_{J} \mathrm{Tr}_J \left(\mathbf{D}^{(J)}(g_0) \mathbf{Y}^{(J)}_{\sigma} \right)
\end{aligned}
$$

where:

$$
\left(\mathbf{Y}^{(J)}_{\sigma} \right)_{\mu,\lambda} \equiv \sum_{\nu} a^{(J)\dagger}_{\nu,\mu;\sigma} a^{(J)}_{\nu,\lambda;-\sigma}
$$

Of course, an analogous relation holds for:

$$
\tilde{Y}_{g_0,\sigma} \equiv \sum_{g \in \mathcal{G}_\omega} a^{\dagger}_{g,\sigma} a^{\dagger}_{g_0 \circ g,-\sigma}
$$

Even though we do not report here the whole algebraic structure, which is much more complex than that corresponding to the *Yangian*, which holds in one dimension (see reference [41] for a detailed analysis of the complete set of relations and tensor representations pertaining to equation 38), we indicate that such algebraic structure is generated by relations of the type:

$$
\begin{aligned}
&[Y_{g_0,\sigma}, Y_{g_1,\sigma'}] = \\
&= \delta_{\sigma,-\sigma'} \textstyle\sum_J \mathrm{Tr}_J \left(\mathbf{D}^{(J)}(g_0) \mathbf{D}^{(J)}(g_1) \mathbf{X}^{(J)}_{\sigma} - \mathbf{D}^{(J)}(g_1) \mathbf{D}^{(J)}(g_0) \mathbf{X}^{(J)}_{-\sigma} \right)
\end{aligned}
$$

Once more a deformation of the Serre relations can be found, generated now not only by the fermionic nature of the representation adopted but also by the intrinsic curvature of the lattice, such that, at fixed filling, the Hubbard hamiltonian defined over $\mathcal{G}_\omega$ (see equation 37) commutes with the whole algebra generated by equation 38. This symmetry has a very high degree of complexity, yet the simplifications in the problem of finding the spectrum of H', to which it leads, are dramatic. To this effect, one should keep in mind that not only does the lattice group introduce partial block diagonalization of the hamiltonian (for example, in the case of $\mathcal{G}_{60}$ the existing representations have dimensions 2, 2, 4, 6, only, so that the maximum number of fermions one has to deal with is not 60, but 36), but that it is the stringent q-symmetry, induced by the generalized Yangian, which factorizes the Hilbert space of states into sectors of reasonably low dimension. This leads us to hope that an exact numerical diagonalization of the Hubbard model for the fullerene molecule may be soon achieved.

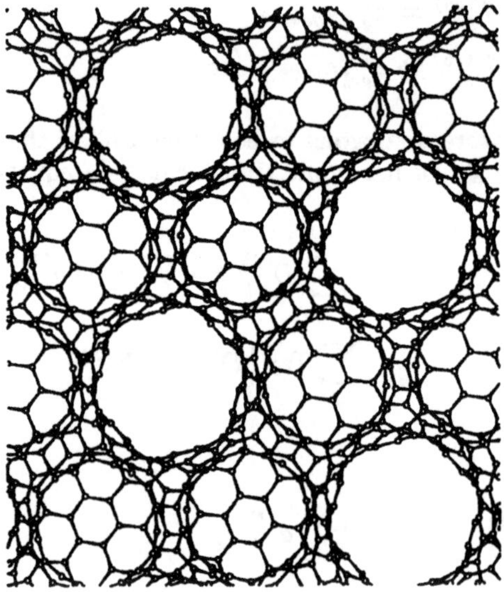

Figure 6. Two layers of the *superdiamond* tubular lattice.

3. Fermionic Linearization and Mott-Hubbard Transition

In this section, we shall show how the free-energy of the fermi – linearized Hubbard model in dimension $d \geq 2$, first expressed as a series expansion in the variational parameter of the theory, can be evaluated explicitly at the MIT point – assuming only that the transition is of the second-order, which is believed to be the case, at least for zero temperature [8, 9, 10] – in terms of 1-D correlation functions. This approach allows us, in principle, to discuss the phase space of the model at any temperature; however, for the sake of simplicity, we shall consider, in the last part of the paper, only the paramagnetic case at $T = 0$. In particular, we shall show that for $d \geq 2$ a transition may occur only at half-filling, in which case $U_c \approx 10$. The system entropy will be shown to play a subtle but extremely relevant role in the transition onset, which justifies the constraint it imposes on filling. When applied to the Hubbard hamiltonian, the *CFLS* amounts to replacing the hopping term between neighbouring clusters of sites, which is bilinear in fermionic operators, with a linear form in the same operators, whose coefficients are variables in a Clifford [42, 43] algebra, anticommuting with the fermionic operators. Explicitly, for a given covering of Λ with non-overlapping clusters Γ:

$$\sum_{<\mathbf{i},\mathbf{j}>} a^{\dagger}_{\mathbf{i},\sigma} a_{\mathbf{j},\sigma} \longrightarrow \sum_{\Gamma \subset \Lambda_n} \left\{ \sum_{<\mathbf{i},\mathbf{j}>\in\Gamma} a^{\dagger}_{\mathbf{i},\sigma} a_{\mathbf{j},\sigma} + \sum_{\mathbf{i}\in\partial\Gamma} \sqrt{q_{\mathbf{i}}} \left(a^{\dagger}_{\mathbf{i},\sigma} \eta_{\mathbf{i},\sigma} + \bar{\eta}_{\mathbf{i},\sigma} a_{\mathbf{i},\sigma} \right) \right\} \quad (39)$$

with $\partial\Gamma$ denoting the boundary of Γ and $q_{\mathbf{i}}$ the number of sites nearest neighbours of $\mathbf{i}$ not in Γ. In equation 39, we introduced the auxiliary fermionic mean field:

$$\eta_{\mathbf{i},\sigma} = \frac{1}{\sqrt{q_{\mathbf{i}}}} \sum_{\substack{\mathbf{j}=n.n.(\mathbf{i}) \\ \mathbf{j} \notin \Gamma}} < a_{\mathbf{j},\sigma} > \tag{40}$$

where $< a_{\mathbf{j},\sigma} >$ stands for equilibrium expectation value of $a_{\mathbf{j},\sigma}$.

The above scheme, in its simplest version, for which Γ coincides with a single site, reproduces [42] the Brinckmann-Rice [7] result for the metal-insulator transition at zero temperature in the Gutzwiller approximation, if the values of $\eta_{\mathbf{k}}$ are chosen to satisfy the fermion algebra, $\{\bar{\eta}_{\mathbf{i},\sigma}, \eta_{\mathbf{j},\sigma'}\} = c^2 \delta_{\mathbf{i},\mathbf{j}} \delta_{\sigma,\sigma'}$, $\{\eta_{\mathbf{i},\sigma}, \eta_{\mathbf{j},\sigma'}\} = 0$, as well as $\{\bar{\eta}_{\mathbf{i},\sigma}, a_{\mathbf{j},\sigma'}\} = 0 = \{\eta_{\mathbf{i},\sigma}, a_{\mathbf{j},\sigma'}\}$. The variational parameter c is to be determined by self-consistently implementing equation 40. In fact, c coincides with the discontinuity in the single-particle occupation number at the Fermi surface and hence, is a natural parameter for describing the MIT. Moreover, the approximation implied by assuming values of $\eta_{\mathbf{j},\sigma}$ to generate a Clifford algebra becomes exact in the limit of infinite dimension of the lattice [43] with the choice $c = 1$.

We shall now study the hamiltonian, given in equation 6, approximating the hopping term between neighbouring chains by equation 39, and keeping it unchanged within each single chain. In other words, we shall use a version of the *CFLS* where the clusters are whole single 1-D chains, $\Gamma \equiv \Lambda_1$. The linearized hamiltonian reads:

$$\begin{aligned} H &= \sum_{\{\Lambda_1 \subset \Lambda\}} \left(H_{Hub}^{(\Lambda_1)} - c H_c^{(\Lambda_1)} \right) \\ H_c^{(\Lambda_1)} &\doteq t' \sum_{j \in \Lambda_1} \sum_{\sigma} (a_{j,\sigma}^{\dagger} \theta_{j,\sigma} + \bar{\theta}_{j,\sigma} a_{j,\sigma}) \end{aligned} \tag{41}$$

where, assuming Λ cubic and therefore, $q_{\mathbf{j}} = 2(d-1)\ \forall \mathbf{j}$ and $t' = t\sqrt{2(d-1)}$. Moreover, we limit ourselves to the paramagnetic phases, setting $\eta_{j,\sigma} \equiv c\theta_{j,\sigma}$, $\forall j \in \Lambda_1$. The new variables, $\theta_{j,\sigma}$, now satisfy the (anti-) commutation relations:

$$\{\bar{\theta}_{j,\sigma}, \theta_{j,\sigma'}\} = \delta_{k,j} \delta_{\sigma,\sigma'} \quad \text{and} \quad \{\theta_{j,\sigma}, \theta_{j,\sigma'}\} = 0$$

The above approximation appears to be particularly well designed to characterize the metal-insulator transition by letting $c \to 0$. In fact, the vanishing of c implies that, to all effects, there is no hopping between different chains, so that the system becomes a sum of non-interacting linear chains with no external field and hence – at half-filling – an insulator. This is necessarily the case in $d = 1$, where $t' = 0$.

We shall assume also that the MIT is second-order, as suggested by numerical calculations, and – according to Landau theory – discuss the transition through the behavior of the free energy f (constrained by the self-consistency condition for c) around the critical point $c = 0$.

In order to evaluate f explicitly, we must first work out the partition function:

$$Z_d = Tr_d \left\{ \exp(-\beta H_{Hub}^{(\Lambda)}) \right\} \Longrightarrow \left(tr \left\{ \exp -\beta H^{(\Lambda_1)} \right\} \right)^{N_c}$$

β being, as usual, the inverse temperature and N_c the number of chains in Λ. The trace, Tr_d, is to be intended over the Fock space pertaining to the whole d-dimensional system, but, due to the approximation in equation 41, it reduces, in fact, to the evaluation of the trace tr over the single-chain Fock space, in which $H^{(A_1)}_{Hub}$ lives, and over the space spanned by the Clifford number-operators $\bar{\theta}_{j,\sigma}\theta_{j,\sigma}$ (assuming values 0, 1). We express Z_d as a series expansion in the variational parameter c, by the customary integral representation of the exponential of a sum of non commuting operators [44, 45]. Besides, from now on, we restrict for simplicity our attention to $d = 2$. Explicitly:

$$\begin{aligned} f &= -\frac{1}{\beta N}\ln Z \\ &= -\frac{1}{\beta N_1}\times \\ &\times \ln\left\{ tr\left[\exp(-\beta H^{(A_1)}_{Hub})\left(1+\sum_{n=1}^{\infty}(\beta c)^n\int_{\Delta_n}\prod_{j=1}^{n} du_j E_1(\beta)E_2(\beta)\dots E_n(\beta)\right)\right]\right\} \end{aligned}$$

with:

$$E_j(\beta) = \exp[\mathrm{ad}(-\beta u_j H^{(A_1)}_{Hub})](H^{(A_1)}_c), \quad N_1 = \frac{N}{N_c}$$

where $\Delta_n = \{\{u_j\}, j = 1,\dots,n \,|\, u_{k-1} \leq u_k \leq 1; k = 1,\dots,n; u_0 = 0\}$, ad denotes the adjoint action and we measure energy in units of t. Direct calculation shows that:

$$\begin{aligned} E_j(\beta) &= \sum_{k\in A_{1,\sigma}} \exp[-\beta u_j(U n_{k,-\sigma} - \mu)] \\ &\times\sum_{s=0}^{\infty} I_s(2\beta u_j)\sum_{\epsilon=\pm 1}\left(a^{\dagger}_{k+\epsilon s,\sigma}\theta_{k,\sigma} + \bar{\theta}_{k,\sigma}a_{k+\epsilon s,\sigma}\right) \end{aligned} \tag{42}$$

with $I_s(z)$ denoting the modified Bessel function of the first kind, of order s. Upon performing the trace over the Clifford variables, $\theta_{k,\sigma}$, one easily recognizes that, in fact, only the terms even in c give a non-vanishing contribution to the free energy. If the MIT is of second-order, the transition is determined by the vanishing of the coefficient $\alpha_2(\beta)$ of c^2 in f. We obtain, after some algebra and using the invariance of Tr under cyclic permutations of its arguments:

$$\alpha_2(\beta) = \frac{1}{\beta}\int_0^{\beta} dx(\beta - x)\sum_{j=-\infty}^{\infty} I_{|j|}(2tx)\left[P_j(x,\mu,U) - 4\delta_{j,0}e^{\mu x}\right] \tag{43}$$

where:

$$\begin{aligned} P_j(x,\mu,U) &= \left[e^{\mu x} - (-)^j e^{-\mu x}\right] C_j \\ &+\left[(e^{-Ux} - 1)e^{\mu x} - (-)^j(e^{Ux} - 1)e^{-\mu x}\right] T_j \end{aligned} \tag{44}$$

Here, denoting the thermodynamic expectation value of any operator, $\bullet$, for the one dimensional Hubbard model by:

$$
\begin{aligned}
<\bullet> &\equiv \frac{1}{Z_1} Tr_1 \left[\bullet \exp(-\beta H_{Hub}^{(A_1)})\right] \\
\mathcal{C}_j &= \left\langle \frac{1}{N_1} \sum_{i,\sigma} \left(a_{i+j,\sigma}^{\dagger} a_{i,\sigma} + a_{i,\sigma}^{\dagger} a_{i+j,\sigma} \right) \right\rangle
\end{aligned}
$$

is the correlation function for two points at distance j, whereas:

$$
\mathcal{T}_j = \left\langle \frac{1}{2N_1} \sum_{i,\sigma} (n_{i+j,-\sigma} + n_{i,-\sigma}) \left(a_{i+j,\sigma}^{\dagger} a_{i,\sigma} + a_{i,\sigma}^{\dagger} a_{i+j,\sigma} \right) \right\rangle
$$

The equation $\alpha_2(\beta) = 0$ – which provides in 2-D, for finite β, the transition critical temperature or, in the limit $\beta \to \infty$, the critical value U_c of U, above which the system is always an insulator – is thus reconducted to evaluating only properties (correlation functions) of the one-dimensional Hubbard model. It should be noted that the equations fixing the system filling and entropy (which will have to be used later on) also reduce to their 1-D version, when $c = 0$. If $T \neq 0$, correlations $\mathcal{C}_j$ and $\mathcal{T}_j$ are not presently known, however, the generalized Bethe Ansatz equations for finite temperature given by Takahashi [46] lead to a recursive scheme from which they can be obtained. Work is in progress along these lines. In particular, the critical equation for infinite temperature $\alpha_2(0) = 0$ can be easily seen to have no solutions.

In what follows, we shall consider, for the sake of simplicity, only the case $T \to 0$, for which the exact solution of Lieb and Wu [6] provides the free energy $f_{Hub}^{(A_1)}$. We shall consider, therefore, exactly those correlations, which can be obtained from such a solution, as derivatives of the free energy. The other correlations entering $\alpha_2(\infty)$ will in turn be approximated by their random phase value:

$$
\mathcal{C}_j = \begin{cases} \mathcal{O}(c) & \text{if } j > 1\,; \\ -\dfrac{\partial f_{Hub}^{(A_1)}}{\partial t} & \text{if } j = 1\,; \\ -2\dfrac{\partial f_{Hub}^{(A_1)}}{\partial \mu} & \text{if } j = 0\,; \end{cases} \qquad \mathcal{T}_j = \begin{cases} \mathcal{O}(c) & \text{if } j > 1\,; \\ n\mathcal{C}_1 & \text{if } j = 1\,; \\ 4\dfrac{\partial f_{Hub}^{(A_1)}}{\partial U} \equiv 4p & \text{if } j = 0\,; \end{cases} \tag{45}
$$

evaluated, of course, at $T = 0$. In equation 45, $j > 1$ is simply the random phase value obtained within the Clifford fermionic linearization, which gives no relevant contribution to α_2, because α_2 is itself the coefficient of the term of order c^2 in the Landau scheme. $\mathcal{T}_1$ is the Hartree random phase value, where n denotes the average electron number per site (*i.e.* the filling). Implementing the equation fixing the filling $-\partial f/\partial \mu = n$, one finds, at $c = 0$, $n = 1/2\mathcal{C}_0$. Moreover, p is the average number of doubly occupied sites.

We now insert equation 45 into equations 43-44 and study the resulting equation $\alpha_2(\infty) = 0$:

$$
\lim_{\beta \to \infty} \int_0^1 (1-x)\Big\{(n-2p)\sinh(\beta\mu x) - 2p\sinh\left(\beta(U-\mu)x\right) + \mathcal{C}_1(1-n)
$$

$$
\cosh(\beta\mu x) + n\mathcal{C}_1 \cosh\left(\beta(U-\mu)x\right) - \frac{1}{2}e^{\beta\mu x}\left[2 - n\left(1 - e^{-\beta U x}\right)\right]\Big\} \frac{e^{2\beta x}}{\sqrt{x}}\, dx = 0 \tag{46}
$$

where μ is to be thought of as given by the filling equation. Equation 46 shows that, depending on whether the value of μ is greater than, equal to or less than $\frac{1}{2}U$, one must analyze three different regimes, corresponding respectively to n greater than, equal to or less than 1. When $n < 1$, equation 46 becomes $p = \frac{1}{2}nC_1$. Resorting to the known results on the limiting behaviour for both $U \to 0$ and $U \to \infty$ of p and C_1 in 1-D [46, 47], one can check that equation 46 has no solution in U, because p turns out to be always $< \frac{1}{2}nC_1$. Analogously, one finds no solution for $n > 1$, in which case equation 46 becomes $p = \frac{1}{2}(n-1)(2-C_1)$ while p is now found to be larger than $\frac{1}{2}(n-1)(2-C_1)$ for any U. This suggests that there may exist a solution only at $\mu = \frac{1}{2}U$.

Indeed, at half-filling, equation 46 leads to a critical value U_c, as one can derive by the following analysis. To begin with, the equation for filling at $T = 0$ determines μ only up to an extra term, linear in T, $\mu \mapsto \mu_0 + sT$. In fact, if ε_k is the spectrum of the system hamiltonian, then:

$$n = \int D_k \frac{1}{1 + e^s e^{-\beta(\varepsilon_k - \mu_0)}} \xrightarrow[\beta\to\infty]{} \int_{\mathcal{D}} D_k$$

for any s. Here:

$$D_k = \frac{d^d k}{(2\pi)^d}$$

is the measure over the reciprocal space and $\mathcal{D}$ denotes the set $\{k|\varepsilon_k - \mu_0 < 0\}$). One can easily check that for $\mu_0 \neq \frac{U}{2}$ the factor s does not affect equation 46, whereas, at $\mu_0 = \frac{1}{2}U$, it changes equation 46 into:

$$2\left(1 + e^{2s}\right)p = 2(n-1)e^{2s} + C_1\left[e^{2s}(1-n) + n\right]$$

This latter equation is naturally to be considered at $n = 1$, where it reduces to:

$$2p\left(1 + e^{2s}\right) = C_1 \tag{47}$$

however, it is instructive to write it, as was done above in equation 47, for generic n, because that form shows explicitly how, for $s \to -\infty$ and $s \to \infty$, it formally reproduces the equations for $n <$ and > 1, respectively.

The important feature in equation 47 is that s is a function of U: in fact, it is simply the entropy at zero temperature. Actually, as the grand-canonical thermodynamic potential Ω now reads:

$$\Omega = -T \int D_k \ln\left(1 + e^s e^{-\beta(\varepsilon_k - \mu_0)}\right) \tag{48}$$

recalling that $S \equiv -\left(\partial\Omega/\partial T\right)_\mu$, one infers:

$$S|_{T=0} = \lim_{T\to 0} -\left(\frac{\partial\Omega}{\partial T}\right)_{\mu_0} = s \tag{49}$$

The entropy S, at $T = 0$, can be obtained resorting, once more, to the solution of Lieb and Wu [6],

$$S = -\left(\int_{-\pi}^{+\pi} dk\varrho(k)\ln\varrho(k) - \int_{-\pi}^{+\pi} dk\varrho_0(k)\ln\varrho_0(k)\right) - \left(\int_{-\infty}^{\infty} d\Lambda\sigma(\Lambda)\ln\sigma(\Lambda) - \int_{-\infty}^{\infty} d\Lambda\sigma_0(\Lambda)\ln\sigma_0(\Lambda)\right) \quad (50)$$

where:

$$\varrho(k) = \frac{1}{2\pi}\left\{1 + 2\cos k\int_0^\infty d\omega \frac{\cos(\omega\sin k)\,J_0(\omega)}{1+\exp\left(\frac{1}{2}\omega U\right)}\right\} \quad (51)$$

$$\sigma(\Lambda) = \frac{1}{2\pi}\int_0^\infty \operatorname{sech}\left(\frac{1}{2}\omega U\right)\cos(\omega\Lambda)J_0(\omega) \quad (52)$$

$$\rho_0(k) \equiv \frac{1}{\pi}\theta\left(\frac{\pi}{2} - |k|\right) \text{ and } \sigma_0(\Lambda) \equiv \frac{1}{2\pi}\frac{\theta(1-|\Lambda|)}{\sqrt{1-\Lambda^2}}$$

These last two equations denote $\rho(k)$ and $\sigma(\Lambda)$ evaluated at $U = 0$; $J_\nu(\omega)$ is the Bessel function of the first kind, of order ν. The entropy in equation 50 is defined so as to be zero at $T = 0$, for the non-interacting system ($U = 0$). Upon evaluating s, by equations 50-52 and recalling that:

$$p = \int_0^\infty d\omega \frac{J_0(\omega)J_1(\omega)}{1+\cosh\left(\frac{1}{2}\omega U\right)}$$

$$C_1 = \int_0^\infty d\omega J_0(\omega)J_1(\omega)\left[\frac{4}{\omega\left(1+e^{\frac{1}{2}\omega U}\right)} + U\frac{1}{1+\cosh\left(\frac{1}{2}\omega U\right)}\right]$$

equation 47 can be solved for U. In order to evaluate U_c, one could perform a purely numerical analysis. We estimate U_c ,instead, by resorting to the asymptotic form of the various factors in equation 47. Solution of the resulting equation gives:

$$U_c = \frac{\pi}{\pi - 2\sqrt{2}} \approx 10.03 \quad (53)$$

which is in remarkable agreement with the value $32/\pi \approx 10.18$ obtained by Brinkmann and Rice [7] and is also comparable with the result $U_c \approx 13.5$ in $d = \infty$ derived in [9, 10] and with the exact value $U_c = 8$ for certain simplified, extended Hubbard models [48, 49]. It is interesting to note that the system entropy, which plays a significant role in the onset of the transition, has to be high. This is not surprising, because the ground state is highly degenerate, when the system is an insulator, for any non-vanishing T.

Summarizing, there may exist a metal-insulator transition for the Hubbard model in $d \geq 2$. The equation derived for the critical point holds, in principle, for generic T and U, assuming that the transition is second-order. Even though so far we have solved it explicitly only at $T = 0$, in which case the critical value is $U_c \approx 10$, we conjecture that this value represents a good estimate of the critical U for the *MIT* transition in $d \geq 2$ in that, in the scheme proposed, c becomes zero in correspondence to such value, implying in two different ways that the metal ($c \neq 0$) becomes an insulator. On the one hand, it corresponds to the vanishing for $U = U_c$, of the discontinuity in the single-particle occupation number at the Fermi surface. On the other hand, when c is identically zero ($U > U_c$), the $d \geq 2$ model is faithfully mapped on the standard 1-D model, which is known to be insulating for any $U \neq 0$ at half-filling. The case $T \neq 0$, which is controlled by the same conceptual scheme, leads to more involved analytical and numerical calculations, but preliminary results of work now in progress [50] are very promising.

References

1. M. Gutzwiller, Phys. Rev. Lett. **10**, 159 (1963)
2. J. Hubbard, Proc. Roy. Soc. A **276**, 238 (1963)
3. *The Hubbard Model*: A Reprint Volume, A. Montorsi, ed., World Scientific Publ. Co., Singapore, 1992
4. J. Hubbard, Proc. Roy. Soc. A **281**, 401 (1964)
5. N.F. Mott, *Metal-Insulator Transition*, Taylor & Francis, London, 1990
6. E.H. Lieb and F.Y. Wu, Phys. Rev. Lett. **20**, 1445 (1968)
7. W.F. Brinkman and T.M. Rice, Phys. Rev. B **2**, 4302 (1970)
8. A. Georges and G. Kotliar, Phys. Rev. B **45**, 6479 (1992)
9. A. Georges and W. Krauth, Phys. Rev. Lett. **69**, 1240 (1992)
10. X.Y. Zhang, M.J. Rozenberg and G. Kotliar, Phys. Rev. Lett. **70**, 1666 (1993)
11. A. Montorsi, M. Rasetti and A.I. Solomon, Int. J. Mod. Phys. B **3**, 247 (1989)
12. C.N. Yang and S.C. Zhang, *Mod. Phys. Lett.* **B4**, 759 (1990)
13. O.J. Heilmann and E.H. Lieb, *Ann. New York Acad. Sci.* **172**, 583 (1971)
14. C.N. Yang, *Phys. Rev. Lett.* **63**, 2144 (1989)
15. F.H.L. Essler, V.E. Korepin and K. Schoutens, *Phys. Rev. Lett.* **67**, 3848 (1991); and *Nucl. Phys.* **384 B**, 431 (1992)
16. F.H.L. Essler, V.E. Korepin and K. Shoutens, *Phys. Rev. Lett.* **68**, 2960 (1992)
17. A.S. Alexandrov, J. Ranninger and S. Robaszkiewicz, *Phys. Rev.* **B 33**, 4526 (1986)
18. A.L. Kuzemsky, A. Holas and N.M. Plakida, *Physica* **122 B**, 168 (1983)
19. T. Holstein, *Ann. Phys.* (N.Y.) **8**, 325 (1959)
20. E. Hückel, *Z. Phys.* **70**, 204 (1931) and **76**, 628 (1932)
21. S. Robaszkiewicz, R. Micnas and J. Ranninger, *Phys. Rev.* **B 36**, 180 (1987)
22. L.G. Lang and Yu.A. Firsov, *Zh. Eksp. Teor. Fiz.* **43**, 1843 (1962) [*Sov. Phys. JETP* **16**, 1301 (1963)]
23. I.M. Gel'fand and D.B. Fairlie, *Commun. Math. Phys.* **136**, 487 (1991)
24. A. Lerda and S. Sciuto, *Nucl. Phys.* **B 401**, 613 (1992)
25. A. Montorsi and M. Rasetti, *Phys. Rev. Lett.* (1993)
26. M. Varenna, *Quantum Groups in Condensed Matter Physics*, in *Quantum Groups in Physics*, L. Castellani and J. Wess, eds.; Editrice Compositori, Bologna, 1995
27. M. Jimbo, *Lett. Math. Phys.* **10**, 63 (1985); **11**, 247 (1986)
28. C.N. Yang, *Revs. Mod. Phys.* **34**, 694 (1962)
29. D.B. Uglov and V.E. Korepin, *The Yangian Symmetry of the Hubbard Model*, Institute for Theoretical Physics, Stony Brook, preprint ITP-SB-93-41, *Phys. Rev. Lett.*, in press
30. V.G. Drinfeld, in: Proceedings of the *International Congress of Mathematicians*, Berkeley 1986, A.M. Gleason, ed.; American Mathematical Society, Providence, RI, 1987; and *Sov. Math. Dokl.* **32**, 254 (1985)
V.G. Drinfeld, *Sov. Math. Dokl.* **27**, 68 (1983) and **32**, 254 (1985)
31. H.W. Kroto, *Science* **242**, 1139 (1988)
32. R.F. Curl and R.E. Smalley, *Science* **242**, 1017 (1988)
33. M. Rasetti and R. Zecchina, *Physica* **A 199**, 539 (1993)
34. F. Klein, *Vorlesungen ueber die Theorie des Ikosaeders*, Teubner, Leipzig, 1888
35. F. Klein and R. Fricke, *Vorlesungen ueber die Theorie der elliptischen Modulfunktionen*, Teubner, Leipzig, 1890
36. W. Magnus, *Non-Euclidean Tesselations and their Groups*, Academic Press, New York, 1974

37. H.S.M. Coxeter and W.O. Moser, *Generators and Relations for Discrete Groups*, Springer-Verlag, Berlin, 1965
38. W. Magnus, A. Karrass and D. Solitar, *Combinatorial Group Theory*, J. Wiley, New York, 1966
39. M. Rasetti and T. Regge, *Rivista Nuovo Cimento* **4**, 1 (1981)
40. S. Cesare and V. Del Duca, *Rivista Nuovo Cimento* **10**, 1 (1987)
41. A. Ceresole, M. Rasetti and R. Zecchina, *Generalized Yangian Symmetries of the Hubbard Model over Non-Euclidean Lattices*, Politecnico of Torino, Preprint POLFIS-TH-36-95, september 1995; submitted to *Physica* **A**
42. A. Montorsi and A. Pelizzola, J. Phys. A: Math. Gen. **25**, 5815 (1992)
43. A. Montorsi, *Mod. Phys. Lett.* **B 9**, 971 (1994)
44. W. Magnus, Comm. Pure Appl. Math. **7**, 649 (1954)
45. M.L. Metha, abd K. Kumar, J. Phys. A: Math. Gen. **9**, 197 (1976)
46. M. Takahashi, Prog. Theor. Phys. **47**, 69 (1972)
47. J. Carmelo and D. Baeriswyl, Phys. Rev. B **37**, 7541 (1988)
48. A. Montorsi and M. Rasetti, Phys. Rev. Lett. **66**, 1383 (1991)
K. Michielsen, H. De Raedt and T. Schneider, Phys. Rev. Lett. **68**, 1410 (1992)
49. A.A. Ovchinnikov, *Metal-Insulator transition in the Generalized Hubbard model*, preprint cond-mat 9412109
50. A. Montorsi, M. Rasetti and T. Segalini, in preparation

Appendix A

Schedule of the 1994 GICC School of Computational Chemistry

19 September

Morning Session

8.30 – 10.30	D. Viterbo:	The Language of Space Group Theory
11 – 12	R. Dovesi:	The Language of Band Theory
12 – 13	C. Roetti:	The HF-LCAO (CRYSTAL) Technique

Afternoon Session.
Use of the International Tables of Crystallography (Viterbo, Ugliengo, Aprà)
Graphical tools for periodic Systems (Harrison)

20 September

Morning Session

8.30 – 10.30	C. Pisani:	Ab-initio Quantum-Mechanical Treatment of Periodic Structures
11 – 12	K. Schwarz:	The DF-LAPW Technique
12 – 13	A. Dalcorso:	The PW-SCF Technique

Afternoon Session.
Guide to the use of the PWSCF and CRYSTAL programs (Dalcorso, Roetti, Orlando, Aprà)

21 September

Morning Session

8.30 – 9.30	V.R. Saunders:	The Coulomb Problem in Periodic Structures; Ewald Technique
9.30 – 10.30	R. Dovesi:	Crystal Energy, Elastic Properties, Phase Transitions
11 – 13	M. Catti:	Lattice Dynamics and Thermodynamic Properties

Afternoon Session.
Choice of Basis Set and Use of Pseudopotentials (Roetti, Dalcorso, Orlando, Aprà)
Estimate of Formation Energies and of Correlation Corrections (Dalcorso, Orlando, Causà, Aprà)

22 September

Morning Session

8.30	–	9.30	M. Causà:	Numerical Integration in LCAO-DFT Techniques
9.30	–	10.30	A. Dalcorso:	Special Point Techniques for Reciprocal Space Integration
11	–	12	N.M.H. Harrison:	Direct SCF and Parallel Algorithms for Crystalline Systems
12	–	13	E. Aprà:	Magnetic Properties of Crystalline Systems

Afternoon Session.
The Bader Analysis of Crystalline Electron Charge Densities (Gatti)
Magnetic Properties of Crystalline Systems (Dovesi, Aprà)

23 September

Morning Session

8.30	–	10.30	M. Weyrich:	Density matrix; Structure Factors, Compton Profiles
11	–	13	R.Resta:	Induced and Spontaneous Dielectric Polarization

Afternoon Session.
X-Ray Structure Factors and the Problem of the Thermal Corrections (Lichanot, Dovesi, Roetti)
Calculation of Compton Profiles (Asthalter)

24 September

Morning Session

8.30	–	9.30	M. Rasetti:	The Hubbard Model and Superconductivity
9.30	–	10.30	V.R. Saunders:	The Electronic Correlation in Crystals
11	–	12	C. Pisani:	Loss of Symmetry: Surfaces and Local Defects

Appendix B

Subject Index

Appendix C

List of Acronyms

AFM – Antiferromagnetic
AO – Atomic Orbital
APW – Augmented Plane Wave
ASA – Atomic Sphere Approximation
ASW – Augmented Spherical Wave
a.u. – atomic units
BCS – Bardeen, Cooper and Schrieffer
BF – Bloch Function
BHS (PP) – Bachelet, Hamann and Schlüter (Pseudo-Potentials)
BO – Born-Oppenheimer (approximation)
BSSE – Basis Set Superposition Error
BZ – Brillouin Zone (first)
CC – Coupled-Cluster
CFLS – Cluster Fermionic Linearization Scheme
CMO – Canonical Molecular Orbital
CO – Crystalline Orbital
CP – Car-Parrinello
CPU – Central Processing Unit
DF(T) – Density Functional (Theory)
DM – Density Matrix
DOS – Density of States
ECP – Effective Core Potentials
EFG – Electric Field Gradient
FFT – Fast Fourier Transform
FLAPW or FP-LAPW – Full-potential Linearised Augmented Plane Wave
FM – Ferromagnetic
FWHM – Full Width Half Maximum
GC – Gradient-Corrected
GFMC – Green-Function Montecarlo
GGA – Generalised Gradient Approximation
GS(ES) – Ground State (Electronic Structure)
GT(O) – Gaussian Type (Orbital)
KS – Kohn and Sham
HC – Host Crystal
HF – Hartree-Fock
IDOS – Integrated Density of States
IR – Irreducible Representation
KKR – Korringa-Kohn-Rostoker
LAPW – Linearised Augmented Plane Wave
LCAO – Linear Combination of Atomic Orbitals
LDA – Local Density Approximation
LMO – Localised Molecular Orbital
LMTO – Linear Muffin-Tin Orbital
LO – Local Orbital
LSDA – Local Spin Density Approximation
MD – Molecular Dynamics
MIT – Metal-Insulator Transition
MO – Molecular Orbital
MT(A) – Muffin Tin (Approximation)
NSG – Natural Spin Geminal
NSO – Natural Spin Orbital
ODLRO – Off-Diagonal Long-Range Order
PDOS – Projected Density of States
PP – Pseudo-Potential
PW – Plane Wave
QM – Quantum Mechanics
RIM – Rigid Ion Model
ROHF – Restricted Open-shell Hartree-Fock
RHF – Restricted Hartree-Fock
RIF – Residual Instrumental Function
SC – Supercell
SCF – Self-Consistent-Field
SIC – Self-Interaction Correction
SM – Shell Model
STO – Slater Type Orbital
TG – Translation Group
UHF – Unrestricted Hartree-Fock
VMC – Variational Montecarlo

Lecture Notes in Chemistry

For information about Vols. 1–29
please contact your bookseller or Springer-Verlag

Editorial Policy

This series aims to report new developments in chemical research and teaching - quickly, informally and at a high level. The type of material considered for publication includes:

1. Preliminary drafts of original papers and monographs
2. Lectures on a new field, or presenting a new angle on a classical field
3. Seminar work-outs
4. Reports of meetings, provided they are
   a) of exceptional interest and
   b) devoted to a single topic.

Texts which are out of print but still in demand may also be considered if they fall within these categories.

The timeliness of a manuscript is more important than its form, which may be unfinished or tentative. Thus, in some instances, proofs may be merely outlined and results presented which have been or will later be published elsewhere. If possible, a subject index should be included. Publication of Lecture Notes is intended as a service to the international chemical community, in that a commercial publisher, Springer-Verlag, can offer a wider distribution to documents which would otherwise have a restricted readership. Once published and copyrighted, they can be documented in the scientific literature.

Manuscripts

Manuscripts should comprise not less than 100 and preferably not more than 500 pages. They are reproduced by a photographic process and therefore must be typed with extreme care. Symbols not on the typewriter should be inserted by hand in indelible black ink. Corrections to the typescript should be made by pasting the amended text over the old one, or by obliterating errors with white correcting fluid. Authors receive 50 free copies and are free to use the material in other publications. The typescript is reduced slightly in size during reproduction; best results will not be obtained unless the text on any one page is kept within the overall limit of 18 x 26.5 cm (7 x $10^1/_2$ inches). The publishers will be pleased to supply on request special stationary with the typing area outlined.

Manuscripts should be sent to one of the editors or directly to Springer-Verlag, Heidelberg.

GPSR Compliance

The European Union's (EU) General Product Safety Regulation (GPSR) is a set of rules that requires consumer products to be safe and our obligations to ensure this.

If you have any concerns about our products, you can contact us on ProductSafety@springernature.com

In case Publisher is established outside the EU, the EU authorized representative is:

Springer Nature Customer Service Center GmbH
Europaplatz 3
69115 Heidelberg, Germany

Batch number: 10704711

Printed by Printforce, the Netherlands